Intuitive Axiomatic Set Theory

Set theory can be rigorously and profitably studied through an intuitive approach, thus independently of formal logic. Nearly every branch of Mathematics depends upon set theory, and thus, knowledge of set theory is of interest to every mathematician. This book is addressed to all mathematicians and tries to convince them that this intuitive approach to axiomatic set theory is not only possible but also valuable.

The book has two parts. The first one presents, from the sole intuition of "collection" and "object", the axiomatic ZFC-theory. Then, we present the basics of the theory: the axioms, well-orderings, ordinals and cardinals are the main subjects of this part. In all, one could say that we give some standard interpretation of set theory, but this standard interpretation results in a multiplicity of universes.

The second part of the book deals with the independence proofs of the continuum hypothesis (CH) and the axiom of choice (AC), and forcing is introduced as a necessary tool, and again the theory is developed intuitively, without the use of formal logic. The independence results belong to the metatheory, as they refer to things that cannot be proved, but the greater part of the arguments leading to the independence results, including forcing, are purely set-theoretic.

The book is self-contained and accessible to beginners in set theory. There are no prerequisites other than some knowledge of elementary mathematics. Full detailed proofs are given for all the results.

José Luis García is Emeritus Professor at the University of Murcia, Spain. He began his study of Mathematics at the University of Granada (Spain). After several years teaching Mathematics in secondary school, he was enrolled by the University of Murcia (Spain) as a member of the group working on algebra, led by Professor Gomez Pardo. He received his doctorate and a permanent position at the same university. He developed his research on Ring and Module Theory. He became a full professor in 1991, and then served for thirty years.

Textbooks in Mathematics

Series editors:
Al Boggess, Kenneth H. Rosen

Classical Vector Algebra
Vladimir Lepetic

Introduction to Number Theory
Mark Hunacek

Probability and Statistics for Engineering and the Sciences with Modeling using R
William P. Fox and Rodney X. Sturdivant

Computational Optimization: Success in Practice
Vladislav Bukshtynov

Computational Linear Algebra: with Applications and MATLAB® Computations
Robert E. White

Linear Algebra With Machine Learning and Data
Crista Arangala

Discrete Mathematics with Coding
Hugo D. Junghenn

Applied Mathematics for Scientists and Engineers
Youssef N. Raffoul

Graphs and Digraphs, Seventh Edition
Gary Chartrand, Heather Jordon, Vincent Vatter and Ping Zhang

An Introduction to Optimization with Applications in Data Analytics and Machine Learning
Jeffrey Paul Wheeler

Encounters with Chaos and Fractals, Third Edition
Denny Gulick and Jeff Ford

Differential Calculus in Several Variables
A Learning-by-Doing Approach
Marius Ghergu

Taking the "Oof!" out of Proofs
A Primer on Mathematical Proofs
Alexandr Draganov

Vector Calculus
Steven G. Krantz and Harold Parks

Intuitive Axiomatic Set Theory
José Luis García

https://www.routledge.com/Textbooks-in-Mathematics/book-series/CANDHTEX-BOOMTH

Intuitive Axiomatic Set Theory

José Luis García

CRC Press is an imprint of the
Taylor & Francis Group, an **informa** business
A CHAPMAN & HALL BOOK

First edition published 2024
by CRC Press
2385 Executive Center Drive, Suite 320, Boca Raton, FL 33431

and by CRC Press
4 Park Square, Milton Park, Abingdon, Oxon, OX14 4RN

CRC Press is an imprint of Taylor & Francis Group, LLC

ISBN: 978-1-032-58120-0 (hbk)
ISBN: 978-1-032-58401-0 (pbk)
ISBN: 978-1-003-44991-1 (ebk)

DOI: 10.1201/9781003449911

Typeset in Latin Modern font
by KnowledgeWorks Global Ltd.

Publisher's note: This book has been prepared from camera-ready copy provided by the authors

To my grandchildren, Rocío,
Rodrigo and Alicia

Contents

Preface

There are so many good books on set theory at all levels, either introductory or advanced, that presenting yet another one requires well-founded reasons. Thus, the most urgent question we must address is this: is there anything in this book that is not to be found in any other?

There is a quick answer to this question: this is a book with an intuitive presentation of set theory, hence with no use of results of mathematical logic, but which includes the independence theorems of Gödel and Cohen on the continuum hypothesis and the axiom of choice, while these results are usually regarded as advanced material, only accessible to students with a good command of formal logic. But this is probably too quick to give a fair idea of the purpose and plan of this book, so I shall try to expand a bit on this issue.

First of all, how should we understand an *intuitive axiomatic* theory? Better than entering a theoretical discussion about this point, let us proceed by example: among other books, the influential texts by Halmos [8] or Hrbacek and Jech [9] develop axiomatic set theory without referring to mathematical logic, except for the symbols and rules that are usual in any other part of mathematics. And some (but few) central concepts of the theory, like *set* or *property*, are introduced hoping that their meaning will be understood after some indications. It is in this same sense that our presentation is intuitive, with our only primitive concepts being *object* and *collection*, and this one with a restricted use. More decisively, we also accept some intuitions about the natural numbers and especially about the idea of finiteness, as it is explained in Section 2.6. Though formal or symbolic presentations of set theory try to obtain the set of natural numbers as a product of the theory, and this try comes very close to its objective, one cannot say that it is completely successful – see Chapter 6. Therefore, intuition may have some role in connection with this problem.

Therefore, our intuitive presentation of the theory is similar to that of the above-mentioned texts, which avoid the use of formal logic, but it is different in that we study and prove the independence theorems of Gödel and Cohen. Like this, to the reasons that justify this approach to the theory and which are given or suggested in those books, we may add the following. One of the things that makes forcing (the principal technique in independence proofs) hard for beginners is that it requires the combination of several ideas and concepts from different domains: Boolean algebras, generic sets, forcing language, Boolean values of formulas, and forcing relation. Through the intuitive approach, however, these ingredients of forcing are reduced to just the first two, and this makes the presentation of generic extensions simpler.

This book is a course on set theory at an introductory level. Its first part contains the basic theory, and the material is approximately standard, covering the axioms, the role of set theory as the basis of most of mathematics and its significance in extending arithmetic to infinity, by means of cardinals and ordinals. The second part of the book deals mainly with the metatheory, studying not just the results that can be proved, but results that cannot be proved, and justifying why they cannot be proved. Jointly, both parts give an introduction to the core of set theory, but leave out the extensive collection of important applications that the reader may find in other more complete and authoritative books.

Set theory has undergone a long history: it has shaped the way of thinking in many classical areas of mathematics; it has convinced (most) mathematicians to be working on solid, as well as comfortable ground, much more so than before the advent of set theory; it has provided a unification of the majority of the branches of mathematics under a common basis. But set theory is not only the universal language and unifying ground for all of mathematics; it has also become indispensable in the development of mathematics in practice, particularly Topology, but also Analysis or Algebra. In fact, many results in those classical fields have been obtained through the use of advanced techniques of set theory. Set theory is currently the foundation of mathematics and at the same time a very useful tool for mathematical research. It follows that knowing the basics of set theory, as presented in the first part of this book or in any other text, is a necessity for any student of Mathematics.

The book can serve as a textbook for a course on set theory for advanced undergraduate or beginning graduate students. If they have already had some training on sets, they can skip most of the first chapters, though a quick review of these would be needed in order to identify the terminology and notation, and get familiar with the basic ideas as they are presented here. But these first chapters are developed at such a slow pace that they are also suitable for most undergraduates, provided they have a good command of elementary mathematics. For these, Part I of the book can be the basis for a first course on set theory, complemented if necessary with the Appendices. In this direction, Appendix B contains a sketchy introduction to first-order logic as applied to set theory, and it could be a helping complement, though it would have to be further developed for those wanting to attain a good knowledge of first-order set theory.

The book has the usual aspect of mathematical books: the text in each chapter simulates a lecture to an imaginary audience who does not the have right to interrupt. The lecturer states what are the problems, the means to solve them, the reasons for following such or such path; and, of course, states and proves the results. The results are labeled as theorems, propositions, lemmas or corollaries; explanations about the use of these names are given in Section 1.5. The end of a proof is marked by the symbol □. The explanations may include examples and remarks, but there are also remarks and examples labeled as such, and normally this happens when they have a greater extension or when they are later referred to. There is also the label "Comment", written with smaller sized characters. Sometimes these comments are addressed more to the instructor than to the student, sometimes they are digressions on some side matter or contain additional information that is not required in the rest of the book. So, they are basically remarks, but with the common feature that they can be omitted by the reader (if she or he is not, in a later chapter, asked to come back and read the comment, but this is rare) without affecting the understanding of the rest of the text. All these labeled items are numbered following the numbers of chapters and sections; thus, Proposition 8.5.7 appears in Section 5 of Chapter 8 and is the seventh numbered item in that section.

A number of exercises are included in each chapter. They range from trivial to challenging. Some or many of them are either taken from other books or are variants of exercises appearing in other books, perhaps sometimes uncounsciously on my side. Instead of indicating in the text the source of each of such exercises, I state a general acknowledgment in this preface. Sometimes an exercise is used in the proof of a proposition; it is to be understood that the student should solve the exercise to have a complete proof of the corresponding result. The bibliography lists only the titles that are cited in the text and a few other books where the reader may examine alternative expositions, frequently with additional material, of the theory presented here.

I thank José Luis Gómez Pardo for reading a preliminary version of this book and sending me several corrections. My sister Pepa, (also a mathematician), read parts of the book and provided me with useful feedback and encouragement. It is also a pleasure to thank the editors, Bob Ross and Beth Hawkins, and the production teams led by Paul Boyd and Meeta Singh, for their suggestions and help in the process of turning a draft into a book.

Part I

The Zermelo-Fraenkel Theory

1

Introduction

1.1 The Beginnings of Set Theory

For determining the number of objects in a given collection, we use counting. Like this, you can find how many trees are there in a park by counting them. But what if the collection is infinite? True, we cannot find material infinite collections in normal life, but for mathematicians, infinity appears so frequently that one is forced to deal with it in some form. It pops up in infinite processes, like studying limits of functions, and it is therefore at the base of calculus. But it also appears through infinite collections, like that of all the points in a finite line, or, of course, in the infinite sequence $0, 1, 2, 3, \ldots$ of all the natural numbers. Would it be possible to use counting with infinite collections as we do with finite ones?

Certainly, we cannot get much by counting the objects of an infinite collection because it has no end. But, counting is not the only way to obtain information about the size of collections. Indeed, there is something about the size of even finite collections that can be assessed without having to count. Imagine that you have a collection of stamps and a collection of coins, and you wish to know whether you have more stamps than coins or vice versa. Of course, you may count them all. But there is another way: you place all your stamps in a row, then you put one of your coins in front of each stamp until you run out of coins, and then you know that there are more stamps than coins, or the other way round. Or perhaps the two collections match perfectly, each coin is paired with one stamp, and you know that there are as many coins as stamps.

What happens if we apply this same procedure to infinite collections? Say we want to compare the collection of all natural numbers $0, 1, 2, 3, \ldots$ and the collection of all even natural numbers $0, 2, 4, 6, \ldots$. Thus suppose that we have all the natural numbers in a row (we have to use our imagination for that, because the row is necessarily infinite); and below each number, we place one of the even numbers, like this:

0	1	2	3	4	5	6	7	8	9	10	11 ...
0	2	4	6	8	10	12	14	16	18	20	22 ...

DOI: 10.1201/9781003449911-1

Surprisingly, this seems to be a perfect match: as far as we proceed, we will always have an even number paired with a natural number, $2n$ corresponding to an arbitrary n. So, in some sense, the collection of even numbers has the same size as the collection of all natural numbers, despite the obvious fact that even numbers are just a part of the collection of all natural numbers and therefore the size of the first collection should be less than that of the second. How can this be?

What is evidenced by this paradox is the need for precise definitions of the concepts we use: we have not defined what is to be understood as the size of infinite collections, so we really do not know what *size* means in this context, and as a result of careless use of the concept, we have arrived to a kind of contradiction, one of the so-called *paradoxes of infinity.*

A German mathematician (though born in Russia), Georg Cantor (1845–1918) was the first to construct a precise definition of something akin to the size of infinite collections, something that he called the *power* of the collection. For example, in 1874 he showed that it was impossible to get a perfect pairing between the collection of all natural numbers and that of all the points in a segment of a straight line (but there is a pairing between all natural numbers and a collection which contains some, but not all, of the points of the segment). In contrast, he also showed later that the collection of all the points of a segment has the same power as that of all the points in the inside of a square. And for the first time, an instrument had been devised to differentiate between infinities, some of them being greater than others.

Comment 1.1.1. Incidentally, the words "collection" and "set" will be used interchangeably in this introduction, but not later. The words used by Cantor for this concept were different and evolved with time, but one of the terms he used, that of *set* (Menge), prevailed as the preferred name for these collections.

After having shown the existence of different sizes (or *powers*) of infinities, namely the power of the collection $\mathbb{N}$ of natural numbers and the power of the set $\mathbb{R}$ of real numbers (which is the same as the power of the set of points of a segment), Cantor tried to find some set whose power would be intermediate between these two. He devoted much effort to this search, studying infinite subsets of $\mathbb{R}$ and determining their size. In all the cases he considered, the set had either the power of $\mathbb{N}$ or the power of $\mathbb{R}$, the *continuum* as it was called. This led him to *conjecture* that there is no set with an intermediate power; that is, he expressed his belief that this is what happens, but without having convincing proof that it is so. Proving or disproving this conjecture, known as the *continuum hypothesis*, remained for decades as the most noteworthy problem in set theory. As for Cantor, he worked in constructing the theory of sets for most of his life, so it is fair to say that the basic concepts and the main part of early set theory is due to him.

1.2 The Antinomies

Another German mathematician, a contemporary of Cantor, was Richard Dedekind (1831–1916). As Cantor, Dedekind had an interest in studying sets, but his reasons were different from those of Cantor. He did not try to compare the sizes of infinite sets; instead, he wanted to use sets to construct a sound foundation for classical mathematics, especially algebra and analysis. By introducing the idea of a set, he sought to construct the natural numbers from logical principles, and through them, all the other classes of numbers, and also functions, limits, algebraic concepts, calculus and much more, could be described just from sets.

Both Cantor and Dedekind felt the necessity of being as accurate as possible in specifying the principles upon which their theory was founded, and especially in clarifying the basic concept of set. They thus gave careful explanations on the idea of a set. Leaving aside some differences between both views, they accepted the existence of objects of our thinking, and considered the reunion of some of those objects as a new object, which would be a set. Like this, we may think of, say, the polynomial functions, and then form the collection of all such functions, which would be a set according to this idea. In a definition of set he gave in 1895, Cantor precised that a set is a reunion of *well determined* objects *into a whole*. Perhaps this was a precaution against the paradoxes which were about to appear in the theory.

We have seen how Cantor had introduced the idea of the power of a set. In the 1880s, he further developed his ideas and considered the powers of sets as *transfinite numbers*, that is, numbers beyond infinity. This amounts to consider these "numbers" as having a reality like natural numbers, and thus the theory is an extension of arithmetic to the infinite. As of 1897 (and in a letter to Dedekind in 1899), Cantor referred to the following paradox. Let us consider sets having all the possible powers of sets, and suppose that all the objects that are elements of those sets can be seen as a whole; that is, they form a set S. What would be the power of S? Clearly, it is not less than any possible power as it includes sets with every possible power. But Cantor had shown that for any transfinite number, there is always a greater one. Hence, a set whose power is greater than the power of S exists, and this contradicts the preceding sentence. But he did not see a contradiction here; for him this simply showed that the transfinite numbers formed a kind of "unfinished" collection that could not be made into a whole. Of course, this interpretation leaves the main problem unanswered, that is, how could we know when a reunion of objects can be made into a whole and become a set?

But other paradoxes caused more damage. The Danish logician and mathematician Gottlob Frege (1848–1925) published in 1893 and 1903 the two volumes of his main work *Grundgesetze der Arithmetik* (The fundamental laws of Arithmetic); one of the basic Frege's laws had the consequence that, given any concept that objects may fulfill or not, there is always a class (a term more or less equivalent to that of "set") containing precisely those objects that fall under the given concept (that is, that satisfy the property expressed by the concept). But in 1902, shortly before the appearance of the second volume, Frege received a letter from the British mathematician Bertrand Russell (1872–1970), showing that if one applies this principle to the concept "class that is not a member of itself", a contradiction follows and the very foundation of Frege's system falls apart.

Russell's paradox not only ruined Frege's system but also threatened the basics of set theory. To see this, let us call *principle of comprehension* to the assertion that follows from Frege's law that has been just mentioned. That is, according to this principle of comprehension, in order to construct or describe a set it suffices to state some property that objects may possess or not: there is a set formed with the objects that have the property. Implicitly, this principle was being used in practice in the beginnings of set theory; Dedekind, for instance, had spoken of the set of "all the things that can be object of my thinking". Not only was it used; it was the main tool for producing sets. If this principle of comprehension is contradictory and we are denied its use, how are we supposed to obtain sets and study their properties?

The first years of the 20th century witnessed a thorough debate, not only about these paradoxes and some others (all of which became to be known as *antinomies*), but also about the principles on which set theory and mathematics could be based, or even whether set theory could have a firm basis or was inevitably inconsistent. Some proposals were presented with the goal of building set theory on solid ground, clarifying its methods and freeing it from antinomies, yet preserving the results obtained by Cantor and Dedekind. One of these proposals was made by Ernst Zermelo(1871–1953) in 1908, and completed around 13 years later by Abraham Fraenkel (1891–1965) and Thoralf Skolem (1887–1963), with a last addition by John von Neumann (1903–1957) in 1925. Zermelo's theory, modified by those additions, is now known as Zermelo-Fraenkel theory and is the most widely used in the study of set theory and its applications. Since this theory is an axiomatic theory, we explain next some generalities about the axiomatic method.

1.3 Axiomatic Theories

Among those that have been preserved to our days, the most ancient book containing a systematic and deductive exposition of mathematics is the *Elements* of Euclid of Alexandria (about 300 B.C.). This extraordinary work is divided into 13 parts called "books". Proposition 47 of Book I reads: *In right-angled triangles, the square on the side subtending the right angle is equal to the squares*[1] *on the sides containing the right angle.* This is the deservedly famous Pythagoras' theorem. Right after stating the theorem, Euclid goes on to give the proof; that is, to convince the reader that this assertion is true. This proof is a reasoning based upon Propositions 4, 14, 41 and 46 of the same Book I. In turn, the proof of these propositions applies Propositions 13, 29, 31, 34 and 37. And these in their turn rest on Propositions 11, 15, 23, 26, 27 and 35. This is an example of the type of reasoning that makes Mathematics a deductive science.

It is important to mention a particular feature of deduction. Still with Pythagoras' theorem, let us consider the drawings below.

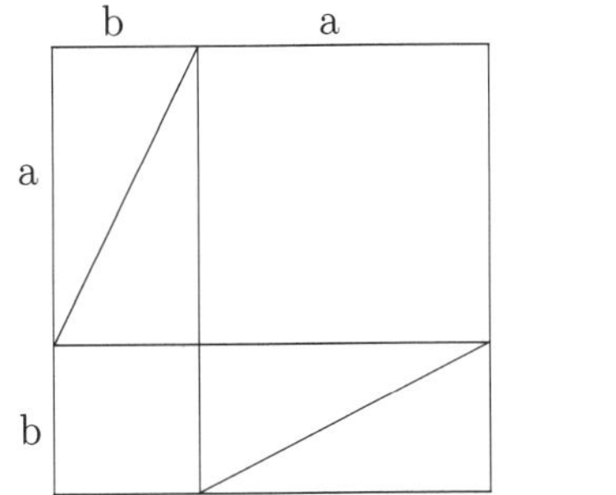

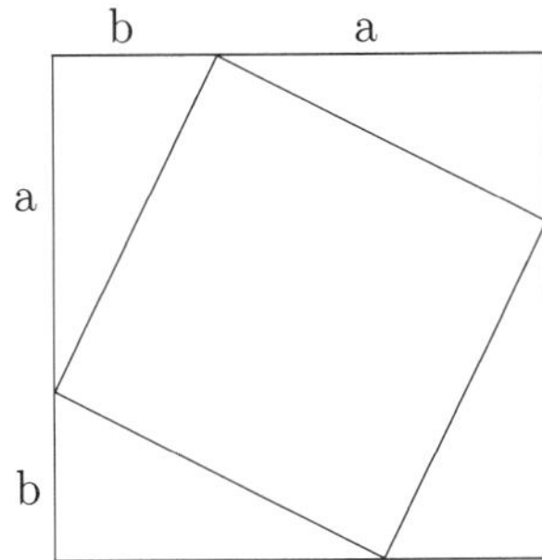

Figure 1.1
Illustrating Pythagoras' theorem.

The figure is almost a visual proof of the theorem. The idea of these drawings is very old, but it does not appear in Euclid. The two squares in the figure are equal and have side $a + b$. In both, there are four equal rectangular triangles whose sides are a, b and the hypotenuse. If we subtract the area of these four triangles from the total square in each of the drawings, the remaining areas must be equal. The remaining area on the left drawing is the sum of the areas of two squares, one with side a, the other with side b; that on the right is the inner square, whose side is the hypotenuse of the rectangular triangles. Therefore, the square on the hypotenuse is the sum of the squares on the sides that form the right angle.

There are reasons why this is not properly a proof: for instance, the fact that the inner figure in the right square is itself a square is suggested by the figure but not really proved. But let us leave this type of remarks aside. The point I am interested in is that the property that we want to show is seen directly from the figure, in just one stroke, so to speak. One **sees**, with only a simple reflection (and overlooking some possible defects of the proof), that the inner square on the right-hand side equals the sum of the two smaller squares on the left-hand side. And this is what someone who has not been trained in

[1]Nowadays, we would say "equal to **the sum of** the squares ...".

deduction would expect of a convincing proof of a property: to see directly the reason why the property has to be true.

But the procedure in the *Elements* is totally different: Pythagoras' theorem is proved after a long chain of propositions. If you are a beginner and have seen no proofs before, this procedure may be unexpected, perhaps strange. Say you are shown a proof of the fact that if some assertion (A) is true, then another assertion (B) must be true; what you expect is something of the type of the visual "proof" of Pythagoras' theorem seen above. Instead, you are asked to follow a chain of deductions: first you see that if (A) is true, then some other property (A1) is true; then that when (A1) is true, (A2) must be true, and so on; the number of such steps may vary, but say that after five steps, you see the final argument that the proposed property (B) is true if the property (A5) is true. It might be the case that you understand each of the steps, and at the end of the proof you are still waiting to see the reason why (B) is true when (A) is true. The series of deductions does not show, in just one movement, a direct connection between the starting assumption (A) and the conclusion (B), and so the immediate link that in your imagination should have tied (B) to (A) simply does no exist. But hopefully, in case this type of problem troubles you, this will be quickly surmounted. You should not expect that a proof be like a revelation. Instead, you will understand, after some training, that chains of correct deductions form an impeccable proof, because deduction transmits truth.

There is another fundamental ingredient in deduction and proof. It has been mentioned above, that Proposition 47 of Euclid's *Elements* (Pythagoras' theorem) is proven by means of arguments where other 15 propositions of Book I are used; and in turn, some of these 15 propositions require of other results for being themselves proved. But it is obvious that the process of founding a result on previous results, these again based on some previous-to-them results, and so on, is a process which must stop somewhere. This will be familiar to anyone who has witnessed the never-ending questions of a little child, asking systematically why? to each reason that he or she is given by grown-ups. Usually, this stops by the grown-up invoking the authority in support of the final reason, for example "this is so because I say so". In mathematical deduction too, the process stops by invoking some final reason. But there is a substantial difference: when proving the results that lead to Pythagoras' theorem (or any other result), the final reasons that may be given belong necessarily to a short explicit list of assertions, which appear at the start of the book, the postulates or *axioms*.

In Euclid's Book I of the *Elements*, the postulates are five relatively simple statements concerning plane geometry. For instance, the first postulate says that it is possible *to draw a straight line from any point to any point*: the third, that it is possible *to describe a circle with any center and any distance.* And the idea is that everything in the book is proved through deductions which are ultimately founded on these axioms or postulates. But there is something more.

Along with the postulates, Euclid includes at the start of the book some other material. Thus there are some rules of logic (called common notions in the book) which will be here and there applied when making a proof. These are assertions like: *Things which are equal to the same thing are also equal to one another* or *If equals be added to equals, the wholes are equal.* Besides, there are 23 definitions. Some of them can be considered as true definitions, identifying positively a certain concept to which a name is given; for instance, *A diameter of a circle is any straight line drawn through the center and terminated in both directions by the circumference of the circle.* But what can we make of other "definitions" like *A point is that which has no part*, or *A line is breadthless length*? These are quite weak as definitions, and if one has no previous idea of what we usually call a line or a point, will not be illuminated by that definition. They have a role however: if the reader of Euclid has already an idea of point and line, these definitions will confirm her or him that the author is going to use the concept in the same sense. And in fact, they give some property of the corresponding notion; thus we learn from the "definition" of point that it is not an object that can be divided. Though not real definitions, they help in setting the scene of what is to come.

So, an axiomatic theory as Euclid's is a complex formed by the definitions, axioms (or postulates) and common notions that are set at the start; and then by the propositions that are obtained from those first principles through deduction. The basic concepts, like point or line are what are called *primitive notions*: some concepts that are not properly and unequivocally defined, but that the author hopes the reader will understand. Likewise, the author hopes that the reader may also share that the axioms state true, or at least plausible facts about those concepts. If these two conditions are met, then the reader will see how the theory is smoothly developed, consisting of a chain of deductions correctly leading to the propositions.

Let us add, finally, that the deductive exposition of the results of a theory does not reflect the way in which those results have been found in practice. As for Pythagoras's theorem, for instance, there is no evidence about how it came to be known, but it is pretty sure that this involved a good deal of discussions, tries, experiments or conjectures, and took quite a long time since the moment that someone noticed the property, probably in some particular situation, until it became common knowledge. Usually, it is only after some property has been established in some form, or, at least, reasonably seen as probably true, that one tries to find an argument to prove it. Mathematicians use deduction in order to convince everybody (and first of all, oneself) beyond any doubt that a certain property is indeed true.

1.4 Intuitive and Formal Axiomatic Theories

The theory in Book I of the *Elements* is what is sometimes called a *material axiomatic theory*, and I name here as *intuitive axiomatic theory.* Let us summarize: there is a field of study, plane geometry in the example; there are some *primitive concepts* pertaining to that field, like line or point in our case; and there are the *axioms*, some assertions about those primitive concepts or about concepts which are defined from the primitive ones. Then, by applying deduction guided by natural logic (*common notions*), one obtains properties about that field of study. The properties so obtained will appear as absolutely certain for everyone who agrees that the axioms are true. Even if someone does not accept the axioms as evident truths, he or she will admit that *if the axioms were true* then the deduced properties would also be true.

There is, however, a problem with intuitive axiomatic theories, and it appears already in the proof of Proposition 1 of Euclid's Book I. In that proof, he uses the following property: given a circle and a point A on its circumference, if we draw the circle with center A and the same radius of the first one, the two circumferences have some point of intersection. This is a property which seems obvious to our intuition of points and lines: the circumference with center A has points in the inside of the first circle and outside it, and thus it seems evident that the two circumferences must cross one another. But this property has not been given as an axiom nor has it been deduced from other propositions; and so it would be inaccurate to declare that all properties can be proved from the axioms alone.

We do not know what Euclid himself could have thought of this. But one possible view of his axiomatic system is the following: we have some intuition about the primitive concepts, even though they are not completely defined. And the above property about the circles is evident from that intuition. Therefore, there is no need to set it as an axiom; everybody having the same intuition about our primitive notions will accept that it is true. The axioms have a different role than that of being evident truths; by setting the axioms, Euclid is asking something of the reader or hearer: accept that I can draw straight lines and circles, accept just two plausible properties, and I will show you all that can be done and learned about plane geometry.

In this view of intuitive axiomatic theories, there are primitive concepts with an intuitive meaning; and properties that are obviously true, not obtained by deduction, about these concepts are automatically accepted. These are some kind of "hidden axioms" in a sense. But the axioms are not necessarily evident truths. In Euclid, some of them are just the starting points of the game: the possibility of drawing. In other cases, they could vary from statements which are justifiable from the basic intuition, to conventions. However, the weakness in these axiomatic systems is that the boundary between "hidden axioms" and explicit axioms may be difficult to be accurately determined.

There is an alternative to intuitive axiomatic theories which is addressed precisely to solve this problem. They are the *formal axiomatic theories*. The most conspicuous example of a formal axiomatic theory was presented in 1899, with the publication of *Grundlagen der Geometrie* (Foundations of Geometry) by David Hilbert (1862–1943). This book is a text on the same subject of

Euclid's geometry, but its aim is to develop the theory without hidden axioms, by incorporating them into the list of explicit axioms.

It is important to remark the consequence this has on the primitive notions. Primitive notions are still needed in formal theories, because they are the objects to which the theory refers. But, since every property that we could intuitively suppose of these concepts has to be included in the axioms or proven from them, the idea we may have of the primitive notions has now no role in the theory; no proof can be founded on the meaning of those concepts, just using the axioms will be enough; and so that meaning is useless. The theory is then developed without giving a meaning to the primitive notions. So, this is what formal axiomatic theories do: there are primitive concepts in them, but they are abstract: they are neither defined nor explained directly. We will only know about their properties through the axioms.

At the start of his book, Hilbert presents his primitive notions: there are three systems of objects: a system of "points", one of "straight lines", one of "planes". But there is no description or information about these objects, they are left as abstractions. They have no meaning, except that of being objects of different groups of entities. He goes on to mention that there are certain relationships between these objects, and these relationships are again undefined. Then the axioms give properties that those relationships are supposed to satisfy. And thus Hilbert's theory is not about points or drawings, it simply obtains consequences of some assumptions about some abstractions. Of course, the reader can imagine an interpretation of the primitive concepts as referring to geometry; if she or he accepts the truth of the axioms under that interpretation, the theorems will also appear as true facts within the same interpretation.

1.5 Axiomatic Set Theory

In the first years of the 20th century, it had been evidenced by the antinomies that set theory was in need of a thorough revision and clarification. At the time there were many discussions about the possible remedies (and even more about the causes of the problems) and, as mentioned above, Zermelo gave in 1908 a presentation of set theory with the aim of providing it with the solid foundation that was needed, avoiding the antinomies, while at the same time preserving essentially what Cantor (and complementarily Dedekind) had constructed. In the form that Cantor and Dedekind had introduced the idea of a set, it was unclear whether the universe of those objects from where sets are constructed is subject to perpetual change because of our action of forming sets and viewing them as new objects; or it has to be considered as definite and invariable so that one cannot change anything, only study the universe as an objective entity. Zermelo decided clearly in favor of this second view:

he supposes a *domain* of objects ("things" is how he call these objects) given from the start, and therefore not subject to change.

Zermelo supposes also that a certain relationship between those things exists, so that two objects a, b may be connected in that relation or not – in the same sense that inequality is a relationship in the domain of numbers, so that, for two arbitrary numbers a, b, it may be that $a < b$ or not. That relationship is left undefined by Zermelo, though we may understand that the intention is that it should represent membership; so that some things will be members of other things, and these will then be sets, formed with objects of the domain. Then, Zermelo gives the axioms concerning the objects of the domain and the relations between them, and thus constructs set theory as an axiomatic theory.

Zermelo leaves that relationship undefined, or abstract, because he is constructing set theory as a formal axiomatic theory. Accordingly, his axioms cannot be said to be evident truths – this is impossible for a formal axiomatic theory, because the primitive notions have no meaning. Yet, if the reader imagines that they refer to the construction of sets, they are reasonable enough to be generally accepted as truths. But above all, they serve the purpose of designing a universe of sets where all the results developed by Cantor and Dedekind, as well as the usual ways in which mathematicians deal with sets, may be justified as following from the axioms.

In this book, we study a presentation of this same theory as an intuitive axiomatic theory. This means that some non-abstract primitive notions will be introduced. Apart from this, we follow the use of almost all mathematics books that develop an axiomatic theory: the results obtained from the axioms are called *propositions*; and each proposition needs a *proof* in which we may use previously proven propositions along with the axioms. Though all proven results are propositions, some receive additionally another name. For instance, propositions that are considered to be of special importance are usually called *theorems*. Calling a certain proposition a theorem is many times a free election: the same result might appear as a theorem in some book and as a proposition in another; to call it a theorem simply stresses the importance that we attribute to the result, but it does not refer to any intrinsic property of that result. But some results, like Pythagoras' theorem, have already earned the title and they are "theorems with a name". Nobody would think of calling it "Pythagoras' proposition".

Another possible term for some propositions is *corollary*. This is usually employed for results which are almost immediate consequences of a previously given proposition (or theorem). By naming a result as a corollary, we emphasize that, though it is just an easy consequence of the proposition to which it is tied, it has some importance of its own and deserves to be highlighted.

Finally, other propositions will be called *lemma*. Normally, a lemma is attached to some proposition or theorem which is more relevant; the lemma is an important step toward proving that theorem, but the interest it has of its own is relatively reduced. Probably, it will be used in proving the theorem which comes after it, but will not be employed in any future proof. But of course, it could be justifiably employed, as it is no other than a proposition after all. These different terms for "proposition" have just the intention to serve as a guide in assessing the value of the results, stressing the importance of some or revealing the secondary role of others.

Still another step in the progressive loss of meaning was undergone by the theory around 1930, leading to see it as a *symbolic axiomatic theory*, specifically a *first-order theory*. But this will be dealt with by the end of the book.

1.6 Logical Symbolism and Truth Tables

Besides the usual symbols in mathematics for numbers, operations, equations and the like, symbols linked to elementary logic will also be used throughout the text. We present here some of these symbols, called *connectives*, and explain the precise meaning we are going to give them.

A type of statement that is constantly employed in mathematical texts is $P \Rightarrow Q$ (which is read 'P implies Q' or 'if P, then Q'); of course, with particular assertions substituted for P and Q. For instance, when dealing with integers, we may assert

$$x > 7 \Rightarrow x > 5$$

and we naturally consider this as a valid statement, that has to be true no matter the number that we substitute for x. This type of statement is a *conditional statement*, also called *implication*. It asserts that if P (*the condition*) is true, then Q (*the conclusion*) must be true as well. A non-symbolic way of asserting $P \Rightarrow Q$ is: 'P is a sufficient condition for Q' – and also 'Q is a necessary condition for P'.

The fact that the conditional of the above example is considered to be true whatever the value of the number x, entails that sentences like $2 > 7 \Rightarrow 2 > 5$ or $6 > 7 \Rightarrow 6 > 5$ must be considered true. The only possibility for a conditional $P \Rightarrow Q$ to be false is that P be true and Q false; for instance $8 > 7 \Rightarrow 8 > 11$ is a false implication. So, the meaning in mathematics of an implication is precisely this: that it is not the case that the condition is true and the conclusion false. This rule may be represented by what is called a *truth table*, as shown next.

P	Q	$P \Rightarrow Q$
T	T	T
T	F	F
F	T	T
F	F	T

What the table gives is the truth value (T for 'true', F for 'false') of $P \Rightarrow Q$ according to the truth values of the condition P and the conclusion Q. The value of $P \Rightarrow Q$ is given in the third column. Thus we see that the conditional is only considered to be false when P is true and Q is false. This is the normal use of the conditional, and one must get accustomed to it. By the way, it is also possible to write $Q \Leftarrow P$ with exactly the same meaning of $P \Rightarrow Q$.

Sentences that offer no problem of interpretation are those of the form 'P and Q'. By saying, for instance '$3^3 = 27$ and $3 < 3^3$' we mean that both assertions hold. The symbol for representing a sentence like the above is $\wedge$; so $(3^3 = 27) \wedge (3 < 3^3)$ is the symbolic form of the sentence (though one does not write the parentheses if there is no risk of confusion). An assertion of this form is called a *conjunction*, and here is its truth table.

P	Q	$P \wedge Q$
T	T	T
T	F	F
F	T	F
F	F	F

That is, a conjunction is considered true when and only when both members of the conjunction are true.

There is also a symbol, $\neg$, for the negation of a proposition P, i.e., for the assertion that P does not hold; this would be $\neg(P)$. The table in this case is even simpler than the previous ones.

P	$\neg(P)$
T	F
F	T

While the meaning of 'and' in mathematics matches quite well that of natural language, the case of 'or' is trickier. It is common in usual language to use 'or' referring to options which exclude each other: for instance 'take it or leave it'; this is the *exclusive* meaning of 'or'. But it is also used in the *non-exclusive* sense: 'It will rain today or tomorrow', where the assertion includes the possibility that it might rain both days. In mathematics, sentences of the type 'P or Q' follow the non-exclusive interpretation of 'or'. This is shown in the truth table below corresponding to the sentences of the form $P \vee Q$ ($\vee$ is the symbol for 'or'), which are called *disjunctions*.

P	Q	$P \vee Q$
T	T	T
T	F	T
F	T	T
F	F	F

That is, $P \vee Q$ is considered true unless both members of the disjunction are false. Thus $3 > 9 \vee 3^2 = 9$ is true; as also $3 < 9 \vee 3^2 = 9$. A false sentence of this form would be $3 > 9 \vee (\neg(3^2 = 9))$.

The symbol $\Leftrightarrow$, which is read 'if and only if', is a combination of the two forms $\Rightarrow$ and $\Leftarrow$ of the implication symbol which we have seen. Accordingly, assertions of the form $P \Leftrightarrow Q$ have the meaning that both conditional statements $P \Rightarrow Q$ and $Q \Rightarrow P$ are satisfied; and so $P \Leftrightarrow Q$ has the same value as $(P \Rightarrow Q) \wedge (Q \Rightarrow P)$. In non-symbolic form: 'P is a necessary and sufficient condition for Q'. This can be applied to get the truth table for $\Leftrightarrow$:

P	Q	$P \Rightarrow Q$	$Q \Rightarrow P$	$P \Leftrightarrow Q$
T	T	T	T	T
T	F	F	T	F
F	T	T	F	F
F	F	T	T	T

The truth tables of $P \Rightarrow Q$ and $Q \Rightarrow P$ appear in the third and fourth columns. Then the last column is obtained by using the truth table of the conjunction $\wedge$ applied to the previous two columns, thus giving the value of $(P \Rightarrow Q) \wedge (Q \Rightarrow P)$. Thus it is the fifth column that gives the table for $P \Leftrightarrow Q$, according to the values of P and Q in the first two columns.

The statement $P \Leftrightarrow Q$ is called a *biconditional* or a *double implication*. $P \Leftrightarrow Q$ is also said to be an *equivalence*, and it is true when P and Q are both true or both false. An example of an assertion of this type which is true for whatever value of x is: $x^2 + 2x - 15 = 0 \Leftrightarrow (x = 3 \vee x = -5)$.

1.7 Permutations

The Gold-bug is a short novel by E. A. Poe. There is in it a buried treasure and a note on a parchment with the instructions for finding it. The beginning of this handwritten note reads: *pdttsdapvvghneibgvetlvetvniaghneisiogavvipn* (I apologize for taking the liberty of changing the original text, which consisted of digits and symbols, and for placing instead these letters. I do this for typographical reasons as well as for convenience. But though the style and the atmosphere created in the novel is shamefully betrayed by this change, the logic of the story is not affected). This, of course, is a ciphered text which one needs to translate to get what is called the *plaintext*, that is, the original text. And for this translation, there is a key that the possessor of the message should have in order to make the decipherment. A substantial part of the book is devoted to explain how the text can be deciphered without knowing the key, just through logic and ingenuity. But we are going to suppose that we have the key. Here it is:

a	b	d	e	g	h	i	l	n	o	p	s	t	v
l	b	g	h	i	n	e	p	t	v	a	d	o	s

Possessing this key, it is child's play to decipher the text. We simply have to substitute each letter of the ciphered text with the letter below it in the table.

Like this, the first three letters of the plain text would be "ago". For Captain Kidd, the pirate who hid the treasure and wrote the note[2], the problem would have been the inverse: he would have known the plain text and wanted to cipher it. For this, the key would have been obtained just by interchanging the rows of the above table – and then reordering the letters so that the upper row appears in alphabetical order.

a	b	d	e	g	h	i	l	n	o	p	s	t	v
p	b	s	i	d	e	g	a	h	t	l	v	n	o

So, to write the first three letters of the note, he would have looked at those letters in the plain text he wanted to transmit, "ago", and written "pdt" on the parchment.

This is an example of what mathematicians call a *substitution* or a *permutation*. It is not essential for the study of substitutions whether we employ letters or whatever kind of objects. In mathematics, substitutions are preferably made with numbers. For instance, the two substitutions written above could be expressed with numbers (with alphabetical order for the letters corresponding to natural order for the numbers), like this:

1	2	3	4	5	6	7	8	9	10	11	12	13	14
8	2	5	6	7	9	4	11	13	14	1	3	10	12

and

1	2	3	4	5	6	7	8	9	10	11	12	13	14
11	2	12	7	3	4	5	1	6	13	8	14	9	10

Besides appearing in many situations, in mathematics or in life, permutations are also important because we can operate with them, as we operate with numbers or functions. For instance, we may consider permutations of five elements, that is, of $1, 2, 3, 4, 5$. Let

$$\sigma = \begin{pmatrix} 1 & 2 & 3 & 4 & 5 \\ 2 & 5 & 4 & 3 & 1 \end{pmatrix}, \quad \tau = \begin{pmatrix} 1 & 2 & 3 & 4 & 5 \\ 5 & 1 & 2 & 4 & 3 \end{pmatrix}$$

We use the notation of functions for referring to the element that is substituted for any index i by some permutation (we call here *index* to any of the numbers appearing in a permutation; in this example, any of the numbers $1, 2, 3, 4, 5$). For instance, $\sigma(1) = 2$, $\sigma(5) = 1$ or $\tau(3) = 2$. We may multiply these permutations by applying first the substitution σ, then the substitution τ to the result. We represent this multiplication below, by presenting the two

[2]Of course, all this is an oversimplification. Captain Kidd would have a key containing the correspondence for the 26 letters of the alphabet. We have used only 14 to keep a reasonable size, and because it would be sufficient for this first part of the message.

successive steps: first, we apply σ; then we apply τ to the result of σ. The permutation on the right gives the product.

$$\overbrace{\begin{pmatrix} 1 & 2 & 3 & 4 & 5 \\ 2 & 5 & 4 & 3 & 1 \end{pmatrix}}^{\sigma} \rightarrow \overbrace{\begin{pmatrix} 2 & 5 & 4 & 3 & 1 \\ 1 & 3 & 4 & 2 & 5 \end{pmatrix}}^{\tau}, \qquad \sigma * \tau = \begin{pmatrix} 1 & 2 & 3 & 4 & 5 \\ 1 & 3 & 4 & 2 & 5 \end{pmatrix}$$

The result of multiplying two permutations σ and τ in this form can be written as $\sigma * \tau$ and also as $\tau \circ \sigma$; and like this we may write $(\tau \circ \sigma)(4) = \tau(\sigma(4)) = \tau(3) = 2$. One may use whatever notation she or he likes: the meaning of $\sigma_1 * \sigma_2$ is the same as that of $\sigma_2 \circ \sigma_1$. But we will employ preferably the second one.

Say ρ is another permutation of five elements,

$$\rho = \begin{pmatrix} 1 & 2 & 3 & 4 & 5 \\ 3 & 5 & 1 & 2 & 4 \end{pmatrix}$$

Then we may calculate $\rho \circ (\tau \circ \sigma)$ and $(\rho \circ \tau) \circ \sigma$. It is readily seen that both computations give the same result:

$$\begin{pmatrix} 1 & 2 & 3 & 4 & 5 \\ 3 & 1 & 2 & 5 & 4 \end{pmatrix}$$

And it is also easy to understand that this is a general property of the product of permutations: for if $\sigma_1, \sigma_2, \sigma_3$ are arbitrary permutations and i is an arbitrary index, $[(\sigma_3 \circ \sigma_2) \circ \sigma_1](i)$ is obtained by applying to the index i the substitution σ_1 first, then $\sigma_3 \circ \sigma_2$ to the result; in all, it is obtained just by applying consecutively the substitutions $\sigma_1, \sigma_2, \sigma_3$ respectively to i first, then to the successive results. And this is the same for $\sigma_3 \circ (\sigma_2 \circ \sigma_1)$. That is, the product of permutations satisfies the *associative* property; which is not to be neglected, because other operations do not have that property: the substraction of numbers is not associative, and $9 - (5 - 2)$ is not $(9 - 5) - 2$.

We had started by considering some permutations of 14 elements, and had seen two permutations that, being one for ciphering, the other for deciphering, were inverse. We can give a precise description of inverse permutations by means of the product: two permutations σ and τ are mutually inverse when $\tau \circ \sigma = \sigma \circ \tau$ is the identity permutation; this is the substitution where each i is substituted for itself. Working again with 5 elements, the inverse of the permutation σ above would be obtained by first switching the rows in σ, then reordering both rows so that the first row has all the elements in the natural order. That gives:

$$\mu = \begin{pmatrix} 1 & 2 & 3 & 4 & 5 \\ 5 & 1 & 4 & 3 & 2 \end{pmatrix}$$

One can check that both products $\mu \circ \sigma$ and $\sigma \circ \mu$ give indeed the identity.

The student probably knows that the number of permutations of 5 elements is $5! = 5 \cdot 4 \cdot 3 \cdot 2 \cdot 1 = 120$; and in general the number of permutations of n

elements is $n! = n \cdot (n-1) \cdot (n-2) \cdot \cdots \cdot 3 \cdot 2 \cdot 1$. The collection of all the permutations of n elements is denoted S_n. And the product of permutations we have just considered is defined for permutations of the same S_n. In the example at the beginning of this section, our permutations were elements of S_{14}.

The only result about permutations that we need in the first part of this book is an easy property of S_3. So, we give the necessary definition and then show the property.

Definition 1.7.1. Let $n \geq 3$. A permutation σ of S_n is called a *cycle* when there are indices $i_1, \ldots, i_r$ so that $\sigma(i_1) = i_2, \sigma(i_2) = i_3, \ldots, \sigma(i_{r-1}) = i_r$ and $\sigma(i_r) = i_1$; and $\sigma(j) = j$ if $j \neq i_1, \ldots, i_r$. The number r must satisfy $1 < r \leq n$ and is called the *length* of the cycle σ.

The permutation σ of this definition is written $(i_1 \ \ i_2 \ \ \ldots \ \ i_r)$. For instance, $(1 \ \ 5 \ \ 4 \ \ 2)$ may denote a cycle of S_6 (also of S_5 or S_9). In S_6,

$$(1 \ \ 5 \ \ 4 \ \ 2) = \begin{pmatrix} 1 & 2 & 3 & 4 & 5 & 6 \\ 5 & 1 & 3 & 2 & 4 & 6 \end{pmatrix}$$

A cycle of length 2 is called a *transposition*. The result we need is the following.

Proposition 1.7.2. *Every permutation of three elements is a product of factors, each of them equal to* $(1 \ \ 2)$ *or to* $(1 \ \ 2 \ \ 3)$.

Proof. S_3 has $3! = 6$ elements. Since obviously the identity is equal to $(1 \ \ 2) \circ (1 \ \ 2)$, it is only necessary to check the remaining three elements. This is shown below:

$$\begin{pmatrix} 1 & 2 & 3 \\ 3 & 2 & 1 \end{pmatrix} = (1 \ \ 2 \ \ 3) \circ (1 \ \ 2), \quad \begin{pmatrix} 1 & 2 & 3 \\ 1 & 3 & 2 \end{pmatrix} = (1 \ \ 2) \circ (1 \ \ 2 \ \ 3)$$

$$\text{and } \begin{pmatrix} 1 & 2 & 3 \\ 3 & 1 & 2 \end{pmatrix} = (1 \ \ 2 \ \ 3) \circ (1 \ \ 2 \ \ 3)$$

□

2

Objects, Collections, Sets

2.1 Introduction

We may see set theory as the study of the mathematical universe, that is, of the collection of all the mathematical objects. Certainly, the assumed objective for the theory in its beginnings was the study of the universe of sets or collections of objects. But it gradually became to be accepted that all (or nearly all) mathematical objects can be seen as sets, so that, for this view, the study of sets includes the study of the mathematical objects. However, it is difficult to identify precisely which are all the mathematical objects, and this is where axiomatic can help. The idea is simple: to define some entity, in this case the mathematical universe, one collects properties of this universe until there is only one entity that satisfies all the collected properties; that will necessarily be our universe and, in this form, it will have been completely identified through its properties. These properties that we want the universe to satisfy are *the axioms*.

But in practice things are not so simple. Our knowledge of the (pretended) universe being far from exact, nothing guarantees that even a careful and experienced selection of the axioms will lead to the desired identification, so it might happen that there are many entities that satisfy the axioms; or it might be that there is no entity that satisfies them. And this is more or less what occurs in set theory: if there is some universe, that is, a collection of objects that satisfies all axioms, then there are many other collections satisfying the same axioms.

This has forced mathematicians to aim at a more modest goal. We have seen in the previous chapter that set theory started with two main uses, following respectively the ideas of Dedekind and Cantor. First, set theory is designed to serve as a frame or common ground to all mathematical theories. It should embrace all of the concepts of Mathematics and every particular mathematical domain should be possibly viewed as included in set theory. Second, set theory is an extension of usual arithmetic to infinite collections: as natural numbers serve to measure the size of finite collections, an extension of natural numbers within set theory serves to introduce a kind of measure of the size of infinite collections. Now, the axioms of set theory are so chosen as to guarantee that every universe that satisfies the axioms is capable of providing these two uses:

DOI: 10.1201/9781003449911-2

whatever the universe, all the usual mathematical objects of everyday life: numbers of any kind, functions, operators, curves, algebraic structures, etc., have a sort of faithful copy inside the universe. Also, every universe which satisfies the axioms contains infinite collections and is able to give measures of the size of those collections. As we will see, however, not all universes are equivalent, and properties of infinities may differ from one universe to another and those familiar mathematical objects of everyday may have different copies (sometimes with different behavior) in different universes.

It seems normal that an axiomatic theory start by explicitly giving the list of axioms, and follow with the propositions deduced from them. But this is not compulsory, and we may instead follow an stepwise procedure. Following Zermelo, we postulate a universe of objects which is the frame where the axioms will be given and the properties of the objects will be obtained. The first concepts or definitions can be presented without assuming any axiom, and very few axioms are necessary to prove the first propositions. Then some more axioms are added to obtain new propositions and so on. In this way, we shall speak, for instance, of 2-universes for those where only axioms $1, 2$ are supposed to hold; or 6-universes for universes satisfying all axioms up to number 6; going on, *ZF-universes* are those universes that satisfy all first 8 axioms; and in *ZFC-universes* all axioms from 1 to 9 are supposed to hold. Like this, the results obtained say for 2-universes, will be valid in 6-universes or in ZFC-universes because they are valid whenever axioms 1 and 2 hold; in this way, the theory will be presented in an accumulative form.

2.2 Membership and Inclusion

We shall always assume (if not explicitly said) that all the concepts and constructions we introduce are referred to and take place in a certain *universe* $\mathcal{U}$ which is a collection of objects. The objects of $\mathcal{U}$ will be called $\mathcal{U}$-objects.

We present set theory as an intuitive axiomatic theory. Our first primitive notion (leaving aside those of universe and $\mathcal{U}$-object) is *$\mathcal{U}$-collection.* A $\mathcal{U}$-collection is a collection consisting of objects of the universe $\mathcal{U}$.

Though we do not give a precise definition of the concept "collection", it is hoped that the reader shares the usual idea of collection. Later, some conventions, examples and axioms will help in showing further characteristics of the concept, making it more precise. When the universe $\mathcal{U}$ is understood, we shall simplify the terminology and speak simply of *object* and *collection* instead of $\mathcal{U}$-object and $\mathcal{U}$-collection. This might lead to ambiguity, since both words, object and collection, have a more general meaning in natural language. But it will be rare that we employ the words in this general sense; and if we do, we shall try to state this explicitly.

Equality is a logical notion that needs no definition. Some clarification will nevertheless be useful, but we defer saying a word about it till later.

Examples 2.2.1. Let us consider the plane of elementary geometry. The collection of all the points in the plane forms a certain universe $\mathcal{U}$, so that the $\mathcal{U}$-objects are the points. The collection of all the points in a straight line, finite or infinite, is a $\mathcal{U}$-collection. But also the vertices of a triangle or the points in a circumference or 1000 points that we choose in whatever way, would form $\mathcal{U}$-collections.

In that example, collections are not objects of the universe. They live in the universe somehow, but are not citizens. But it is possible (and usual) that a collection be an object at the same time. In fact, the objects of interest when we study the plane are not only the points; also the lines, polygons, circles or regions of the plane should be seen as full right citizens. So, we may consider the following simplified situation.

In the same plane, suppose we form a different universe that we shall call $\mathcal{U}_0$. The objects of $\mathcal{U}_0$ will now be all the points and all the finite straight lines, where we understand that *line* means the collection of all the points of the line. Like this, if A, B are two points, the line AB is the $\mathcal{U}_0$-collection formed by all the points between A and B; and at the same time it is a $\mathcal{U}_0$-object. Here, points are objects but not collections, lines are collections and objects, while the collection of all the points in a circle or the collection formed by the three sides of a triangle are collections but not objects.

If C is a collection of objects, those objects are called the *elements* of C. If u is one of the elements of C, we say that u *belongs to* C and write $u \in C$. The contrary (that is, u is an object which is not an element of C) is written as $u \notin C$. Do not pay attention to the fact that some letters are capital and some are not. This use does not hide a meaning, just provides an opportunity for having more symbols at hand. It is true that usually capital letters will denote collections and small case is used for objects. But this is not a established convention, we will not be attached to any rule in this, and moreover such a rule could hardly be meaningful, since some collections may be objects and conversely.

The use of symbols (like $u \in C$ above) is, as always in mathematics, very useful: first, it is a compact, short notation which is hence easy to read, but it is also intended to avoid ambiguity and to exclude referring to the same fact by different means, unlike what happens with natural languages. For instance, the relation $u \in C$ can be expressed by saying, as above, "u belongs to C", but it is also common to refer to it by saying "u is an element of C", "C contains u", "u is a member of C" or even "u is in C". On the contrary, $u \in C$ is the only symbolic notation to refer to that relationship[1]. The relation itself is called the *membership* relation. It should be added that the assertion $u \in C$ includes that u is an object and C is a collection (in fact, it is to be understood that this says that u is one of the objects that form the collection C). Since $\mathcal{U}$ is a $\mathcal{U}$-collection, $u \in \mathcal{U}$ means that u is an object.

[1]Not entirely: $u \in C$ is sometimes written $C \ni u$.

Collections may be identified by giving the entire list of its elements inside curly brackets. Like this, assuming that natural numbers are (hopefully) objects of our universe $\mathcal{U}$, $\{1, 7, 2\}$ denotes the collection that contains precisely three elements, the numbers $1, 2$ and 7.

Obviously, this procedure for identifying collections is not available when the collection contains infinitely many objects; in such case, a possibility is to introduce some special symbol to denote the collection: $\mathbb{Z}$ identifies the collection of all integers, as $\mathbb{N}$ is the collection of all natural numbers (i.e., non-negative integers). Frequently, curly brackets are also used to identify infinite collections, whose elements are described between the brackets, though of course not listed one by one. For instance, $\{x|x \in \mathbb{Z} \text{ and } x > 2\}$ (read: all x *such that* x belongs to $\mathbb{Z}$ and $x > 2$) denotes the collection whose members are all integers greater than 2. The same collection could be also written as $\{x|x \in \mathbb{N} \wedge 3 \leq x\}$. The use of variables as above coupled with the notation $\{\ \ |\ \ \}$ is an efficient way to describe many collections – infinite or not: $\{x|x \in \mathbb{Z} \wedge 2 \leq x < 100\}$ is the collection of all natural numbers from 2 to 99.

Definition 2.2.2. Let A, B be two collections (distinct or not). We say that A *is included in* B (or that A is a *subcollection* of B) when every element of A is also an element of B. This is written $A \subseteq B$.

This relationship is called *inclusion*. To indicate that A is not included in B we write $A \nsubseteq B$. The condition for having $A \subseteq B$ can also be written in symbolic language, thus:

$$x \in A \Rightarrow x \in B$$

According to the meaning of implication shown in the previous chapter, $x \in A \Rightarrow x \in B$ expresses the fact that if x is any object belonging to A, then it belongs to B; and this defines the property $A \subseteq B$.

Examples 2.2.3. If C is any collection, then $C \subseteq C$, because the implication $x \in C \Rightarrow x \in C$ is obviously true. Thus any collection is a subcollection of itself.

Assuming that $\mathbb{N}$ (formed with the natural numbers) is a $\mathcal{U}$-collection, we have $\mathbb{N} \subseteq \mathbb{N}$, but also $E \subseteq \mathbb{N}$ if E is the collection of all even numbers, or $P \subseteq \mathbb{N}$ with P the collection of all prime numbers. Similarly, if $\mathbb{R}$ (the real numbers) is a $\mathcal{U}$-collection, the elements of the interval $(0, 1)$ form a subcollection of $\mathbb{R}$.

Let $\mathcal{U}_0$ be the universe introduced in Examples 2.2.1. Imagine that $ABCD$ is a square, and let P be the collection whose elements are all the points on the sides of the square. Then the side AB is included in P because each element of AB is a point of the square. However, we cannot say that the point A is included in P, because A is not supposed to be a collection, and the inclusion relation occurs exclusively between collections, as given in the definition. Instead, the collection $\{A, B\}$ consisting of those two points is indeed a subcollection of P; and also a subcollection of AB.

Since we are presenting an intuitive theory, the primitive concept of $\mathcal{U}$-collection has some meaning; and so, properties of $\mathcal{U}$-collections that are obviously true from the meaning of the concept might be used in arguments (though, in that case, we shall call attention to the fact). Besides this, there are some features of the notion of $\mathcal{U}$-collection that are not so evident, but which will also be assumed. These serve to complete the idea of how we are going to use the concept of collection, and to make it more precise.

We thus introduce some assumptions about the primitive idea of a $\mathcal{U}$-collection. The first one has already been supposed in the previous discussion when talking of infinite collections. In mathematics we do need and use infinite collections, like that of the natural numbers or of the points in a line. Thus, we must admit that $\mathcal{U}$-collections may be infinite.

A collection can be seen as a plurality of objects; that is, several objects which we unite in our imagination, forming a new entity. It seems therefore that a collection must consist of, at least, two objects. But in our theory we will need collections with less than two objects. It would otherwise be uncomfortable when trying to consider, for instance, the $\mathcal{U}$-collection of all natural numbers between 10 and 20 which are perfect squares (i.e., they are the square of a natural number) and having to admit that there is no such collection because it consists of just one object. So, we adopt the convention that for each object u of the universe, there is a collection $\{u\}$ consisting of just this object. For an analogous reason we accept also the existence of a $\mathcal{U}$-collection with no objects, an *empty collection*. An example is the collection of all natural numbers between 50 and 60 which are perfect squares, if natural numbers are $\mathcal{U}$-objects.

Certainly, these conventions that we accept but are not absolutely evident, play the role of an axiom. But in this, we are going to substitute this name, axiom, with "principle". This gives our first principles.

Principle of infinite collections. It is possible for a $\mathcal{U}$-collection to be infinite.

Principle of singletons. If u is any $\mathcal{U}$-object, there is a $\mathcal{U}$-collection $\{u\}$ which contains u and no other object – a collection with just one object is called a *singleton* whence the name for the principle.

Principle of the empty collection. There is some $\mathcal{U}$-collection which has no object.

Comment 2.2.4. There are reasons for calling these *principles* instead of *axioms*, though we agree that we may use these principles in arguments as we do with the axioms. First, these principles do not give directly properties of objects, while the usual axioms try to determine the universe and thus they refer to the $\mathcal{U}$-objects. However, this reason is not conclusive because the principles speak indirectly of objects, since we have not excluded that $\mathcal{U}$-collections may be $\mathcal{U}$-objects. Thus, properties of collections may have consequences as properties of objects.

The main reason for the substitution is practical. Later in this same chapter, we state our first two axioms about objects; and these two axioms imply our principles of singletons and of the empty collection. Also, axiom 6 has the existence of infinite collections as a consequence. Thus these principles are not really needed in the long run; we just use them for making the idea of collection more precise in our first steps. And so it would not be reasonable to let them appear as axioms, only to be discarded soon because they do not form part of the final list of axioms.

Our last assumption about collections can be exemplified as follows. Let us suppose that natural numbers are objects of the universe $\mathcal{U}$. Then we may consider the collection, say A, of all nonzero natural numbers up to 100 which are perfect squares; and the collection B of all natural numbers up to 100 that can be obtained by the addition of consecutive odd numbers starting with 1; that is, $B = \{1, 1+3, 1+3+5, 1+3+5+7, \dots\}$. Even if the reader does not know the property behind this, he or she may easily be convinced that A and B have the same elements. In such a situation, we will agree that A and B are the same collection, i.e. $A = B$. That is to say, a collection is completely determined by its elements. We shall refer to this as the *principle of extension.*

Comment 2.2.5. In connection with this, we say a word about equality. An object or collection is only equal to itself and is not equal to any other object or collection. We have identified the collections of natural numbers given right above by describing them, and by naming them as A or B. Of course, A and B are just names for the collections, not the collections themselves. But even if making this distinction is sometimes necessary, it would be cumbersome and pedantic to accomodate our speaking to such distinctions. So, we will normally say "the collection A" or "the collection B" (and even "two collections A and B", while not implying that they are indeed two different collections, just two different names) and the meaning is unproblematic for everybody. And it turns out that, as observed above, A and B are (if we identify the names and the collections) one and the same. So, we may write $A = B$ and again the meaning is clear: we are not saying that the names are equal, but that the collections named by them are "equal" (that is, the same). This is the sense in which we will use equality: generally, $x = y$ asserts that x and y denote the same object.

Proposition 2.2.6. ***(Principle of extension)*** *If A, B are given collections, then $A = B$ if and only if $A \subseteq B$ and $B \subseteq A$.*

This, of course, says that two collections are the same when they have exactly the same elements. Let us illustrate this with an example.

Example 2.2.7. Let us consider again the universe $\mathcal{U}_0$ of Example 2.2.1. Let $ABCD$ be a rectangle and let P be the collection of all the points of the rectangle (that is, the points on the border of the figure). Another $\mathcal{U}_0$-collection S consists of the four sides of the rectangle, AB, BC, CD, DA, which are $\mathcal{U}_0$-objects too. These two $\mathcal{U}_0$-collections are different: S has just four members, P is infinite and thus they cannot be equal by the principle of extension.

Note also in this example that a point A is an $\mathcal{U}_0$-object that is not a $\mathcal{U}_0$-collection (A is not the same as the $\mathcal{U}_0$-collection $\{A\}$, since A has no elements); the $\mathcal{U}_0$-collection $\{A, B\}$ with these two points is not an object.

Something important is to be learned from this example. It is understandable that a beginner look collections as if they had some material reality; and then distinguishing P and S above seems unnatural. Because both collections, seen as something material, can be identified to the same rectangle, so how is it that they are different? Well, they are different because one should not look at collections as corresponding to something physical; instead, they should be seen as lists. In the example, S is a list of four objects, because sides are, by convention, $\mathcal{U}_0$-objects. It does not matter how those objects, the sides, are formed. In fact, if X is one point in the side AB, then X is not one of the four elements of S; thus X is not an element of S, but X is an element of P (certainly, X is an element of an element of S). The idea that a collection can be seen as a list is the consequence of the principle of extension; though it is true that this is just an approximation: there is no proper list describing an infinite collection, but the idea of an infinite collection is closer to that of an infinite list than to any other idea. The principle makes our idea of a $\mathcal{U}$-collection more concrete and usable.

To say it all, a collection can be seen as a list, but it is more like a shopping list. A list is normally understood as an ordered list. But if we go shopping and have a list of ten items that we must buy, the order in which these items appear in the list is totally irrelevant. What one has to do is to get all ten articles. And this is also how we see collections: they consist of objects of the universe, independently of any order.

Comment 2.2.8. As with the other principles, we might ask why do we not declare this principle as an axiom. First, the principle refers to collections and not directly to objects. More importantly, here too there is a strong connection between this principle and an axiom which will be given later, axiom 7. But it is not the case that axiom 7 implies the principle of extension nor the other way round. Hence we are forced to state both results as basic assertions, and calling this a principle underlines its difference with axiom 7 or *axiom of extension*. In practice, however, this principle of extension works as an axiom, but it does not refer only to the objects of the universe as all axioms do.

Remark 2.2.9. The principle of extension is the basis in proofs showing that two collections are equal. For example, it implies that there is only one empty collection, since all possible empty collections have the same elements, none. We may thus speak of ***the*** *empty collection*, which will be denoted as $\emptyset$. Note that $\emptyset$ is a subcollection of every $\mathcal{U}$-collection C, because the implication of Definition 2.2.2, $x \in \emptyset \Rightarrow x \in C$ is always true according to the convention on the meaning of implications we gave in Chapter 1, since $x \in \emptyset$ is always false – in such situations, i.e., when the condition in some implication is always false, we say that the implication is *vacuously true*.

It may also be seen from the principle of extension that an $\mathcal{U}$-object u does not, in general, coincide with $\{u\}$. Because $\{u\}$ has just one element, while u

may be any object; for instance, we could take $u = \{2, 3, 4\}$, so u has three objects. However, we still cannot prove that $u \neq \{u\}$ holds for every $\mathcal{U}$-object u.

A collection may be an object of the universe $\mathcal{U}$ (for instance, we hope that $\mathbb{N}$ or $\mathbb{Z}$ will be objects), or may be not (as is the case of $\mathcal{U}$ itself, as we shall see later). When a, b are collections which are objects, then the relations $a \in b$ or $a \subseteq b$ may be satisfied or not, but they are independent. Beginners fail sometimes to rightly appreciate this difference; and this does not improve when one uses the same word to denote both situations, as is the case when we say "a contains b" both for $b \subseteq a$ and $b \in a$. It is for this reason that, as artificial as it can be, we will stick to the notation that we explain now: we shall say that a *contains* u (or u is contained in a) to indicate $u \in a$, and use the term a *includes* u (or u is included in a) for the inclusion relation $u \subseteq a$.

Example 2.2.10. Let us suppose that A, B, C below are collections of objects of the universe.

$$A = \{3, \{2,3\}, -2, \{\pi, 7\}, \mathbb{Z}, \begin{pmatrix} 1 & i \\ 0 & -i \end{pmatrix}, 2\}, \quad B = \{3, 2\}, \quad C = \{3, -2\}$$

$B \in A$ and $B \subseteq A$ too, though by different reasons. Also, $C \subseteq A$, but $C \notin A$. On the other hand, $\mathbb{Z} \in A$, but $\mathbb{Z} \nsubseteq A$.

2.3 Intersections and Unions

Definition 2.3.1. If A, B are two $\mathcal{U}$-collections, the *intersection* of A and B (written $A \cap B$) is the collection formed with all the objects that belong both to A and to B.

That is,

$$A \cap B = \{x | (x \in A) \text{ and } (x \in B)\}$$

or, equivalently,

$$x \in A \cap B \Leftrightarrow (x \in A \wedge x \in B)$$

For instance, with reference to Example 2.2.10, $B \cap C = \{3\}$ and $\mathbb{Z} \cap A = \{3, 2, -2\}$.

Note that we are using here an obvious property of collections: that the objects that belong to two given collections form a collection. As mentioned above, we do not call this property an axiom, because we see it as absolutely evident. However, there would have been a problem if we had not added the principle of the empty collection (and even the principle of singletons). For example, suppose that all integers are objects of $\mathcal{U}$, and D is the collection

$$D = \{x \in \mathbb{Z} | x > 5\}$$

Then $A \cap D = \emptyset$, A being the collection of Example 2.2.10. And we know that this is indeed a collection by the principle of the empty collection.

The symbolic description of D above has a difference with previous uses of brackets. To be in accordance with those uses, we should have written here $D = \{x | (x \in \mathbb{Z}) \wedge (x > 5)\}$, which is correct. But the new notation is self-explanatory and its use in subsequent situations will not cause any confusion.

The collections A, D in the example have an empty intersection because they have no elements in common. In general, two collections A, B such that $A \cap B = \emptyset$ are called *disjoint*.

Definition 2.3.2. If A, B are collections, the *union* of A and B (written $A \cup B$) is the collection formed with all the objects that belong to at least one of the collections A, B. It will be denoted $A \cup B$.

Once again, we write this in symbols,

$$x \in A \cup B \Leftrightarrow (x \in A \vee x \in B)$$

or

$$A \cup B = \{x | x \in A \vee x \in B\}$$

Referring to the objects mentioned in Example 2.2.10, we may see that $B \cup C = \{2, -2, 3\}$. As for $A \cup \mathbb{Z}$ it should be noted that it is not equal to $\mathbb{Z}$: though it includes $\mathbb{Z}$ (and contains $\mathbb{Z}$ as well), $A \cup \mathbb{Z}$ has exactly four objects which do not belong to $\mathbb{Z}$.

2.4 Differences

Definition 2.4.1. Let A, B be two collections. We call *difference A minus B* (written $A \setminus B$) to the collection whose objects are the elements of A which do not belong to B.

In symbols,

$$A \setminus B = \{x \in A | x \notin B\}$$

Proposition 2.4.2. *If A, B are collections, then $A \setminus (A \setminus B) = A \cap B$.*

Proof. We use the principle of extension (Proposition 2.2.6): to prove the equality of two collections $X = Y$, we show $X \subseteq Y$ and $Y \subseteq X$. Since this is our first proof of a proposition, we try to be very careful and show all the details.

Suppose $x \in A \cap B$. Then $x \in A$ by Definition 2.3.1. If we had $x \in A \setminus B$, then $x \notin B$ by Definition 2.4.1. But, since $x \in B$ from the hypothesis, $x \in A \setminus B$ is not possible, hence $x \notin A \setminus B$. Thus, $x \in A$ and $x \notin A \setminus B$, whence $x \in A \setminus (A \setminus B)$ by the same definition. This proves the implication $x \in A \cap B \Rightarrow x \in A \setminus (A \setminus B)$, that is, $A \cap B \subseteq A \setminus (A \setminus B)$ by Definition 2.2.2.

Conversely, suppose $x \in A \setminus (A \setminus B)$. Then $x \in A$ and $x \notin A \setminus B$, once more by Definition 2.4.1. This entails that $x \in B$ by the same definition, and that $x \in A \cap B$ by Definition 2.3.1. So $A \setminus (A \setminus B) \subseteq A \cap B$. We now apply the principle of extension and deduce the equality in the statement. □

When $B \subseteq A$, the difference $A \setminus B$ is also called the *complement of* B *with respect to* A. In particular, the complement of a collection B with respect to the universe $\mathcal{U}$ (i.e., $\mathcal{U} \setminus B$) is called *the complement* of B and denoted B^c, so that $B^c = \{x | x \notin B\}$. And, even though this might lead to ambiguity, we sometimes speak of the *complement of* B to mean "the complement of B with respect to A", in a context where the collection A may be understood without risk. If this is so, we may even write B^c to denote $A \setminus B$. By using this notation, it follows from Proposition 2.4.2 that if $B \subseteq A$, then $(B^c)^c = A \setminus (A \setminus B) = A \cap B = B$, thus giving $(B^c)^c = B$.

Examples 2.4.3. (1) If A, B, C are the collections of Example 2.2.10, then $B \setminus C = \{2\}$, $B \setminus A = \emptyset$, $A \setminus \mathbb{Z} = \{\{2,3\}, \{\pi, 7\}, \mathbb{Z}, \begin{pmatrix} 1 & i \\ 0 & -i \end{pmatrix}\}$ and $\mathbb{Z} \setminus A$ contains all integers except $2, 3, -2$.

(2) In the universe $\mathcal{U}_0$ of Example 2.2.1, if A is the collection of all the finite lines, then $\mathcal{U} \setminus A$ is the collection of all the points.

Some properties of complements are worth noting.

Proposition 2.4.4. *(De Morgan laws) Given collections A, B, C such that $A \subseteq C$ and $B \subseteq C$, and considering the complements with respect to C, the following relations hold.*

(1) $(A \cup B)^c = A^c \cap B^c$.

(2) $(A \cap B)^c = A^c \cup B^c$.

Proof. (1) Let $x \in (A \cup B)^c$. By definition, $x \notin A \cup B$, so that $x \notin A$ and $x \notin B$. Since $x \in C$, we see that $x \in A^c$ and $x \in B^c$, from which $x \in A^c \cap B^c$. This shows $(A \cup B)^c \subseteq A^c \cap B^c$.

Now, let $x \in A^c \cap B^c$, so that $x \in A^c$ and $x \in B^c$. Therefore $x \notin A$ and $x \notin B$. From Definition 2.3.2 we see that $x \notin A \cup B$, but $x \in C$ and thus $x \in (A \cup B)^c$. This justifies the inclusion $A^c \cap B^c \subseteq (A \cup B)^c$ and completes the proof of (1) by the principle of extension.

(2) Exercise. □

We mention other properties concerning intersections and unions. They are known as the *distributive* properties.

Proposition 2.4.5. *Let A, B, C be collections. The following relations hold.*

(1) $A \cup (B \cap C) = (A \cup B) \cap (A \cup C)$.

(2) $A \cap (B \cup C) = (A \cap B) \cup (A \cap C)$.

Proof. (1) We use again the principle of extension and prove the double inclusion. Let us first show that $x \in (A \cup B) \cap (A \cup C) \Rightarrow x \in A \cup (B \cap C)$. To this end, we will use for the first time here a very useful method of proof, the *reductio ad absurdum*[2] (RAA). It works as follows: in order to prove a given property P, one makes the supposition that P does not hold; then we reason to obtain from that supposition that something impossible must occur. Then we may safely conclude that P does hold, because the contrary possibility is untenable.

Back to our matter, if we assume that the conditional proposition given above is not valid, then this means that there is some object u for which the implication fails, and hence $u \in (A \cup B) \cap (A \cup C)$ is true, but $u \in A \cup (B \cap C)$ is false. In particular, $u \notin A$; since $u \in (A \cup B) \cap (A \cup C)$, we see that $u \in B$ and $u \in C$, hence $u \in B \cap C$. But this contradicts the fact that $u \notin A \cup (B \cap C)$. By RAA, this shows that the starting implication is valid and our inclusion $(A \cup B) \cap (A \cup C) \subseteq A \cup (B \cap C)$ is proved.

We may prove the reverse inclusion by more direct means. Suppose $x \in A \cup (B \cap C)$. Either $x \in A$ or $x \notin A$. In the first case, $x \in A \cup B$ and $x \in A \cup C$, hence $x \in (A \cup B) \cap (A \cup C)$. If $x \notin A$, then $x \in B \cap C$, and thus $x \in B$ and $x \in C$. It follows that $x \in A \cup B$ and $x \in A \cup C$, and we get the same conclusion: $x \in (A \cup B) \cap (A \cup C)$ in any case.

(2) Exercise. □

Remark 2.4.6. The method RAA for proofs is related to a rule of elementary logic, the *contrapositive rule*. This says that an implication "$P \Rightarrow Q$" is valid if and only if the implication "$\neg(Q) \Rightarrow \neg(P)$" is valid. This can be deduced by constructing the truth table of the double implication $(P \Rightarrow Q) \Leftrightarrow ((\neg(Q)) \Rightarrow (\neg(P)))$, and observing that it gives the value True in every case; so that one of the implications is true if and only if the other is true.

[2]That is, reduction to absurd.

As for the link between this rule and the RAA, when using the RAA to prove P, we suppose that $\neg P$ holds and prove from this that a contradiction, say C, can be deduced. Hence we show $(\neg P) \Rightarrow C$. By the above rule (and bearing in mind that $\neg(\neg P)$ asserts the same as P), we see that we may deduce $(\neg C) \Rightarrow P$. But, since C is impossible, $\neg C$ must be true, from which it follows that P is indeed true, the same that we infer by using RAA.

2.5 The First Axioms

It is now time to introduce the term to which set theory owes its name.

Definition 2.5.1. A *set* is an object of the universe $\mathcal{U}$ which is a $\mathcal{U}$-collection. When a is a set and A is a collection such that $a \subseteq A$, we say that a is a *subset* of A.

Examples 2.5.2. Consider the universe $\mathcal{U}_0$ of Example 2.2.1, where the objects are the points and the (finite) lines in the plane. The sets in this universe are the lines, because they are $\mathcal{U}_0$-collections and $\mathcal{U}_0$-objects. However, the universe $\mathcal{U}$ in the same example has no sets.

Remark 2.5.3. In the pre-axiomatic times of set theory (roughly, the last 30 years of the 19th century) the concept of set was not dependent on an explicitly assumed universe, as in the preceding definition. But certainly a set was considered as a collection and as an object. And it was generally agreed upon that if a collection of objects was formed, then this collection was in turn an object, hence a set. As mentioned in Chapter 1, the *principle of comprehension* which could be derived from the axioms in Frege's theory, was a kind of formalization of this assumption; stating basically that, given a property that objects could satisfy or not, there is a set consisting of those objects that satisfy the property. However, this principle led to contradictions or *antinomies.* The most simple of these is Russell's antinomy: one considers the collection A of all sets x having the property $x \notin x$ and by the comprehension principle, A is a set. On the other hand, the definition of A means that for any set x, one has $x \in A \Leftrightarrow x \notin x$; applying this to the set A, one sees that $A \in A \Leftrightarrow A \notin A$; therefore none of the possibilities $A \in A$ or $A \notin A$ can happen, which is a contradiction. This and other paradoxes gave historically a motivation for the axiomatization of the theory, with explicit assumptions which restricted the concept of set.

Comment 2.5.4. Historically, the first steps of set theory were devoted to the study of infinite sets and to compare them as to their sizes. This seems to have been the main motivation for Cantor, the creator (with Dedekind) of set theory; and its development led to a general theory of infinities which can be seen as an extension of ordinary arithmetic. But besides having the power of taming the infinite, set theory has also the power of unifying mathematical concepts. This is done by considering the common nature of mathematical objects: surely, what they have in common is that they are objects and this concept of object is the first basic notion of the theory. But this concept alone is not enough to achieve unification of objects so different as lines, numbers, functions, surfaces, etc. So, there is a second necessary step to the unifying process, since it is impossible to pretend that all these are really homogeneous beings. The solution has been to create an artificial universe where all the mathematical objects have *replicas*, imitations that behave (hopefully) exactly as the true beings, but they are all homogeneous, because they are sets. So, we recover inside set theory: numbers as sets (though we know that numbers are not sets, this is just a simulation), functions as sets, points as sets and so on. This unification is indeed rewarding. Compare for example the pain taken by 17th or 18th century mathematicians to establish a clear concept of function, and the easiness of the concept obtained with set theory, though it might appear at first sight as artificial. In fact, the universe of set theory is a kind of duplicate for the real universe of mathematics.

In Zermelo's construction of set theory [25], the concept of set is similar to the one given here. Zermelo postulates a universe or domain of objects, but instead of considering the membership relation, he supposes an unspecified, abstract relation, denoted ϵ, between objects of the universe – this abstract character of the relation makes his theory a formal axiomatic theory. Then set is defined (with just one exception, the empty set) as any object x of the universe with the property that there is some object y of the universe such that $y \epsilon x$. On the other hand, the reader will note that our definition of set depends on the universe, and so it is far from being a definition which tries to capture the "true nature of the concept of set", if this exists.

Most of the axioms of set theory give information about the objects of the universe, and in particular, show which collections are sets. This is the case for the first two axioms which we now present.

Axiom 1 (of the empty set). The empty collection is an object of $\mathcal{U}$.

And thus it is a set by definition. As mentioned after Proposition 2.2.6, it will be written $\emptyset$ and called the *empty set*.

Axiom 2 (of pairs). If a, b are objects of the universe $\mathcal{U}$, distinct or not, then the collection $\{a, b\}$ formed with those objects is an object of $\mathcal{U}$.

And therefore a set. Since we may take $a = b$, this axiom gives in particular that $\{a\}$ is a set if a is an object. The set $\{a, b\}$ is called *a pair*, even if it contains only one element when $a = b$.

Examples 2.5.5. (1) We consider again the universe $\mathcal{U}_0$ of Examples 2.2.1, whose objects are points and lines. This universe does not satisfy axiom 1, since the empty collection has not been declared as an object. But we may simply add the object $\emptyset$ to $\mathcal{U}_0$ and get a new universe $\mathcal{U}_1$ whose objects are all the points and lines, plus $\emptyset$. This will be a 1-universe.

$\mathcal{U}_1$ is not a 2-universe: if A, B are two points, $\{A, B\}$ is not a point nor one line, so it is not a $\mathcal{U}_1$-object. We might add to the universe $\mathcal{U}_1$ all those pairs $\{A, B\}$, all the objects $\{l_1, l_2\}$ for l_1, l_2 lines, all the objects $\{l, A\}$ for l a line, A a point, all the pairs $\{\emptyset, A\}$, etc. But this would not be sufficient. Because we would have a collection, say $\{A, \{B, C\}\}$ which should be an object if axiom 2 has to be satisfied. This results in an endless process if we wanted to extend $\mathcal{U}_1$ to a 2-universe.

(2) Let us now start with an empty universe and call it $\mathcal{U}_0$ (not the same $\mathcal{U}_0$ as in the previous example). Now, $\mathcal{U}_0$ does not satisfy axiom 1, so let us call $\mathcal{U}_1$ to the universe which has only one object, $\mathcal{U}_1 = \{\emptyset\}$. This satisfies axiom 1, but not axiom 2. We may try by setting $\mathcal{U}_2 = \{\emptyset, \{\emptyset\}\}$, but still other collections with two objects are not sets, so axiom 2 is not fulfilled. What about $\mathcal{U}_3 = \{\emptyset, \{\emptyset\}, \{\{\emptyset\}\}, \{\emptyset, \{\emptyset\}\}\}$? But this is again insufficient: $\{\emptyset, \{\{\emptyset\}\}\}$ is not an object of $\mathcal{U}_3$. This series of universes is insatiable: the more pairs you add, the more it needs.

For the coming results and definitions, we assume that $\mathcal{U}$ is a 2-universe; i.e., axioms $1, 2$ are satisfied in the universe $\mathcal{U}$. We will use frequently *ordered pairs*, a concept which we now define.

Definition 2.5.6. Let $a, b \in \mathcal{U}$, distinct or not. The set $\{\{a\}, \{a, b\}\}$ is called the *ordered pair* of a and b, and will be denoted as $\langle a, b\rangle$.

This definition was proposed by Kuratowski. The idea is to identify which are the elements of the pair and which one comes first. That $\{\{a\}, \{a, b\}\}$ is a set can be deduced by applying axiom 2 three times. The order in the writing of the ordered pair $\langle a, b\rangle$ is significant: a is said the first element of the ordered pair $\langle a, b\rangle$, b is the second one.

Example 2.5.7. Suppose $a = b$ is an $\mathcal{U}$-object. Then $\langle a, b\rangle = \{\{a\}, \{a, a\}\} = \{\{a\}, \{a\}\} = \{\{a\}\}$, since $\{x, x\} = \{x\}$ for any object x by the principle of extension. So we see that an ordered pair can consist of just one object, $\{\{a\}\}$.

Ordered pairs is our first example of a phenomenon which is present at many points in set theory and was mentioned in Comment 2.5.4: we construct inside the theory a kind of copies of the objects and concepts of mathematics. We are surely accustomed to use ordered pairs in the obvious sense, as two mathematical objects given in an order, like the cartesian coordinates that identify points in the plane. But we need to view that concept as a set in order to make it homogeneous with all the objects of the universe. Thus $\langle a, b\rangle$ is not properly our usual idea of what an ordered pair ought to be, but it will be seen that it behaves inside the theory as a "normal" ordered pair. In fact, it satisfies the characteristic property of usual ordered pairs, as we see now.

Proposition 2.5.8. *Let a, b, c, d be arbitrary objects of $\mathcal{U}$. We have $\langle a, b\rangle = \langle c, d\rangle$ if and only if $a = c$ and $b = d$.*

Proof. If $a = c$ and $b = d$, then it is obvious that $\{\{a\}, \{a, b\}\} = \{\{c\}, \{c, d\}\}$ by the principle of extension. Conversely, let $\langle a, b\rangle = \langle c, d\rangle$. Suppose first that $a \neq c$. Then $\{a\} \neq \{c\}$ by extension, hence $\{a\} = \{c, d\}$ and $c \in \{a\}$, whence $c = a$, absurd.

So we conclude $a = c$. If $a = b$, then $\{a, b\} = \{a\}$ and $\{c, d\} = \{a\}$ so $d = a = b$. If $a \neq b$ then $\{a, b\} \neq \{c\}$ and thus $\{a, b\} = \{c, d\}$ so that $b \in \{c, d\}$ and $b = d$ in any case. □

As a consequence, observe that the ordered pairs $\langle a, b\rangle$ and $\langle b, a\rangle$ are different unless $a = b$.

By the axiom of pairs, the ordered pair $\langle a, b\rangle$ exists (i.e., is an object of $\mathcal{U}$). Though we cannot, for the moment, justify that the collection formed by three objects whatever is a set, three objects form always a new object called an *ordered triple*.

Definition 2.5.9. Let a, b, c be objects, distinct or not. The ordered pair $\langle\langle a, b\rangle, c\rangle$ is called the *ordered triple* (or, simply, triple) of a, b, c and is written $\langle a, b, c\rangle$. It is an object by the axiom of pairs.

It follows easily from Proposition 2.5.8 that two ordered triples $\langle a, b, c\rangle$ and $\langle a', b', c'\rangle$ are the same if and only if $a = a', b = b'$ and $c = c'$.

Remark 2.5.10. Given three objects a, b, c, there is also the object $\langle a, \langle b, c\rangle\rangle$, and this we will call a *pseudo-triple*. The pseudo-triple $\langle a, \langle b, c\rangle\rangle$ will be written as $[a, b, c]$. It might happen that a pseudo-triple is equal to a triple. This is precisely the case if we have an ordered pair of the form $\langle\langle a, b\rangle, \langle c, d\rangle\rangle$: it is both the triple $\langle a, b, \langle c, d\rangle\rangle$ and the pseudo-triple $[\langle a, b\rangle, c, d]$.

2.6 Some Remarks on the Natural Numbers

Natural numbers have already appeared in this text, for instance when we introduced the axiom of pairs by saying that **two** objects whatever form a set; or when an ordered **triple** has been defined. In these cases, numbers are not seen as objects of the universe. Instead, they are used in their intuitive, ordinary sense, as a tool in our reasoning about the universe. But it is important to make clear which are the properties of the ordinary natural numbers that we assume for its use as a reasoning tool.

In this sense, we admit that we understand what it means that a certain collection includes five objects or has more than twenty objects. Even very great numbers can be assumed as results of our basic intuition that any natural number n has a next one, namely $n + 1$. Besides, natural numbers have an obvious order: we start with $0, 1, 2, 3, \ldots$ and can always think of the next number to any given number. Along with this idea, we accept also that numbers can be used to order finite collections of objects, and thus a (finite) sequence of objects is again an intuitive concept, which may be identified by writing natural numbers as subscripts: $x_1, x_2, x_3, \ldots, x_n$.

An important issue concerns the relations between natural numbers and the pair of concepts finite-infinite. As said, we may imagine the natural numbers ordered in an endless sequence; and this sequence is *potentially infinite*, which means that it is impossible to reach its end. But, in our use of natural numbers as a tool, we do not think of the collection of all natural numbers as a complete entity, that is, as an *actual infinity*. On the other hand, infinite sets may appear from the axioms as objects of the universe, but the objective domain of the universe and the domain of our activity in studying the properties of the universe are different domains. In this second sense, we do not imagine that we may carry out infinitely many reasonings, constructions or, generally, actions.

We understand that a collection is *finite* when its objects can, in principle, be given in some order and the list reaches eventually an end. And thus each natural number n is finite in the sense that, starting with 0, then 1, then 2, and going on by adding one unit at a time, we arrive to the number n. Disregarding philosophical or epistemological objections (which may be justified), when we consider natural numbers in our reasoning activity, we will follow the usual uncritical view of numbers which is shared by almost everybody.

We finally add that, once a certain copy of the natural numbers will be introduced from the axioms as an object of our universe, we may imitate and even extend the above intuitive notions (sequences, number of objects of a collection, finite sets) by means of precise constructions inside the universe $\mathcal{U}$.

2.7 Cartesian Products

Definition 2.7.1. Let A, B be collections. Then the collection formed with all ordered pairs $\langle a, b\rangle$ such that $a \in A$ and $b \in B$, is called the *cartesian product* of A and B, and denoted $A \times B$.

Symbolically, $A \times B = \{\langle x, y\rangle | x \in A \text{ and } y \in B\}$. As with unions or differences, the fact that the cartesian product $A \times B$ is a collection is considered to be obvious from the intuitive concept of collection.

Example 2.7.2. (1) Let us assume that the real numbers are objects of $\mathcal{U}$; and $\mathbb{R}$ denotes the $\mathcal{U}$-collection of all real numbers. Ordered pairs of real numbers are the crucial device which allows us to identify points in the plane, as done in analytic geometry or *cartesian geometry*, as it is usually called. Thus inside set theory, the cartesian plane can be seen as the collection $\mathbb{R} \times \mathbb{R}$: each point is determined by the ordered pair of its coordinates, which are real numbers hence elements of $\mathbb{R}$.

(2) Let us consider again the example of the universe $\mathcal{U}_0$ in Examples 2.5.5, and recall that we called $\mathcal{U}_1$ to $\mathcal{U}_0 \cup \{\emptyset\}$. Suppose that by adding pairs of elements of $\mathcal{U}_1$, then pairs of the new elements, pairs of these new elements, and so on, it is possible to create a 2-universe $\mathcal{U}_\infty$ which contains $\mathcal{U}_0$. Call P to the collection of all points, L to the collection of all lines. We may describe a certain $\mathcal{U}_\infty$-collection C by writing $C = \{\langle l, p\rangle \in L \times P | p \in l\}$; this means that the elements of C are the ordered pairs consisting of one line and one point (this is the collection $L \times P$) so that the point is on the line ($p \in l$). The difference $(L \times P) \setminus C$ contains all ordered pairs $\langle l, p\rangle$ such that p is not on the line l.

As obvious properties of cartesian products we mention that $A \times \emptyset = \emptyset \times A = \emptyset$, whatever the collection A. If A, B are not empty and $A \neq B$ then $A \times B \neq B \times A$.

Another operation related to ordered pairs is given in the next definition. Since a new symbol appears in it, we first explain its use. The expression $\exists y$ introduced below (read "there exists y such that ..."), coupled with the condition that follows it, has the meaning of asserting that some object y satisfies the condition that is given thereafter, a condition usually written between parentheses for clarity. That is, with $\exists y\ (y \cdot y = 2)$ we assert that some object y satisfies the equation $y \cdot y = 2$. This would be a false assertion if our domain consists of the natural numbers, because no natural number is a square root ot 2, but it would be true in the domain of real numbers: there exist in fact two real numbers whose square is 2.

Definition 2.7.3. Let C be a collection. Then $\mathrm{D}(C) = \{x | \exists y\ (\langle x, y\rangle \in C)\}$ and $\mathrm{cD}(C) = \{y | \exists x\ (\langle x, y\rangle \in C)\}$. $\mathrm{D}(C)$ is called the *domain* of C and $\mathrm{cD}(C)$ is the *codomain* of C.

So, this says that the domain $\mathrm{D}(C)$ (respectively[3], the codomain $\mathrm{cD}(C)$) is formed with all the objects that appear as the first (resp., the second) element in some of the elements of C which is an ordered pair.

[3]This adverb is used in this and similar contexts with the intention of making possible two parallel and valid readings: one may read the sentence forgetting all the parentheses beginning with "respectively" – which may also appear abbreviated as "resp."; or alternatively one may read the sentence including only the items inside the parentheses.

Example 2.7.4. Let us consider the universe $\mathcal{U}_\infty$ mentioned in Example 2.7.2, and the collection C of that example. $\mathrm{D}(L \times P)$ is the collection L of all lines; because for every line l there is some point p on it; and therefore $\langle l, p\rangle \in C$ and $l \in \mathrm{D}(C)$. By similar reasons, $\mathrm{cD}(C) = P$.

In the same universe, let $l = AB \in \mathcal{U}_\infty$ be the line limited by the points A, B. Then $\{l\} \times P$ will be the collection of all ordered pairs $\langle AB, p\rangle$, p being any point. Then $H = (\{l\} \times P) \cap C$ contains all pairs $\langle AB, p\rangle$ where p is a point on the line $l = AB$. Thus $\mathrm{cD}(H)$ consists of all the points of the line l, that is, $\mathrm{cD}(H) = l$ and $H = \{l\} \times l$.

The symbol $\exists$ (which is always followed by a variable, like in $\exists x$ or $\exists y$) is the *existential quantifier*. We shall also employ the *universal quantifier* $\forall$; thus $\forall x$ (read "for all x"), has the corresponding meaning: every object x satisfies the condition that comes after it. Both quantifiers are frequently used limiting the domain where the variable is to take values: like this, "$\exists x \in C$", when C is a given collection, has the meaning that some element of C satisfies the condition which is given subsequently. The same is applied in the case of a universal quantifier "$\forall x \in C\ (P)$" says: every element of C satisfies the condition (P).

We have already defined ordered triples. It should be clear how we could also define ordered quadruples, and how this same process can be continued. We summarize this construction by indicating that once we will have defined ordered n-tuples, we may define ordered $n + 1$-tuples by stating that the ordered $n + 1$-tuple of the objects $a_1, \ldots, a_{n+1}$ is the ordered pair $\langle\langle a_1, a_2, \ldots, a_n\rangle, a_{n+1}\rangle$, where $\langle a_1, \ldots, a_n\rangle$ denotes of course the ordered n-tuple of those n objects. For completion, we also define a 1-tuple as just one object, $\langle u\rangle = u$.

We know that the cartesian product of the collections A and B is written $A\times B$. Similarly, the analogous collection of ordered triples taken from A, B, C will be written $A\times B\times C$ $(= (A\times B)\times C)$, and so on. The cartesian product $A\times A$ can be represented as A^2, and the notation A^3, A^n serves for the collections $A \times A \times A$ or $A \times A \times \cdots \times A$. Note that $A^3 = A^2 \times A$ and $A^{n+1} = A^n \times A$. But in general $A^3 \neq A \times A^2$.

2.8 Exercises

1. All the objects that are referred to in the items that follow are supposed to be objects of the universe $\mathcal{U}$. Describe in symbolic form (i.e., by using the symbol {.....}) the following collections:

   a) All the natural numbers that are greater than 9 and divisors of 55.

   b) All the numbers of the leap years from 1701 till today.

c) All the natural numbers which are powers of 3 with an integer exponent.

d) All the real numbers in the interval $(-6, 10)$ which are integer multiples of π (The collection of all real numbers will be denoted as $\mathbb{R}$).

e) All the natural numbers that divide 1000 and are multiples of 15.

In the list that follows, it is assumed that all the objects and collections defined are elements of the universe $\mathcal{U}$. Exercises 2 to 8 refer to these objects.

$U = \{2, 4, 6, 8, \dots\}, \quad A = \{1, 3, 5, 7, \dots\}, \quad B = \{5, 10, 15, 20, \dots\}$

$X = \{2, \{2, 3\}, \{2, 3, 4\}, \{2, 3, 4, 5\}, \dots\}, \quad C = \{6, 12, 18, 24, 30, \dots\}$

$Y = \{2, \{2, 4\}, \{2, 4, 6\}, \dots\}, \quad D = \{\{1, 3\}, \{3, 5\}, \{5, 7\}, \dots\}$

2. Explain which of the relations that follow are valid and which are not:

$$2 \notin U, \quad 2 \in Y, \quad 5 \notin X, \quad C \in U, \quad 1200 \notin C, \quad \{3, 5\} \in A$$

3. Same question for the following:

$$2 \in X \wedge 4 \in Y, \quad 5 \in A \wedge 120 \notin A, \quad \{15, 17\} \in D \vee 15 \in D$$

4. Same question for the following:

$$X \subseteq U, \quad Y \nsubseteq U, \quad Y \subseteq X, \quad B \nsubseteq A, \quad C \subseteq Y, \quad D \subseteq A$$

5. Explain which of the following assertions are valid and which are not:

$$6 \in C \Rightarrow 135 \in B, \quad \{2, 4\} \in X \Rightarrow \pi \in A, \quad 25 \in B \Rightarrow 25 \in C$$

6. Same question for the following:

$$x \in A \Rightarrow x \notin B, \quad x \in B \Rightarrow x \in Y, \quad x \in Y \Rightarrow x \notin X$$

7. Same question for the following:

$$(x \in X \wedge x \notin \{2, 4\}) \Rightarrow x \in Y, \quad x \in C \Rightarrow x \in U$$

8. Describe the following collections:

$$A \cap B, \quad A \cup B, \quad B \cap C, \quad A \cup C, \quad A \cap X, \quad U \setminus Y, \quad Y \setminus U$$

9. Assuming that A, B are $\mathcal{U}$-collections, find the collection $[(A \setminus B) \cup (B \setminus A)] \cap A$.

10. For the collection A defined in Example 2.2.10, identify the collection $(A \cup \mathbb{Z}) \setminus \mathbb{Z}$.

11. Let $\mathcal{U}$ be a universe where the usual natural numbers are $\mathcal{U}$-objects. Let A be the collection of all numbers that are multiple of 5 and B the collection of all even numbers which are ≥ 20. Give the explicit list of all the elements of each of the sets $A \cup B, A \cap B, A \setminus B$ and $B \setminus A$ which are ≤ 35.

12. Suppose that the real numbers are $\mathcal{U}$-objects. Let $A = \{x \in \mathbb{R} | \sin x = 1\}$, and $B = \{x \in \mathbb{R} | \cos x = 0\}$. Find $A \cap B, A \cup B, B \setminus A$.

13. Construct the truth table that has been mentioned in Remark 2.4.6.

14. Suppose that $x \in \mathcal{U}$ and $\mathcal{U}$ is a 2-universe. Prove that $x \notin x \Rightarrow x \neq \{x\}$, and $x \neq \{x\} \Rightarrow \{x\} \notin \{x\}$.

15. In the universe $\mathcal{U}$ of Example 2.2.1, suppose we have a coordinate system, so that every point is given by its coordinates. Let A be the collection of all the points whose x-coordinate is an even integer, B the collection of those whose y-coordinate is an odd integer. Describe the grid formed by the points of $A \cap B$. How is this collection of points related to the collection $A \times B$? How many points of $A \cap B$ are there in the interior of a square centered at the origin and with side equal to 10?

16. Let $\mathcal{U}_0$ be the universe of Example 2.2.1, and consider a rectangle of sides AB, BC, CD, DA. If x is the collection of all the points of the rectangle and y is the collection whose elements are the two lines that are the diagonals AC, BD, what is $x \cap y$? Is the center of the rectangle an element of $x \cup y$? What is the intersection of the $\mathcal{U}_0$-collections x and AC?

17. Let A, B be collections. Prove that $A \cap (B \cup A) = A = A \cup (B \cap A)$. Infer from this that if A, B, C are collections such that $A \cap B = A \cap C$ and $A \cup B = A \cup C$, then $B = C$.

18. Suppose that $A, B, C \subseteq \mathcal{U}$ are collections. Complements are taken with respect to $\mathcal{U}$. Prove the following properties:

    a) $A \cup B = B \Leftrightarrow A \cap B = A$.

    b) $A^c \cup B = (A \cap B^c)^c$.

    c) $A \subseteq (B \cup C)^c \Rightarrow A \cap B = A \cap C = \emptyset$.

19. Prove item (2) of Proposition 2.4.4 in two ways: first, deduce it from item (1) by using the property $(X^c)^c = X$, for any collection $X \subseteq C$. Then, give a direct proof similar to the proof of (1).

20. Suppose that all real numbers are objects and consider the collections of real numbers

$$A = \{x \in \mathbb{R} | x^2 > 4\}, \quad B = \{x \in \mathbb{R} | 1 \leq |x| \leq 3\}$$

Identify these collections as intervals or unions of intervals of the real line. Do the same for the collections $A \cap B, A \cup B, A \setminus B, B \setminus A$.

21. Prove item (2) of Proposition 2.4.5 in two ways: first, show the property $X^c = Y^c \Rightarrow X = Y$ and deduce (2) from (1) and this property. Then, give a direct proof similar to that of (1).

22. Suppose A, B, C are collections. For each of the assertions that follow, prove it if it is true and give a counterexample (i.e., an example where the assertion is not true) if it is false.

    (i) $A \cap (B \setminus C) = (A \cap B) \setminus (A \cap C)$.

    (ii) $A \cup (B \setminus C) = (A \cup B) \setminus (A \cup C)$.

    (iii) $(A \cup B) \setminus C = (A \setminus C) \cup (B \setminus C)$.

    (iv) $(A \times B)^c = A^c \times B^c$, where complements are taken with respect to the class of all ordered pairs.

23. Let us define the *symmetric difference* of two collections A, B, as follows:

$$A \triangle B = (A \setminus B) \cup (B \setminus A)$$

Suppose A, B, C are collections included in a collection U. Prove that:

    (i) $(A \triangle B)^c = (A \cap B) \cup (A^c \cap B^c)$.

    (ii) $A \triangle (B \triangle C) = (A \triangle B) \triangle C$.

    (Complements are taken with respect to U).

24. Suppose A, B, C are collections with finitely many elements, and let us write $|X|$ to denote the number of elements of the collection X. Prove:

    (i) $|A \cup B| = |A| + |B| - |A \cap B|$.

    (ii) $|A \cup B \cup C| = |A| + |B| + |C| - |A \cap B| - |A \cap C| - |B \cap C| + |A \cap B \cap C|$.

25. There are 40 students in a class. Each of them has to choose at least two courses from the following three: Latin, Greek, Arabic. There are 33 students in the Latin course and 30 in the Greek course. Also 9 students follow all three courses. How many students follow the Arabic and Greek courses simultaneously?

26. Prove that for non-empty collections A, B, $A = B$ if and only if $A \times B = B \times A$.

27. For $\mathcal{U}$-collections A, B, C, D explain when the equality $A \times B = C \times D$ holds.

28. Justify the assertion after Definition 2.5.9.
29. Write a triple $\langle a, b, c \rangle$ as a set, giving its elements explicitly as sets (i.e., do not write an element as $\langle a, b \rangle$, for instance). Which are the elements of the set $\langle a, a, a \rangle$? Justify that $\langle a, a, a \rangle \neq [a, a, a]$, assuming that $a \neq \{\{a\}\}$.
30. It has been seen in the text that with 3 objects a, b, c, one can form a triple $\langle a, b, c \rangle$ and a pseudo-triple $[a, b, c]$, which are two different ordered pairs with the same elements in the same order. In how many ways can 4 objects a, b, c, d be arranged as different ordered pairs while keeping the same order between them?
31. More generally, if we write a_n to denote the number of ways in which n objects may be arranged as different ordered pairs as in the preceding exercise, find a formula that gives a_{n+1} in terms of the values a_k for $k \leq n$.

3

Classes

Throughout this chapter, $\mathcal{U}$ is assumed to be a 2-universe.

3.1 The Formation of Classes

3.1.1 Class sequences

We all know that the Equator is an imaginary line in the Earth's surface (assuming that the Earth is a sphere, it would be the great circle which is everywhere equidistant from both poles). It is not a real component of the Earth itself, but a concept scholars have created for the study of the Earth. Likewise, in set theory we accept that the objects of the universe $\mathcal{U}$ are the real entities we want to study, but for this, we may create other entities which assist in studying our subject. These entities will be a certain type of $\mathcal{U}$-collections that are called *$\mathcal{U}$-classes* (*classes* when the universe $\mathcal{U}$ is understood).

Approximately, classes are those collections that have a concrete description. Classes are constructed in an explicitly defined way, and this makes it possible to handle them effectively. The main idea about constructing classes is to guarantee that the basic operations with collections that have been seen in Chapter 2 (unions, intersections, cartesian products, differences, domain, codomain) produce classes when applied to classes. It turns out that if three of these operations (products, differences, domains) have that property of producing classes from classes, the others will also have that same property. Before giving the formal definition, let us see some examples of how the construction of classes can be carried out. First, a definition.

Definition 3.1.1. The following collections are called *basic classes*:
The collection $\mathcal{M}$ of all ordered pairs $\langle a, b\rangle$ of $\mathcal{U}$ such that $a \in b$.
The collection $\mathcal{E}$ of all ordered pairs $\langle a, b\rangle$ of $\mathcal{U}$ such that $a = b$.

$\mathcal{M}$ is called the *membership class* and $\mathcal{E}$ is the *equality class*.

Example 1. The universe $\mathcal{U}$ is a class: it can be constructed from the basic classes by applying the above mentioned operations.

(1) $\mathcal{E}$
(2) $\mathrm{D}(\mathcal{E})$. This is the universe: for every object x, $\langle x, x\rangle \in \mathcal{E}$ and $x \in \mathrm{D}(\mathcal{E})$.

DOI: 10.1201/9781003449911-3

Example 2. The collection of all ordered pairs; that is, $\{\langle x, y\rangle | x, y \in \mathcal{U}\}$.
(1) $\mathcal{E}$
(2) $\mathrm{D}(\mathcal{E}) = \mathcal{U}$
(3) $\mathcal{U} \times \mathcal{U}$

Example 3. Suppose A, B are classes; that is, they are collections and there exist (finite) constructions giving A and B. Then $A \cup B$ can also be constructed by a sequence of classes.
(1) There will be some lines which end with the class A, by the hypothesis.
(2) Some more lines arrive to the class B.
(3) There is also a construction, as we have seen, for the universe $\mathcal{U}$.
(4) $\mathcal{U} \setminus A$. Let us denote this class as X.
(5) $\mathcal{U} \setminus B$. Let us call it Y.
(6) $X \setminus Y$. Let us call it Z.
(7) $X \setminus Z$. Let us call it W.
(8) $\mathcal{U} \setminus W$.
This class is $A \cup B$, as the reader may check. As a suggestion for the checking, imagine an element not belonging to $A \cup B$, and see through the construction whether it belongs to X, Y, Z, W and then to $\mathcal{U} \setminus W$. Analogously for an element of A, then for an element of B not belonging to A.

The main idea in the construction of classes is simple: you start with the two basic classes and then apply repeatedly (but finitely many times) any of the three operations we mentioned: difference, cartesian product, domain. However, there are still three facts to be taken into account to fully understand the definition that will follow.

- Classes are certain $\mathcal{U}$-collections, and sets are $\mathcal{U}$-collections too – and also $\mathcal{U}$-objects. We want sets to be classes because we want to be able to consider operations involving sets and other classes. One of the rules for defining classes will establish this, i.e., that sets are classes.

- The domain of a class is a class. Obviously, we want a similar property for the codomain. The simplest way to do so is by accepting the rule that changing the order of the elements in a class of ordered pairs will also produce a class. In fact, it will be easier to obtain this property from related properties for ordered triples; and thus two of the rules will refer to such changes in ordered triples.

- It will be practical to attach a kind of code for each step in the formation of a class. The code will include which of the rules has been applied in the step, and to which class already constructed has that rule been applied. Hence the code will consist of one digit identifying the rule, and two more digits identifying the (possibly) two classes on which the operation has been executed – because difference and product apply to two classes. From the codes, it will be possible in many cases to reconstruct the steps. For instance, consider our Examples above. In Example 1, the code will have to give: (1) identification of the basic class $\mathcal{E}$; (2) identification of the operation "domain" applied to line 1. In Example 2, we would have: (1) and (2) same as in Example 1; (3) cartesian product applied to lines $2, 2$.

A *class sequence* is the sequence of the terms in the construction of a class.

Definition 3.1.2. A *class sequence* is a finite sequence of collections $C_1, C_2, \dots, C_n$ together with a label $(e(k), u(k))$ for each $k = 1, 2, \dots, n$ where $e(k)$ is a positive integer ≤ 7, and the following conditions are satisfied for each k.

(i) If $e(k) = 1$, then C_k is a set, and $u(k) = (0, 0)$.

(ii) If $e(k) = 2$, then C_k is either $\mathcal{M}$ or $\mathcal{E}$, and $u(k) = (1, 0)$ or $u(k) = (0, 1)$, respectively.

(iii) If $e(k) = 3$, then $u(k) = (i, j)$ with $i, j < k$ and $C_k = C_i \setminus C_j$.

(iv) If $e(k) = 4$, then $u(k) = (i, j)$ with $i, j < k$ and $C_k = C_i \times C_j$.

(v) If $e(k) = 5$, then $u(k) = (i, i)$ for some $i < k$ and C_k is the domain of the collection C_i, $C_k = \mathrm{D}(C_i)$ – see Definition 2.7.3.

(vi) If $e(k) = 6$, then $u(k) = (i, i)$ for some $i < k$ and C_k is the collection of all triples $\langle a_1, a_2, a_3 \rangle$ such that $\langle a_2, a_1, a_3 \rangle \in C_i$.

(vii) If $e(k) = 7$, then $u(k) = (i, i)$ for some $i < k$ and C_k is the collection of all pairs $\langle a_1, \langle a_2, a_3 \rangle \rangle$ such that $\langle a_1, a_2, a_3 \rangle \in C_i$ (such pairs were called pseudo-triples in Chapter 2; and written $[a_1, a_2, a_3]$).

A *class* is a collection that is one of the terms of some class sequence. If $e(k) = j$ we say that C_k has been obtained by applying rule j.

If A_1, A_2 are classes and $A_1 \subseteq A_2$, we say that A_1 is a *subclass* of A_2. Note that a class may be constructed through different class sequences. As a trivial example, $\emptyset$ is a class that is obtained through the one-line sequence $\emptyset, (1, (0, 0))$; and also through the lines $1 : \mathcal{E}, (2, (0, 1))$, $2 : \mathcal{U}, (5, (1, 1))$, $3 : \emptyset, (3, (2, 2))$. As another example, the intersection can be constructed as a class by applying the identity $A \cap B = A \setminus (A \setminus B)$; finding a different class sequence giving $A \cap B$ will be one of the exercises at the end of the chapter. On the other hand, Definition 3.1.2 suggests that classes are those collections that are capable of being described. We will see later how this observation can be made more precise.

The fact that the construction of a class may include any set as a term is an exception to our observation that classes are collections that we may describe. This is forced because we want every set to be a class, but there is also some interest in considering classes whose construction does not include that type of terms.

Definition 3.1.3. A class sequence is called a *neat class sequence* when rule 1 is not employed in any line of the sequence. A *neat class* is a class that is one of the terms of a neat class sequence.

We remark that a neat class sequence is determined by the sequence of the labels of its terms: knowing these labels it is possible to faithfully reconstruct the sequence. The three-steps sequence above for $\emptyset$ is an example; and the same applies for Examples 1, 2, 3 given before Definition 3.1.2.

3.1.2 Obtaining new classes

Many collections can now be shown to be classes, and we next give several instances where it is shown that certain operations on classes give classes. The following is an immediate consequence of Definition 3.1.2.

Proposition 3.1.4. *If A, B are classes, then $A \setminus B, A \times B$, $D(A)$ are classes.*

Proof. By hypothesis, there are class sequences leading to A and to B. We may form a new class sequence containing them both, and continue the sequence by rule 3, 4 or 5, to obtain the classes in the statement. □

Proposition 3.1.5. *If A, B are classes, then $A \cap B$, $A \cup B$ and A^c (the complement of A with respect to the universe $\mathcal{U}$) are classes.*

Proof. $A \cap B = A \setminus (A \setminus B)$ by Proposition 2.4.2. Therefore the same argument of the preceding proof shows that, applying rule 3 twice, we obtain that $A \cap B$ is a class.

Rule 3 gives that $A^c = \mathcal{U} \setminus A$ is a class if A is, because $\mathcal{U}$ is a class as seen in the examples above. Finally, we know from the De Morgan laws, that $A \cup B = (A^c \cap B^c)^c$. As before, we may glue together the class sequences leading to A and to B and continue, getting $A^c, B^c, A^c \cap B^c$ (by the first part of this proof) and $(A^c \cap B^c)^c = A \cup B$. □

Proposition 3.1.6. *The collection of all ordered pairs is a class.*

Proof. It is $\mathcal{U} \times \mathcal{U}$. □

Proposition 3.1.7. *The collection of all ordered triples and the collection of all pseudo-triples are classes. More generally, the collection of all n-tuples is a class for $n \geq 3$.*

Proof. The collection of all ordered triples (respectively, pseudo-triples) is $(\mathcal{U} \times \mathcal{U}) \times \mathcal{U}$ (resp., $\mathcal{U} \times (\mathcal{U} \times \mathcal{U})$), hence it is a class by rule 4 and Proposition 3.1.6. Then, one can prove that the collection of all 4-tuples is a class by seeing it as the cartesian product $C \times \mathcal{U}$ where C is the class of triples; and similarly the property may be proven for any n we want. □

Proposition 3.1.8. *If C is a class, then $C' = \{\langle y, x\rangle | \langle x, y\rangle \in C\}$ is a class.*

Proof. Since C is a class and $\{\emptyset\}$ is a set, the collection of all triples $\langle x, y, \emptyset\rangle$ such that $\langle x, y\rangle \in C$, is a class by rules 1 and 4. Then rule 6 entails that the triples $\langle y, x, \emptyset\rangle$ with $\langle x, y\rangle \in C$ form a class. The domain of this class is C' and it is a class by rule 5. □

Corollary 3.1.9. *If C is a class, then its codomain $cD(C)$ is a class.*

Proof. Let C' be the class obtained from C as in Proposition 3.1.8. Then $\mathrm{cD}(C) = \mathrm{D}(C')$ is a class by rule 5. □

Proposition 3.1.10. *The collection of all sets is a class, that will be denoted as $\mathcal{S}$.*

Proof. We start with the basic class $\mathcal{M}$, and apply Proposition 3.1.8 and then rule 5 to obtain the class C of all sets which are not empty. Then $C \cup \{\emptyset\}$ is a class by the axiom of pairs and Proposition 3.1.5, and it is the class of all sets. □

We know from rule 1 that every set is a class, but we have not yet addressed the question whether there are classes that are not sets. In this connection, the ancient Russell's paradox is now Russell's example:

Proposition 3.1.11. *Let $\mathcal{S}_1 = D(\mathcal{E} \cap ((\mathcal{S} \times \mathcal{S}) \setminus \mathcal{M}))$. Then $\mathcal{S}_1$ is a class that is not a set.*

Proof. It is clear from Proposition 3.1.10 that $\mathcal{S}_1$ is a class. Also, for any set s the equivalence $s \in \mathcal{S}_1 \Leftrightarrow s \notin s$ follows from the definition of $\mathcal{S}_1$. Now, if $\mathcal{S}_1$ is a set, we conclude that $\mathcal{S}_1 \in \mathcal{S}_1 \Leftrightarrow \mathcal{S}_1 \notin \mathcal{S}_1$. But this entails that both $\mathcal{S}_1 \in \mathcal{S}_1$ and $\mathcal{S}_1 \notin \mathcal{S}_1$ are impossible. Since some of these two possibilities must be true, the conclusion follows by RAA. □

Classes that are not sets, like $\mathcal{S}_1$, are called *proper classes.*

Example 3.1.12. Once we have constructed a proper class, namely $\mathcal{S}_1$, we may construct many more. For instance, since $\emptyset \notin \emptyset$, $\emptyset \in \mathcal{S}_1$ and the class $\mathcal{S}_1 \setminus \{\emptyset\} \neq \mathcal{S}_1$. It happens that this new class is in fact a proper class, though the proof is a bit longer than that of Proposition 3.1.11. Going on, the principle of extension tells us that $\emptyset \neq \{\emptyset\}$, hence $\{\emptyset\} \notin \{\emptyset\}$ and $\{\emptyset\} \in \mathcal{S}_1$. As before, $\mathcal{S}_1 \setminus \{\{\emptyset\}\}$ is a new class, different from $\mathcal{S}_1$ and from $\mathcal{S}_1 \setminus \{\emptyset\}$, and again it turns out that it is a proper class. And we may proceed in this way to find (potentially) infinitely many different proper classes.

Comment 3.1.13. Let us pause here to ponder some aspects of the results obtained so far about classes. For instance, compare Corollary 3.1.9 and the following property of sets, easily proven from the axiom of pairs: if s is a set, then $\{\{s\}\}$ is a set. Though the sentences conveying these properties have the same form, they are totally different in their function and significance. The second one is a simple assertion about the universe. It states what we may call *a fact* concerning the objective level of the objects of the universe. On the contrary, Corollary 3.1.9 does not give any information about the universe, since classes are not objects of the universe, just our creations. It concerns our activity of forming classes, and thus it expresses a property of that activity: if we form a certain class A, then also the collection $\mathrm{cD}(A)$ can be formed as a class. So, properties about the universe and properties about our activity of forming classes give two different and, for now, basically unrelated levels of our discourse. It is true that we obtain the material for our construction of classes from the universe, but there is no feedback in the other direction, and, in what concerns the results given in this section, our activity has no impact on the properties of the universe. Only after the introduction of new axioms will we be able to obtain some knowledge about the objective level of the universe from properties of the level of our action about classes.

The difference between these two levels of our study carries with it a difference in our understanding of the results of each level. In that of the universe, we try to observe and prove facts about the objects; and we accept, as a fact, that there may be infinitely many objects. However, when we come to the study of properties of classes, we cannot accept the actual existence of infinitely many classes, because this would mean that our action has created infinitely many entities, while this can only be understood potentially: we may form more and more classes, but we can never put an end to this process. As a rule, when we consider a result like that of Corollary 3.1.9, we will understand it as hiding a potential infinity of concrete results: if we have obtained a certain class A, then we may deduce that $\mathrm{cD}(A)$ is a class. But in practice this can only be done finitely many times. This applies to all the results in this section which state, apparently, a property shared by all classes – of course, in Propositions 3.1.6 or 3.1.7, there is only one particular collection that is shown to be a class, and there is no question about infinity.

Rules 6 and 7 show that from a class of triples, it is possible to obtain the corresponding class of pseudo-triples; or inverting the first two terms and obtain again a class. These rules are enough to show that it is also possible to extend these properties to n-tuples for any $n > 3$ that we need; and also that different manipulations with a class of n-tuples produce another class. The proofs of these results are a bit technical, but quite useful. Besides, what is really necessary to understand in a first reading is: the idea of constructing classes, the instances of classes seen above and a few more, and the fact that any formula identifies (basically) a class.

Rule 6 is self-inverse: if rule 6 is applied twice in succession to a given class C, the result will be the class consisting of all triples of the initial class C. We see next that rule 7 has also an inverse rule.

Proposition 3.1.14. *Let C be a class and call $C^* = \{\langle a_1, a_2, a_3\rangle | [a_1, a_2, a_3] \in C\}$. Then C^* is a class.*

Proof. Note that the pseudo-triples of C form a class by Propositions 3.1.7 and 3.1.5. We show how C^* can be obtained from the class of pseudo-triples of C through transformations allowed by rules 6,7 (and their consequences), by describing the effect of those transformations on a general pseudo-triple $[a_1, a_2, a_3]$ of C.

$\langle a_1, \langle a_2, a_3\rangle\rangle \rightarrow \langle\langle a_2, a_3\rangle, a_1\rangle \rightarrow \langle a_3, a_2, a_1\rangle \rightarrow \langle a_3, \langle a_2, a_1\rangle\rangle \rightarrow \langle a_2, a_1, a_3\rangle \rightarrow \langle a_1, a_2, a_3\rangle$, where we have applied successively Proposition 3.1.8, rules 6, 7, Proposition 3.1.8 again, and rule 6. □

Proposition 3.1.15. *Let C be a class, and let σ be any permutation of three elements, $\sigma \in S_3$. Then the collection*

$$C_\sigma = \{\langle a_{\sigma(1)}, a_{\sigma(2)}, a_{\sigma(3)}\rangle | \langle a_1, a_2, a_3\rangle \in C\}$$

is a class.

Proof. By Propositions 3.1.7 and 3.1.5, the triples of C form a class, so we may assume that C is a class of triples. The transposition $\tau = (1\ \ 2)$ takes C to C_τ, which is a class by rule 6. We look now for other permutations ρ with the same property; for this, we show how C_ρ can be obtained from C through transformations using rules 6, 7, as we did in the proof of Proposition 3.1.14.

$\langle a_1, a_2, a_3\rangle \rightarrow \langle a_1, \langle a_2, a_3\rangle\rangle \rightarrow \langle a_2, a_3, a_1\rangle$, where successively rules 7 and Proposition 3.1.8 have been applied. We thus see that the cycle $\rho = (1\ \ 2\ \ 3)$ gives a class C_ρ. By Proposition 1.7.2, every permutation of $\sigma \in S_3$ is a product of $(1\ \ 2)$ and $(1\ \ 2\ \ 3)$. This entails that C_σ can be obtained from C by successive applications of these two operations, thus giving a class. □

Example 3.1.16. Since $\mathcal{U}$ is a class, $\mathcal{U} \times \mathcal{M}$ is a class, according to Proposition 3.1.4. It is a class of pseudo-triples, so that the corresponding ordered triples form a class by Proposition 3.1.14. This is the class $\{\langle x, y, z\rangle | y \in z\}$. The intersection of this class and $\mathcal{M} \times \mathcal{U}$ gives the class $\{\langle x, y, z\rangle | x \in y \wedge y \in z\}$. By Proposition 3.1.15, the collection $\{\langle x, y, z\rangle | x \in z \wedge z \in y\}$ is a class. It is obtained from the preceding class by the transposition $(2\ \ 3)$.

Extensions to n-tuples of Proposition 3.1.15 will be presented in the exercises. On the other hand, a similar extension for Proposition 3.1.14 is readily found. For this, let us define pseudo-n-tuples (for $n > 3$) by letting $[x_1, x_2, \ldots, x_n] = \langle x_1, [x_2, \ldots, x_n]\rangle$.

Lemma 3.1.17. *For $n \geq 3$, let C_1 be a collection of n-tuples and let $C_2 = \{[x_1, x_2, \ldots, x_n] | \langle x_1, x_2, \ldots, x_n\rangle \in C_1\}$. Then C_1 is a class if and only if C_2 is a class.*

Proof. For $n = 3$ this follows by rule 7 of Definition 3.1.2 and Proposition 3.1.14. We now fix any $n \geq 4$, and assume C_1 is a class. C_1 is a class of triples $\langle\langle x_1, \ldots, x_{n-2}\rangle, x_{n-1}, x_n\rangle$. By rule 7 of Definition 3.1.2, $C_1' = \{\langle\langle x_1, \ldots, x_{n-2}\rangle, \langle x_{n-1}, x_n\rangle\rangle\}$ (for the $\langle x_1, \ldots, x_n\rangle \in C$) is a class, which in turn we see as a class of triples $\{\langle\langle x_1, \ldots, x_{n-3}\rangle, x_{n-2}, \langle x_{n-1}, x_n\rangle\rangle\}$. We apply again rule 7 and obtain that (again for the elements $\langle x_1, \ldots, x_n\rangle \in C$) $C_2' = \{\langle\langle x_1, \ldots, x_{n-3}\rangle, \langle x_{n-2}, \langle x_{n-1}, x_n\rangle\rangle\rangle\} = \{\langle\langle x_1, \ldots, x_{n-3}\rangle, [x_{n-2}, x_{n-1}, x_n]\}$ is a class. By continuing in this way, we finally show that C_2 is a class.

The converse follows analogously, by using Proposition 3.1.14 instead of rule 7 of Definition 3.1.2. □

This allows us to obtain the following technical, but useful result.

Lemma 3.1.18. *Let C be a class of ordered n-tuples (with $n \geq 1$). Then the following collections of $(n+1)$-tuples are classes: (a) $\{\langle x_1, \ldots, x_n, y\rangle | \langle x_1, \ldots, x_n\rangle \in C\}$; (b) $\{\langle y, x_1, \ldots, x_n\rangle | \langle x_1, \ldots, x_n\rangle \in C\}$; (c) $\{\langle x_1, \ldots, x_{n-1}, y, x_n\rangle | \langle x_1, \ldots, x_n\rangle \in C\}$; (d) $\{\langle x_1, y, x_2, \ldots, x_n\rangle | \langle x_1, \ldots, x_n\rangle \in C\}$. Moreover, the collections $C_1 = \{\langle x_1, \ldots, x_n\rangle | \exists y\, (\langle y, x_1, \ldots, x_n\rangle \in C)\}$ and $C_2 = \{\langle x_1, \ldots, x_n\rangle | \exists y\, (\langle x_1, \ldots, x_n, y\rangle \in C)\}$ are classes if C is a class of $(n+1)$-tuples.*

Also, if C is a class of pseudo-n-tuples, the same operations as above on the elements of C give classes of pseudo-$(n+1)$-tuples.

Proof. Case (a) follows from rule 4 of Definition 3.1.2 by taking $C \times \mathcal{U}$. For (b) and (d), we first convert C into the corresponding class C' of pseudo-n-tuples by Lemma 3.1.17 and take $\mathcal{U} \times C'$. The class in (b) is obtained by converting $\mathcal{U} \times C'$ into a class of $(n+1)$-tuples by the same lemma. For (d), we see $\mathcal{U} \times C'$ as a class of pseudo-triples, apply the analogous of rule 6 for pseudo-triples (pseudo-triples become triples by Proposition 3.1.14, we apply rule 6, then come again to pseudo-triples), then convert the class to one of triples again by Lemma 3.1.17. (c) is obtained from the class $C \times \mathcal{U}$ by considering it as a class of triples and applying Proposition 3.1.15 to change the order of the last two elements. C_2 is the domain of the class C so that it is arrived at by rule 5. Finally, we obtain C_1 as a class by converting the class C to the corresponding class of pseudo-$(n+1)$-tuples, taking its codomain by Corollary 3.1.9, and then converting it to a class of n-tuples by Lemma 3.1.17. All these operations can be equally carried on pseudo-tuples, or we may apply Lemma 3.1.17 to get the corresponding results. □

Remark 3.1.19. Lemma 3.1.18 may be extended and give that it is possible to obtain classes as in items (b) and (d), by inserting the element y in any position of the sequence of the x_i. This will be seen in the exercises.

3.2 Classes and Formulas

3.2.1 Formulas describe classes

In the preceding section, we have proved that several collections are classes, by showing that there exist class sequences that lead to those collections. But there is a procedure for proving a collection to be a class that is frequently easier to devise and to implement. This procedure is connected to the symbolic description of collections already presented in Chapter 2: many collections may be identified in the form $C = \{x | P(x)\}$ where $P(x)$ is some condition that the elements of C must satisfy. The general idea we are to present now is that collections described in this form are indeed classes.

Proposition 3.2.1. *Let $a \in \mathcal{U}$, $n \geq 1$ a natural number, and $1 \leq i \leq n$. The collections $A = \{\langle x_1, \ldots, x_n\rangle \in \mathcal{U}^n | P\}$ are classes when P is of any of the following forms:*

$$a = x_i, \quad x_i \in a, \quad a \in x_i$$

Proof. The first collection is obtained from the class $\{a\}$ by repeated application of Lemma 3.1.18, items (a) and (b); by that lemma, we may add components on the left and on the right of any given class, and obtain a class. If a is a set, then the same procedure applied to the class a gives the second collection. When a is not a set, then the collection is obviously the empty class. Finally, we use the membership class $\mathcal{M}$ and take the codomain of the intersection $\mathcal{M} \cap (\{a\} \times \mathcal{U})$, along with Lemma 3.1.18 to obtain the third collection as a class. □

Proposition 3.2.2. *Let $n \geq 2$ a natural number, and $1 \leq i \neq j \leq n$. The collections $A = \{\langle x_1, \ldots, x_n\rangle \in \mathcal{U}^n | P\}$ are classes when P is of any of the following forms:*

$$x_i \in x_j, \quad x_i = x_j$$

Proof. These are obtained from the classes $\mathcal{M}, \mathcal{E}$ and $\{\langle u, v\rangle | v \in u\}$ (a class by Proposition 3.1.8), and applications of Lemma 3.1.18, items (a), (b), (c), (d) – see also Remark 3.1.19. □

We call *simple formulas* to the given descriptions, that is, the expressions P of the classes in Propositions 3.2.1 and 3.2.2. They are constructed with the symbols $\in, =$, and with variables or names of objects, so that each simple formula contains at least one variable. Other classes, as we shall see, admit a description which is given by composing simple formulas by means of some connective (these are $\wedge, \vee, \Rightarrow, \Leftrightarrow$ and $\neg$) or by the use of quantifiers.

We shall be more explicit and construct formulas (for describing classes) stepwise. All simple formulas constitute the formulas of level 1. All the variables x_i appearing in simple formulas are considered *free variables* in those formulas. Then, assuming that we have defined which are the formulas of level $n \geq 1$, and what it means for a variable to be free in a formula of level n, we define the same determinations for the level $n + 1$:

(1) If A is a formula of level n, then its negation $\neg(A)$ is a formula of level $n + 1$. The free variables of A are the free variables of $\neg(A)$.

(2) If A, B are formulas of level $\leq n$ and at least one of them has level n; and any free variable of A which appears in B is also free in B and conversely, then $(A) \wedge (B)$, $(A) \vee (B)$, $(A) \Rightarrow (B)$ and $(A) \Leftrightarrow (B)$ are formulas of level $n + 1$ – the meaning of these *connectives* was shown in Chapter 1. The free variables of those formulas are the variables that are free variables of A or of B.

(3) If A is a formula of level n which has at least two free variables, and x_i is a free variable of A, then $\exists x_i(A)$ and $\forall x_i(A)$ are formulas of level $n+1$. The free variables of A other than x_i are the free variables of the new formulas – the meaning of these *quantifiers* was shown in Chapter 2.

It is usual not to write all the parentheses in formulas obtained from these rules, when it is clear where they should be placed. A variable which appears in a formula and is not free, is called a *bounded variable* of that formula. It follows from the above definitions that a variable x in the formula A is bounded when at some point in the process of constructing A, a quantifier $\forall x$ or $\exists x$ has been introduced, so the variable x has been *quantified.* The meaning of the opposition free–bounded variables can be understood from the following informal idea: if a formula A contains, for example, three free variables x, y, z, then A is asserting a property that each triple of objects may fulfill or not. Like this, in arithmetic we have the formula $x = y + y$, an equation that some pairs of numbers satisfy and some do not. When we write $\exists y\ (x = y + y)$, the only free variable is x, and the formula is giving a property that a number x may satisfy or not; specifically, it is the property of being an even number. In turn, y is now a bounded variable and it is hence mute: the formula is not presenting any property for numbers that could be represented by y.

In what follows, a formula A having some free variables will be written as $A(x_1, x_2, \dots, x_n)$ with the following meaning: (1) the free variables of A are among $x_1, \dots, x_n$; and (2) if the variable x_i (for $i = 1, \dots, n$) is not free in A, then x_i does not appear in the formula A. In this way, if, e.g., x_1, x_2 are the only free variables of the formula A, then we can write A also as $A(x_1, x_2)$ or $A(x_1, x_2, x_3, x_4)$ if x_3, x_4 do not appear in A.

Example 3.2.3. To ease the writing, we represent variables with the letters x, y, z and others, sometimes with subscripts. So, $x \in y$ or $y \in a$ (if a represents a concrete set) are formulas of level 1. The formulas $(x \in y) \vee (x = y)$, $\neg(x \in a)$ or $\exists x\ (y \in x)$ have level 2. A formula of level 7 is (we suppose $a \in \mathcal{U}$):

$$\exists x\ ((a \in y) \wedge (\forall z\ ((z \in y) \Rightarrow ((\neg(a = z)) \Rightarrow (z = x)))))$$

Indeed, $\neg(a = z)$ has level 2; the implication $(\neg(a = z)) \Rightarrow (z = x)$ has then level 3; in turn, it is part of the implication starting with $z \in y$, and that implication has level 4; the formula starting with the universal quantifier has level 5, and it is part of a conjunction of level 6. The total formula is given by the existential quantifier, so that it has level 7. The only free variable in the formula is y, so we may represent the formula as $P(y)$; and the objects of the collection $\{y|P(y)\}$ are those sets which contain the element a and, possibly, just one more element.

Remark 3.2.4. In a sense, the definition we have given of formulas is not a true definition. It explains, hopefully in a precise way, which are the formulas of level n provided one knows which are the formulas of level smaller than n. So, the reader can understand how the formulas of level 6 are constructed because she or he knows which are the simple formulas, which are the formulas of level 2, and from this which are the formulas of level 3, etc. The infinite collection of formulas is of the potential type, their construction is an open, never-ending process.

Proposition 3.2.5. *Let* $n \geq 1$ *and* $P_1(x_1, \dots, x_n), P_2(x_1, \dots, x_n)$ *be formulas having some free variables. Consider the collections (when* $n = 1$*, we* *understand that $\langle x_1 \rangle$ denotes x_1):*

$$C_1 = \{\langle x_1, \dots, x_n \rangle | P_1(x_1, \dots, x_n)\}, \quad C_2 = \{\langle x_1, \dots, x_n \rangle | P_2(x_1, \dots, x_n)\rangle\}$$

If C_1, C_2 are classes, then the collections $\{\langle x_1, \dots, x_n \rangle | Q(x_1, \dots, x_n)\}$ are also classes when Q is a formula with some of the following forms:

$$\neg(P_1), \quad P_1 \wedge P_2, \quad P_1 \vee P_2, \quad P_1 \Rightarrow P_2, \quad P_1 \Leftrightarrow P_2$$

Proof. These are, respectively, $\mathcal{U}^n \setminus C_1$, $C_1 \cap C_2$, $C_1 \cup C_2$, $(\mathcal{U}^n \setminus C_1) \cup C_2$, $(C_1 \cap C_2) \cup ((\mathcal{U}^n \setminus C_1) \cap (\mathcal{U}^n \setminus C_2))$. So, these are all classes by Proposition 3.1.5 and rule 3 of Definition 3.1.2. □

Proposition 3.2.6. *Let $C = \{\langle x_1, \dots, x_n \rangle | P(x_1, \dots, x_n)\}$ be a class and $C_1 = \{\langle x_1, \dots, x_{i-1}, x_{i+1}, \dots, x_n \rangle | \exists x_i(P)\}$ with P having at least two free variables (one of them x_i). Then C_1 is a class and the same holds when $\exists x_i(P)$ is replaced with $\forall x_i(P)$.*

Proof. The collection using $\forall x_i(P)$ is also described by the formula $\neg(\exists x_i(\neg(P)))$. Thus it will be enough to prove that C_1 is a class, because the other case follows then from this and Proposition 3.2.5. Now, the argument of Lemma 3.1.17 shows that $C' = \{\langle\langle x_1, \dots, x_i\rangle, [x_{i+1}, \dots, x_n]\rangle\}$ (with the $\langle x_1, \dots, x_n \rangle \in C$) is a class; as a class of triples $\{\langle\langle x_1, \dots, x_{i-1}\rangle, x_i, [x_{i+1}, \dots, x_n]\rangle\}$, it can be reordered by Proposition 3.1.15 giving the class of tuples $\{\langle\langle x_1, \dots, x_{i-1}\rangle, [x_{i+1}, \dots, x_n], x_i\}$. Taking its domain and using again the proof of Lemma 3.1.17, we obtain the class C_1. □

Corollary 3.2.7. *Let $n \geq 1$ and let $P = P(x_1, \dots, x_n)$ be a given formula having at least a free variable. Then the collection $\{\langle x_1, \dots, x_n \rangle | P(x_1, \dots, x_n)\}$ is a class.*

Proof. If P is a simple formula, then the property follows from Propositions 3.2.1 and 3.2.2. Now, suppose we have already proved that the result is true for formulas of level up to k, and let P have level $k+1$. Therefore $P(x_1, \dots, x_n)$ can be written for instance as $P_1(x_1, \dots, x_n) \wedge P_2(x_1, \dots, x_n)$ with P_1, P_2 formulas of level $\leq k$, because $x_1, \dots, x_n$ satisfy the conventions (1)-(2) stated before Example 3.2.3 with respect to the members of this conjunction. The assertion of the proposition in this case follows by Proposition 3.2.5.

If P is $\exists y(Q)$ or $\forall y(Q)$ where Q is a formula of level k which describes a class of $(n+1)$-tuples with variables $y, x_1, \dots, x_n$, then P describes a class of n-tuples by Proposition 3.2.6. □

Example 3.2.8. Let $\mathcal{I}$ be the collection of all ordered pairs $\langle x, y \rangle$ such that x, y are sets and $x \subseteq y$; $\mathcal{I}$ is a class by Corollary 3.2.7, since $\mathcal{I} = (\mathcal{S} \times \mathcal{S}) \cap \{\langle x, y \rangle | \forall z\ (z \in x \Rightarrow z \in y)\}$. In fact, we can use $x \subseteq y$ as an abbreviation of the formula $\forall z\ (z \in x \Rightarrow z \in y)$ and hence we might speak (abusing language) of $x \subseteq y$ or, say, $x \subseteq a$ (for a given set a) as a formula. $\mathcal{I}$ is called the *inclusion class*.

There are other common abbreviations of formulas that may in practice be used as if they were formulas. For instance, $x = \{y, z\}$ will abbreviate $\forall u\,(u \in x \Leftrightarrow (u = y \vee u = z))$; and $x = \langle u, v\rangle$ or $x = \langle u, v, w\rangle$ are also abbreviations of formulas, and the student will be asked in the exercises to find such abbreviations. When C is a particular class described by some formula $P(x)$, it is also possible to consider "$x \in C$" as abbreviating the formula $P(x)$.

3.2.2 The tree of a formula

Though we cannot give a precise definition, a *tree* is a collection of objects, called the *nodes* of the tree, which has a structure similar to that of an ideal tree: it has a *root* and grows from bottom to top; a typical node has a *branching*, with some nodes above it; but some nodes are *terminal nodes*, they have no continuation. The defining feature of a tree is that going up from a non-terminal node can be done normally in different ways, but there is only one way down from any node. It turns out that every formula can be represented as a tree, and the tree of a formula reveals information about the class sequence leading to the class described by the formula. Let us see an example. For easier understanding, we put the tree upside down.

The rules for the construction of the tree of a formula are:

- When the formula in a node is of the type $\forall x(P)$, then the node has exactly one node above it (remember that the tree of the figure is upside down) which is P. The same for the existential quantifier and for formulas $\neg(P)$.
- When the formula in a node is of the type $(P) \wedge (Q)$, there are exactly two nodes above it, one with P, one with Q. The same for the other binary connectives $\vee, \Rightarrow, \Leftrightarrow$.
- The nodes with a simple formula are terminal nodes.

It is easy to construct a class sequence from the tree of a formula. In the example of Figure 3.1 (see next page), we may want to consider the classes corresponding to the terminal nodes as classes of triples. One may find then which are the classes corresponding to each node, from the results of Section 3.2. For instance, the class in $\mathcal{U}^3$ represented by the formula $a = z$ is $C_1 = \{\langle x, y, a\rangle\}$; for $z = x$ we have $C_2 = \{x, y, x\rangle\}$, and for the formula $(\neg(a = z)) \Rightarrow (z = x)$ we see that the class defined is $C_1 \cup C_2$.

Remarks 3.2.9. (1) The proof of Corollary 3.2.7 is of a special type. It shows two things: (1) if the assertion of the corollary is true for formulas of level k, then it is also true for formulas of level $k + 1$; (2) the assertion is true for formulas of level 1. What we get then is a series of partial proofs: from (1) and (2) one gets directly that the assertion is true for formulas of level ≤ 2; then from this and (1) it follows that it holds for formulas of level up to 3; then we prove that it holds for level 4, etc. So, what is shown is that, for any given formula, there is a proof that the assertion is true for that formula. We

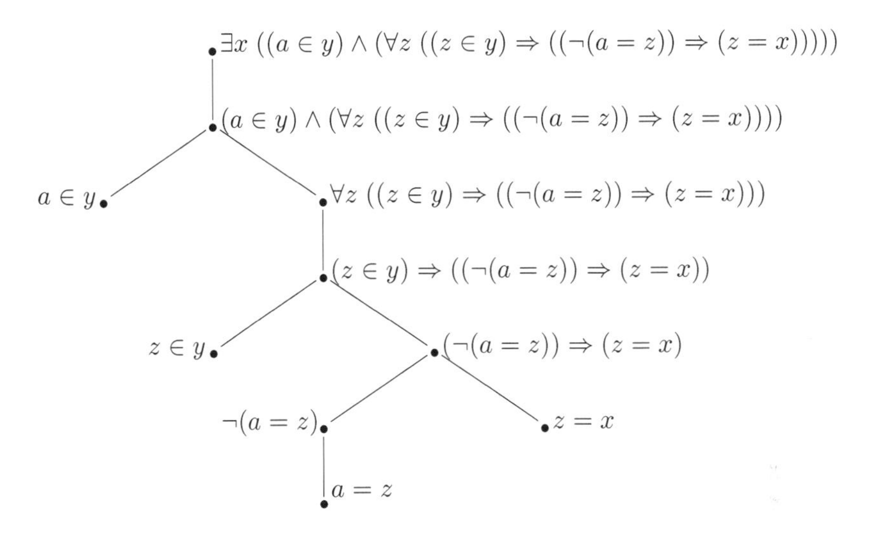

Figure 3.1
Tree of a formula.

cannot speak of all formulas at the same time (as it happens also with classes, see Comment 3.1.13) because the process of constructing formulas is never finished. This type of proof, for properties of formulas or properties of classes, could be called "finite induction".

(2) The class described by a formula is not unique. If P has three free variables, we may represent it as $P(x_1, x_2, x_3)$, but also as $P(x_1, x_3, x_4, x_5, x_2)$, for instance. It may then give a class in $\mathcal{U}^3$ and a class in $\mathcal{U}^5$.

(3) Any class C can be described in the form $C = \{x|P(x)\}$ for some formula P. This is clear for the basic classes; for instance, $\mathcal{M} = \{x|\exists y(\exists z\ (x = \langle y, z\rangle \wedge y \in z))\}$. Then for the constructions of Definition 3.1.2, there is some connective or quantifier giving the corresponding formula; for instance, if C is a class with $C = \{x|P(x)\}$, then $\mathrm{D}(C) = \{y|\exists x(\exists z\ (P(x) \wedge x = \langle y, z\rangle))\}$. More about all these observations is to be found in the exercises.

3.2.3 Other operations with classes

For any class C, we may define its *union* $\cup C$ as follows

$$\cup C = \{x|\exists u\ ((u \in C) \wedge (x \in u))\}$$

and $\cup C$ is a class by Corollary 3.2.7 and the last observation in Example 3.2.8.

Example 3.2.10. Let us consider the 2-universe $\mathcal{U}_\infty$ of Example 2.7.2, let $ABCD$ be a rectangle in the plane of that universe. Suppose that the four sides AB, BC, CD, DA form a class $\mathcal{C}$ – and recall that each of these sides is the set of the points on that line. Then the elements of $\bigcup\mathcal{C}$ are all the points of the rectangle; because they are the objects that belong to the elements of the class, and that is the definition of the union.

We have already used the term "union" in Definition 2.3.2, in a sense which is apparently different. But there is a strong relation: in the same universe $\mathcal{U}_\infty$, consider the set s consisting of two lines, say $s = \{AB, BC\}$. Then $\bigcup s$ contains the points of these two lines, and this is precisely $AB \cup BC$ following Definition 2.3.2. Generally, if a, b are sets in a 2-universe and $s = \{a, b\}$, then $a \cup b = \bigcup s$.

Another important operation with classes consists in taking all the subsets of a given class C:

$$\mathbb{P}(C) = \{u | u \in \mathcal{S} \wedge u \subseteq C\}$$

Here $\mathcal{S}$ is the class of all sets (Proposition 3.1.10) and $\mathbb{P}(C)$ is a class by Corollary 3.2.7. It is called the *power class* of C.

Example 3.2.11. We use again the 2-universe $\mathcal{U}_\infty$. Now, take a triangle ABC, and suppose that the three sides form a class $\mathcal{C} = \{AB, BC, AC\}$. We know that $\emptyset, \{AB\}, \{BC\}, \{AC\}, \{AB, BC\}, \{AB, AC\}, \{BC, AC\}$ are sets by axioms 1, 2; and they are subsets of $\mathcal{C}$. We do not know whether $\mathcal{C}$ is a set. If it is not, the above 7 objects form the power class of $\mathcal{C}$; If $\mathcal{C}$ is a set, then $\mathbb{P}(\mathcal{C})$ would contain 8 objects, the above plus $\{AB, AC, BC\} = \mathcal{C}$.

3.3 Exercises

1. Given two classes A, B, construct a class sequence from A, B leading to $A \cap B$, starting with the line $A \times B$, and ending with $\mathrm{D}(X)$, X being the class in the preceding line.
2. Construct a class sequence for $(\mathcal{U} \times \mathcal{U} \times \mathcal{U}) \cap \mathcal{E}$, and describe this class directly or by means of some formula.
3. Describe class sequences which lead to the following collections, thus proving that they are classes:

   $\{\langle x, y, z\rangle | x \in y \wedge z \notin y\}$, $\bigcup C$ (we assume that C is a class, so a part of the sequence can be supposed to lead to C).
4. Prove the converse direction of Lemma 3.1.17.

5. Let a be a set, so $\{a\}$ is a set by axiom 2. Construct a class sequence, giving explicitly the labels, that leads to the class $\mathrm{D}(\mathcal{M} \cap (\mathcal{U} \times \{a\}))$. Identify the elements of this class.

6. Describe a class sequence which leads to the class $\mathcal{I}$, the inclusion. You may assume that $\mathcal{S}$ (the class of all sets) is a class (Hint: it is advisable to construct a class sequence for the class of those ordered pairs of sets such that $x \not\subseteq y$ first; and then take the complement in $\mathcal{S} \times \mathcal{S}$).

7. Let A, B be classes, so that there is a class sequence which includes A and B; say that A appears in line i and B appears in line $j > i$. Let us continue that class sequence from line j, adding the lines whose labels are shown below:

   $j + 1 : (2, (0, 1))$

   $j + 2 : (4, (i, j))$

   $j + 3 : (3, (j + 1, j + 2))$

   $j + 4 : (3, (j + 1, j + 3))$

   $j + 5 : (5, (j + 4, j + 4))$

   Specify which are the classes corresponding to each of those lines, and identify explicitly the class obtained in the last line.

8. Let $\mathcal{S}_1$ be the class of Russell's example (Proposition 3.1.11). Show that $\emptyset \in \mathcal{S}_1$, and call $\mathcal{S}_2$ to the class $\mathcal{S}_2 = \mathcal{S}_1 \setminus \{\emptyset\}$. Prove that $\mathcal{S}_2$ is a proper class (Hint: observe that, for any set s, $s \in \mathcal{S}_2 \Leftrightarrow s \in \mathcal{S}_1 \wedge s \neq \emptyset$. Use this fact and the property that $\mathcal{S}_2 \neq \emptyset$ to construct an argument inspired in that of Proposition 3.1.11).

9. We know that, writing elements as ordered pairs, a triple $\langle a_1, a_2, a_3 \rangle$ is $\langle\langle a_1, a_2\rangle, a_3\rangle$; and a pseudo-triple $[a_1, a_2, a_3]$ is $\langle a_1, \langle a_2, a_3\rangle\rangle$. Write as triples or pseudo-triples the elements $\langle a_1, a_2, a_3, a_4, a_5\rangle$, $[a_1, a_2, a_3, a_4]$ and $[a_1, a_2, a_3, a_4, a_5]$. Can a pseudo-$n$-tuple be written as a triple?

10. By Proposition 3.1.14, classes formed by pseudo-triples, give classes when transformed into triples. Use this result to prove that if C is a class of pseudo-triples and $\sigma \in S_3$ is any permutation, then the collection $C' = \{[x_{\sigma(1)}, x_{\sigma(2)}, x_{\sigma(3)}] | [x_1, x_2, x_3] \in C\}$ is a class.

11. Imagine that rule 6 had been given only for the case when C_i is a class all of whose elements are ordered triples. That is, instead of rule 6, we would have the following (rule 6′):

    If $e(k) = 6$, then $u(k) = (i, i)$ for some $i < k$ and C_k is the collection of all triples $\langle a_1, a_2, a_3\rangle$ such that $\langle a_2, a_1, a_3\rangle \in C_i$, and all the elements of C_i are ordered triples.

    Show that if C_k is obtained from C_i by rule 6, then it could also be obtained by rule 6′ (and the rest of rules, except rule 6).

12. Prove that, if C is a class of ordered n-tuples and $1 \leq i \leq n$, then the collection $\{\langle x_1, x_2, \ldots, x_n, x_i \rangle | \langle x_1, x_2, \ldots, x_n \rangle \in C\}$ is a class.
13. Prove the following statement, similar to item (d) in Lemma 3.1.18: If C is a class of ordered n-tuples, then the following collection is also a class: $\{\langle x_1, x_2, \ldots, x_i, y, x_{i+1}, \ldots, x_n \rangle | \langle x_1, \ldots, x_n \rangle \in C\}$ (Hint: Use the proof of Lemma 3.1.17 to obtain from C the class of pairs of the form $\langle \langle x_1, \ldots, x_i \rangle, [x_{i+1}, \ldots, x_n] \rangle$, and then apply Lemma 3.1.18).
14. Let $n \geq 3$, $1 \leq i \leq n$, and let C be a class of ordered n-tuples. Prove (without using any result about formulas) that the collection $\{\langle x_1, \ldots, x_{i-1}, x_{i+1}, \ldots, x_n \rangle | \exists x_i \langle x_1, \ldots, x_{i-1}, x_i, x_{i+1}, \ldots, x_n \rangle \in C\}$ is a class (Hint: Consider first the cases $i = n$ and $i = 1$. Then follow the ideas in the proof of Lemma 3.1.17).
15. Give a detailed proof of Proposition 3.2.2.
16. Write formulas which can be abbreviated as $x = \langle u, v \rangle$ or $x = \langle u, v, w \rangle$. That is, write down formulas which describe the classes $\{\langle x, u, v \rangle | x = \langle u, v \rangle\}$ and $\{\langle x, u, v, w \rangle | x = \langle u, v, w \rangle\}$. Writing first a formula describing the class $\{\langle x, u, v \rangle | x = \{u, v\}\}$ could be helpful. For the second formula, you may use the abbreviation found for the first one. Explain how this can be extended for describing the classes whose abbreviated formula would be $x = \langle x_1, \ldots, x_n \rangle$.
17. If $P(x_1, \ldots, x_n)$ describes a certain class C of n-tuples, find a formula $Q(y)$ with just one free variable which describes the same class C.
18. Prove that for any class C there exists some formula $P(x)$ such that $C = \{x | P(x)\}$ (Observation: since this is a result about classes, what one must show is: first, that the basic classes have that property; second, that if A, B are classes having the property and C is obtained from A, B or just from A by some of the rules of Definition 3.1.2, then C has the property. Like this, one would be sure that given any particular class it could be proven that this class has the property).
19. Show that if C is a class of n-tuples (with $n \geq 1$), then there is a formula $P(x_1, \ldots, x_n)$ such that $C = \{\langle x_1, \ldots, x_n \rangle | P(x_1, \ldots, x_n)\}$.
20. Extend Proposition 3.1.15 to classes of n-tuples for arbitrary $n > 3$.
21. Prove that a variable x appearing in a formula P is free in P if and only if $\forall x$ or $\exists x$ do not appear in the formula P. Note that the proof has to be of the same type of that of Corollary 3.2.7, the type that we called "finite induction" in item (1) of Remarks 3.2.9.
22. Construct the tree of the formula $\forall z \, ((z \in y) \Rightarrow (\forall x \, ((y \in x) \Rightarrow (z \in x))))$, and describe the class of those y that satisfy the formula.

4

Relations

4.1 Relations and Operations with Relations

The following concept could be given for arbitrary collections, but we will not have use for it with that generality.

Definition 4.1.1. A class C is called a *binary relation* in case all its objects are ordered pairs. More generally, for $n \geq 2$, an *n-ary relation* is a class which consists of ordered n-tuples.

We shall normally refer to binary relations by calling them simply *relations*. This is without harm insofar we do not use the term "relation" in any other technical sense. When R is a relation and x, y are objects such that $\langle x, y \rangle \in R$, we usually denote this by writing xRy (x is *related to y in the relation R*).

By the definition above, a relation is any subclass of $\mathcal{U} \times \mathcal{U}$. In particular, if A is any class then $A \times A$ is a relation.

Examples 4.1.2. Equality, membership and inclusion, i.e., classes $\mathcal{E}, \mathcal{M}$ and $\mathcal{I}$ are relations. If real numbers are $\mathcal{U}$-objects, then $\{\langle x, y \rangle | x, y \in \mathbb{R} \wedge x \leq y\}$ is a relation if it is a class. The same holds for $\{\langle x, y \rangle | x, y \in \mathbb{R} \wedge x > y\}$ or for $\{\langle x, y \rangle | x, y \in \mathbb{Z} \wedge x | y\}$ (recall that $x|y$ is the notation for "x is a divisor of y" or "x divides y").

A trivial example of a relation is the empty set $\emptyset$. It satisfies the condition that all its elements are ordered pairs because it has no elements. If we write the condition as an implication this property will be clearly seen: R is a relation when $x \in R \Rightarrow (\exists a, b\ (x = \langle a, b \rangle))$. When R is empty, the implication is vacuously true. $\emptyset$ is thus a relation which we call the *empty relation*.

A relation between points in the plane is: $A \ R \ B$ (that is, $\langle A, B \rangle$ belongs to the relation R) if and only if the line AB passes through the origin or $A = B$. If we identify the points and their real coordinates, we also have the following relation S: $(a, b)\ S\ (c, d) \Leftrightarrow (\exists r \in \mathbb{R}\ (ra = c \wedge rb = d))$. Note that it may be that two points A, B with coordinates $(a, b), (c, d)$ respectively, satisfy ARB but (a, b) is not related to (c, d) by the relation S.

DOI: 10.1201/9781003449911-4

Another example: if there is a class E containing all polynomial with real coefficients, and $\mathbb{C}$ is the class of all complex numbers, then $\{\langle z,p\rangle \in \mathbb{C} \times E | p(z) = 0\}$ is a relation that links polynomials and their roots.

Note that we have described particular relations by two means: sometimes we give the class R of pairs which is, by definition, the relation; sometimes we define R by giving a formula "$xRy \Leftrightarrow P(x,y)$". But $P(x,y)$ is a formula (or, most commonly, an abbreviation of a formula: in the example above, we would need to define the product ra through other formulas to obtain really $ra = c$ as a formula) with two variables, hence it determines the class R of ordered pairs, as we saw in Chapter 3. This second form of referring to a relation is probably more intuitive.

Remark 4.1.3. The definition of relation is another example of the ability of set theory in transforming difficult concepts into easy ones. It would be hard to define precisely the abstract idea of "a relationship", be it in mathematics or in normal life. How could one integrate into one notion things so different like "kinship", "being greater or smaller", "being divisible by" and so many? But by making everything an object in the universe, we get a useful copy for every concrete relation under a general concept.

By Corollary 3.1.9 and rule 5 of Definition 3.1.2, the domain $\mathrm{D}(R)$ and the codomain $\mathrm{cD}(R)$ of a relation R are classes. Their union $\mathrm{Fd}(R) = \mathrm{D}(R) \cup \mathrm{cD}(R)$ is also a class by Proposition 3.1.5 and is called the *field* of the relation R.

Proposition 4.1.4. *Let R be a binary relation. Then*

$$R^{-1} = \{\langle x,y\rangle | \langle y,x\rangle \in R\}$$

is a relation, called the inverse relation *of the relation R.*

Let F, G be binary relations, Then

$$G \circ F = \{\langle x,y\rangle | \exists z\ (\langle x,z\rangle \in F \ and\ \langle z,y\rangle \in G)\}$$

is a relation, called the composite relation *(or* composition*) F* followed by *G (Note the order in the writing of these relations, which may come unexpected. Later we will see a reason for this anomaly).*

Proof. That R^{-1} is a relation follows from Proposition 3.1.8.

By item (3) of Remark 3.2.9, $F = \{x|P(x)\}$ and $G = \{y|Q(y)\}$ for certain formulas $P(x), Q(y)$. If we consider $\langle u,v\rangle \in F$ (respectively, $\langle u',v'\rangle \in G$) as an abbreviation of the formula $\exists x\ (P(x) \wedge x = \langle u,v\rangle)$ (resp., $\exists y\ (Q(y) \wedge y = \langle u',v'\rangle)$, then $G \circ F$ is the class described by the following formula $H(x_1,x_2)$: $\exists z\ (\langle x_1,z\rangle \in F \wedge \langle z,x_2\rangle \in G)$. Then $G \circ F$ is a relation by Corollary 3.2.7. □

For a relation R and a class A, we say that $R \cap (A \times A)$ is the *restriction to A* of the relation R, and denote it as R_A. On the other hand, $R \cap (A \times \mathcal{U})$ is another relation, called the *restriction to the domain A* of the relation R. It will be denoted $R|_A$. The codomain of $R|_A$ is written $R[A]$. It is hence a class and thus $R[A] = \{u | \exists x \in A\ (\langle x, u\rangle \in R)\}$. It is easy to see that, employing this notation, $\mathrm{cD}(G \circ F) = G[\mathrm{D}(G) \cap \mathrm{cD}(F)]$ and $\mathrm{D}(G \circ F) = F^{-1}[\mathrm{D}(G) \cap \mathrm{cD}(F)]$.

Examples 4.1.5. (i) Assume that the numbers $2, 3, 4, 5$ are objects of the universe. Then

$$S = \{\langle 2,2\rangle, \langle 2,3\rangle, \langle 3,3\rangle, \langle 3,4\rangle, \langle 4,3\rangle, \langle 5,3\rangle\}$$

is a relation, by the axiom of pairs and Proposition 3.1.5. Its domain $\mathrm{D}(S)$ is $\{2, 3, 4, 5\}$ and its codomain, $\{2, 3, 4\}$. Here the field $\mathrm{Fd}(S)$ coincides with $\mathrm{D}(S)$.

(ii) With similar assumptions, let

$$R = \{\langle 1,6\rangle, \langle 2,2\rangle, \langle 6,2\rangle, \langle 2,6\rangle, \langle 3,4\rangle\}$$

Then $R^{-1} = \{\langle 6,1\rangle, \langle 2,2\rangle, \langle 2,6\rangle, \langle 6,2\rangle, \langle 4,3\rangle\}$. Also, $R \circ S = \{\langle 2,2\rangle, \langle 2,6\rangle, \langle 2,4\rangle, \langle 3,4\rangle, \langle 4,4\rangle, \langle 5,4\rangle\}$.

(iii) Assume now that all the integers are objects of $\mathcal{U}$, and that $\mathbb{N}$ and $\mathbb{Z}$ are classes, as well as the collections of ordered pairs defined through a general condition like $x > y$; that is, $R = \{\langle x, y\rangle | (x, y \in \mathbb{Z}) \wedge (x > y)\}$ and $S = \{\langle x, y\rangle | (x, y \in \mathbb{N}) \wedge (x > y)\}$ are relations. For the first one, its domain and codomain are the same: $\mathbb{Z}$. In the second case the domain is $\mathbb{N} \setminus \{0\}$. On the other hand, $R_{\mathbb{N}} = S$ but $R|_{\mathbb{N}} \neq S$; and in fact $R|_{\mathbb{N}}$ contains infinitely many pairs which do not belong to S.

4.2 Order Relations

4.2.1 Some types of relations

We specify certain properties that a relation may have in connection with another class.

Definition 4.2.1. Let R be a relation and A a class. We say that:

(1) R is *reflexive in* A when each element $x \in A$ satisfies xRx; i.e., $\langle x, x\rangle \in R$. Symbolically, the condition can be written: $\forall x \in A\ (xRx)$.

(2) R is *anti-reflexive in* A when xRx does not hold for any $x \in A$. Symbolically, $\forall x \in A\ (\neg(xRx))$.

(3) R is *symmetric in A* when the following holds:

$$\forall x \in A\ (\forall y \in A\ (xRy \Rightarrow yRx))$$

In the symbolic statement above, there are two consecutive universal quantifiers. In this situation, it is common to abbreviate the writing by using only once the symbol $\forall$ and listing together all the variables which should been preceded by that symbol. Like this, the statement above is more usually written as

$$\forall x, y \in A\ (xRy \Rightarrow yRx)$$

(4) R is *anti-symmetric in A* when (for $x, y \in A$) xRy and yRx cannot both hold if $x \neq y$. In symbols,

$$\forall x, y \in A\ ((x \neq y) \Rightarrow (\neg(xRy \wedge yRx)))$$

Note that we have applied here the convention about the use of the universal quantifier stated just before. Of course, the same convention is applied with the existential quantifier.

(5) R is *transitive in A* when the following holds.

$$\forall x, y, z \in A\ ((xRy \wedge yRz) \Rightarrow xRz)$$

There is a more practical way of stating the anti-symmetric property, by using the contrapositive rule mentioned at the end of Section 2.4, about the equivalence between a conditional proposition $P \Rightarrow Q$ and the proposition $\neg Q \Rightarrow \neg P$. Following that rule, the statement for the property can be given thus:

$$\forall x, y \in A\ ((xRy \wedge yRx) \Rightarrow (x = y))$$

and this way of giving the property has the advantages of using affirmative assertions instead of negative, and equalities instead of inequalities. Accordingly, this is the more usual form for the property and for checking it in concrete examples.

Lemma 4.2.2. *Let R be a relation, $B \subseteq A$ classes. If R is reflexive, symmetric, transitive, anti-reflexive or anti-symmetric in A, the same holds respectively for those properties in B. The converse does not hold in general. However, the converse (except for reflexivity) holds when $Fd(R) \subseteq B$.*

Proof. Exercise. □

Examples 4.2.3. (1) Suppose all integers are objects of the universe $\mathcal{U}$, and the collections $\mathbb{N}, \mathbb{Z}$ of natural or integer numbers are classes. If R is a relation which consists of all ordered pairs $\langle x, y\rangle$ of integers such that x is a divisor of y (denoted as $x|y$), then R is reflexive and transitive in $\mathbb{Z}$, but it is not anti-symmetric in $\mathbb{Z}$: $3|-3$ and $-3|3$ but $3 \neq -3$. On the other hand, R is anti-symmetric in $\mathbb{N}$.

(2) Using again the same classes $\mathbb{N}, \mathbb{Z}$, assume that R is the relation given by $xRy \Leftrightarrow x < y$ with $x, y \in \mathbb{Z}$. Then R is anti-reflexive and transitive both in $\mathbb{Z}$ and in $\mathbb{N}$. It is anti-symmetric in both sets as well, since $(xRy$ and $yRx) \Rightarrow x = y$ holds because the condition in the implication is always false.

(3) Assume that the collection of all ordered pairs $\langle x, y\rangle$ of natural numbers such that x and y have a common divisor which is not 1, is a class, hence a relation R. Then R is symmetric in $\mathbb{N}$ but it has none of the other four properties.

(4) Anti-symmetric is not the contrary of symmetric. A relation may satisfy both properties at the same time, as is the case of the identity relation $\mathcal{E}_A$ in any class A. Of course, it is also possible for a relation to be neither symmetric nor anti-symmetric in a class A, as we have seen for the divisibility in $\mathbb{Z}$. Similarly, anti-reflexive is not the contrary of reflexive. However, here the only example of a relation which is both reflexive and anti-reflexive in A is obtained when $A = \emptyset$.

4.2.2 Order relations

Definition 4.2.4. Let A be a class. A relation R is an *order in A* when R is reflexive, transitive and anti-symmetric in A.

When R is an order in A, we also say that A is *ordered* by R or that R *orders* A. As seen above, the divisibility relation R of Examples 4.2.3 (1) orders $\mathbb{N}$, provided the hypotheses in that example are fulfilled. Similarly, the relation

$$\{\langle x, y\rangle | x, y \in \mathbb{R} \text{ and } x \leq y\}$$

is an order in $\mathbb{R}$. In spite of the fact that this relation has quite special properties, it is the archetypical example of an order relation.

There is a remarkable difference between these two examples. Let us say, with reference to an arbitrary relation R, that two objects x, y are *comparable* if xRy or yRx – as always, a non-exclusive disjunction. In the divisibility relation, the numbers 35 and 45 are not comparable. In the inequality relation in $\mathbb{R}$ just mentioned, any two elements of $\mathbb{R}$ are comparable.

Definition 4.2.5. Let R be an order in the class A. R is a *total order in* A, and we say that A is *totally ordered by* R, if any two elements of A are comparable under the relation R. Total orders are also called *linear orders* in many texts.

To emphasize the distinction, an order in the sense of Definition 4.2.4 is frequently called a *partial order*. Again, if R is an order (respectively, a total order) in A and $B \subseteq A$, then R is also an order (resp., a total order) in B, by Lemma 4.2.2 and the analogous consideration about comparability of elements. Another modification of the name "order" is assigned to another kind of relations, which in fact are not orders.

Definition 4.2.6. Let A be a class and R a relation. R is called a *strict order in* A if R is anti-reflexive and transitive in A.

There is a tight relationship between partial orders and strict orders. In fact, they are two sides of the same coin.

Proposition 4.2.7. *If R is a (partial) order in A, then the following is a strict order in A:*

$$R_0 = \{\langle x, y\rangle | xRy \text{ and } x \neq y\}$$

Conversely, if R is a strict order in A, then the following is an order in A:

$$R_1 = \{\langle x, y\rangle | xRy \text{ or } x = y\}$$

Proof. Let R be an order in A. Clearly, R_0 is anti-reflexive in A. Transitivity of R_0 in A is also obvious except if we had $x, y, z \in A$ with xR_0y and yR_0z but $x = z$. But this is impossible because the anti-symmetry of R would give $x = y$, a contradiction with xR_0y. For the second part, suppose R is anti-reflexive and transitive in A. Reflexivity of R_1 is obvious. If xR_1y and yR_1z, then xR_1z by transitivity of R in case $x \neq y$ and $y \neq z$. In other cases, the relation xR_1z is trivial. Finally, suppose $x \neq y$ in A, and xR_1y. Then xRy; since yRx is impossible because the transitivity of R would imply xRx, we see that $\neg(yR_1x)$ holds, and hence R_1 is anti-symmetric in A. □

The strict order relation which corresponds (according to Proposition 4.2.7) to a total order in A, is called a *strict total order*. It is thus anti-reflexive and transitive, and any two different elements of A are comparable.

4.2.3 Special elements in ordered classes

In this paragraph, we shall always assume that we are given a relation which is a (partial) order in a class A. We shall denote the relation as $\preceq$. The reason for this notation is not that we suppose that the relation is precisely a numerical inequality; rather, we use it because of its suggesting power. As in the preceding section, $\langle x, y\rangle \in \preceq$ will be written $x \preceq y$.

Definition 4.2.8. An element $u \in A$ is called *maximum of A* (with respect to the given relation $\preceq$) when the following holds

$$\forall x \in A \ \ (x \preceq u)$$

It may occur that there is no maximum element for a given order. But if there is a maximum, then it is unique.

Proposition 4.2.9. *If $u, u' \in A$ are maximum elements, then $u = u'$.*

Proof. If both u, u' are maximum elements, then $u \preceq u'$ and $u' \preceq u$, hence $u = u'$ by anti-symmetry. □

The maximum element of an ordered class A will be denoted $\max(A)$, if the relation to which it refers may be understood. The definition of a minimum element is analogous (denoted $\min(A)$), and the foregoing observations apply also to the minimum.

Definition 4.2.10. An element $u \in A$ is called *minimum of A* (with respect to the given relation $\preceq$) when the following holds

$$\forall x \in A \ \ (u \preceq x)$$

Remarks 4.2.11. The reader is probably familiar with the idea of maxima or minima of functions. But the situation there is a bit more complex. Because when speaking of, say, maxima of the real function $f(x) = -x^2$, two relations are considered: the function itself; that is, the relation consisting of the ordered pairs $\langle x, -x^2 \rangle$; and the inequality relation $\leq$ of $\mathbb{R}$ restricted to $\mathrm{cD}(f)$. Then f has a maximum at $x = 0$ means that $f(0)$ is the maximum (with respect to the inequality) of $\mathrm{cD}(f)$, the set of values of the function.

The next definition is not linked to such a well-known concept as that of maximum, so it is advisable to present an example first. Consider the function $f(x) = cos\ x$, and define the relation: $x \prec y$ when $cos\ x < cos\ y$, for $x, y \in \mathbb{R}$. This is a strict order in $\mathbb{R}$, as is easily seen. The corresponding order will be denoted $\preceq$ and we have $x \preceq y \Leftrightarrow (cos\ x < cos\ y) \vee (x = y)$. There is no maximum in $\mathbb{R}$ for this relation: for instance, $cos\ 0 = 1$, but also $cos\ 2\pi = 1$, so 0 is not the maximum, as $2\pi \not\preceq 0$ (nor the other way round). But 0 has still an important property: there is no x such that $0 \prec x$, that is, 0 is not below any other element in that relation. This is a form of maximality, and it is the concept that we now define.

Definition 4.2.12. An element $u \in A$ is called *maximal in A* (with respect to the given order relation $\preceq$) when the following holds

$$\forall x \in A \ \ (u \preceq x \Rightarrow u = x)$$

Unlike maximum elements, a class may have several maximal elements for an order (or none). But if max(A) exists, then it is the unique maximal element:

Proposition 4.2.13. *If $u = max(A)$, then u is maximal in A. Furthermore, if $u = max(A)$ and $w \in A$ is maximal, then $w = u$.*

Proof. Let $u = max(A)$. If $x \in A$, $x \preceq u$ by definition; hence $u \preceq x \Rightarrow u = x$ and hence u is maximal. Next, suppose $w \in A$ is maximal. Since $w \preceq u$, $w = u$ because w is maximal. □

Definition 4.2.14. An element $u \in A$ is called *minimal in A* (with respect to the given relation $\preceq$) when the following holds

$$\forall x \in A \ \ (x \preceq u \Rightarrow u = x)$$

The above result about maximal elements can be applied analogously to minimal elements.

Example 4.2.15. As in previous examples, we admit that divisibility of natural numbers and inequality of real numbers are classes, hence relations. $(0, 1)$ is the open interval, containing all real numbers x such that $0 < x < 1$.

Divisibility is an order relation in $\mathbb{N}$ with a minimum and a maximum element of $\mathbb{N}$; they are respectively, 1 and 0[1]. The same relation is still an order in $A = \mathbb{N} \setminus \{0, 1\}$ but has neither maximum nor minimum. All prime numbers are minimal elements of A, but there are no maximal elements.

The open interval $(0, 1) \subseteq \mathbb{R}$ has the order given by the usual inequality $\leq$: for this, there are no maximal and no minimal elements, and hence neither maximum nor minimum of the interval.

The coming definitions refer also to a relation $\preceq$ which is an order in a class A, but the concepts are relative not only to A but also to a given subclass $B \subseteq A$.

Definition 4.2.16. Let B be a subclass of the class A, $\preceq$ an order relation in A. An element $u \in A$ is an *upper bound* of B in A when

$$\forall x \in B \ \ (x \preceq u)$$

The element $m \in A$ is a *lower bound* of B in A when

$$\forall x \in B \ \ (m \preceq x)$$

In the hypotheses of this definition, let ub(B) be the collection of all the elements of A that are upper bounds of B. ub(B) is a class because $\preceq, A, B$ are classes, and the definition of the upper bounds can then be given by a formula: ub(B) is the class identified through that formula.

---

[1]Yes, 0 is the maximum element for that order relation: $x|0$ because $0 = x \cdot 0$.

Definition 4.2.17. Let B be a subclass of the class A, $\preceq$ an order relation in A. The minimum (if it exists) of the class $\mathrm{ub}(B)$ of upper bounds of B, is called the *supremum* of B in A and written $\sup_A(B)$ (or $\sup(B)$ if A is understood).

Note that $u = \sup_A(B)$ does not imply that $u \in B$. On the other hand, the uniqueness of the minimum shows that the supremum of B in A, if it exists, is unique. The condition that $u = \sup(B)$ can be represented symbolically as a double property:

$$(\forall x \in B \ \ (x \preceq u)) \quad \text{and} \quad [\forall y[(y \in A \text{ and } (\forall x \in B \ \ (x \preceq y))) \Rightarrow (u \preceq y)]]$$

The analogous definition for the maximum of the collection of lower bounds of B in A gives the *infimum* of B, $\inf_A(B)$ or $\inf(B)$.

Example 4.2.18. We use the same conventions and notations of Example 4.2.15. In particular, we consider $\mathbb{R}$ as an ordered class with the usual inequality $\leq$. We already saw that the real interval $(0, 1)$ has neither maximum nor minimum element. But 1 is the supremum and 0 is the infimum of the interval in $\mathbb{R}$. The same is true for the closed interval $[0, 1]$ but in this case 1 is also the maximum and 0 is the minimum.

Consider now $S = \mathbb{N} \setminus \{0\}$ with the divisibility relation. For any $X \subseteq S$ (with $X \neq \emptyset$), the lower bounds of X in S are all the natural numbers that are common divisors of all the elements in X. In turn, the upper bounds are all the non-zero numbers that are multiples of each number in X. Thus, for $X = \{4, 8, 12, 10\}$, there is no maximum nor minimum element of X for divisibility. But 2 is the infimum and 120 is the supremum of X in S.

4.3 Functional Relations

4.3.1 Definitions and notations

Definition 4.3.1. A relation R is said to be *functional* in case the following holds:

$$\langle x, y\rangle, \langle x, y'\rangle \in R \Rightarrow y = y'$$

Forcing an adjective to be used as a noun, we shall say that R is a *functional* when it is a functional relation. Like this, a functional is a relation F with the property that for any element x of the domain, there is a unique element y in the codomain such that $\langle x, y\rangle \in F$. This unique element y is called the *image* of x (under the functional F) and is written usually as $F(x)$. So $y = F(x)$ has exactly the same meaning as $\langle x, y\rangle \in F$, when F is a functional.

Examples 4.3.2. (i) Assume that the real numbers are objects of the universe, and that the collections $\mathbb{R}$ (of all real numbers) and $\mathbb{Z}$ are classes. Then, let $I = \mathbb{R} \times \mathbb{R}$ and assume that the following is a relation

$$R = \{\langle\langle a, b\rangle, n\rangle \in I \times \mathbb{Z} | a < n < b\}$$

The codomain $\mathrm{cD}(R)$ consists of all integers, since each $n \in \mathbb{Z}$ satisfies, for instance, $n - \frac{1}{2} < n < n + 3$. The domain is formed with all pairs $\langle a, b\rangle$ such that $a < b$ and the open interval (a, b) contains some integer. But the relation is not functional: for example, $\langle\langle 2, 6\rangle, 3\rangle$ and $\langle\langle 2, 6\rangle, 5\rangle$ belong to R.

(ii) Let us admit that the following are relations

$$\{\langle x, y\rangle \in \mathbb{Z} \times \mathbb{N} | y = x^2\}, \quad \{\langle x, y\rangle \in \mathbb{N} \times \mathbb{Z} | x = y^2\}, \quad \{\langle x, y\rangle \in \mathbb{N} \times \mathbb{N} | x = y^2\}$$

Call them F, G, H respectively. F is a functional (each number has a unique square) with domain $\mathbb{Z}$ and codomain K formed with all square natural numbers. G has domain K and codomain $\mathbb{Z}$ but is not functional; for example both $\langle 9, 3\rangle$ and $\langle 9, -3\rangle$ belong to G. Obviously, G is the inverse relation (see Proposition 4.1.4) of F. H is functional with domain K.

(iii) For any class A, the diagonal relation $\mathcal{E}_A$ (which is the restriction of the identity relation $\mathcal{E}$ to A) formed with all ordered pairs $\langle x, x\rangle$ for $x \in A$ is a functional, called the *identity functional of* A. Also, the empty relation is functional, with $\mathrm{D}(\emptyset) = \mathrm{cD}(\emptyset) = \emptyset$.

We discuss next some other examples and notations.

Proposition 4.3.3. *Let $n \geq 2$ and let R be a n-ary relation (i.e., $R \subseteq \mathcal{U}^n$ is a class). If $1 \leq i \leq n$, the i-th* projection *of R is $\{\langle x, y\rangle | x = \langle x_1, \ldots, x_n\rangle \in R \wedge y = x_i|\}$. Any projection of R is a functional class.*

Proof. The projection is a class by Corollary 3.2.7 and it is obviously functional. □

In particular, if R is a binary relation, it has two projections which we call the *projection onto the first (or second) component.*

If F is a functional with domain D and $X \subseteq D$ is a subclass, then $F|_X$ is again a functional (the notation $F|_X$ was introduced just before Example 4.1.5). Obviously, if $u \in X$, then $(F|_X)(u) = F(u)$. We usually say that $F|_X$ is the *restriction of the functional F to X*, instead of "restriction to the domain X" which would be in accordance with the definition of the restrictions for a general relation.

Definition 4.3.4. Let R be a relation and A a class. We say that R is *functional on* A when

$$\forall x \in A\, (\langle x, y\rangle, \langle x, y'\rangle \in R \Rightarrow y = y')$$

As an example, the relation $\{\langle x, y\rangle \in \mathbb{R} \times \mathbb{R} | y^2 - 2y + x = 0\}$ is not functional (for instance, there are two solutions of the equation $y^2 - 2y = 0$; i.e., for $x = 0$), but it is functional on the elements of the interval $[1, \infty)$: for these values of x, the corresponding equation either has no solution or has just one solution (when $x = 1$). It is clear that a relation R is functional on A if and only if $R|_A$ is a functional. Also, a relation R is functional if and only if it is functional on $\mathrm{D}(R)$.

We saw in Chapter 2 the notation $\{-|-\}$ to describe collections. There is a related notation that uses functionals for describing classes. Though the same idea works with arbitrary relations instead of functionals, using functionals is more practical. For example, we may write $\{x^3 | x \in \mathbb{Z}\}$ to identify the collection of all numbers that are cubes of integer numbers. Indeed, this is precisely the codomain of a functional with domain $\mathbb{Z}$, the functional F given by $F(x) = x^3$ (we admit for now that this is indeed a class, hence a functional). Like this, $\{F(x) | x \in D\}$ denotes the codomain of a given functional F whose domain includes D; and this codomain is a class by Corollary 3.1.9: it is the class of all the values of the functional F when the argument x ranges over D.

As an example of this notation, if F is a functional and $X \subseteq \mathrm{D}(F)$, then $F[X] = \{F(u) | u \in X\}$ (the notation $R[X]$ was introduced after Definition 4.1.4).

4.3.2 Inversion and composition of functionals

Inversion and composition of binary relations have been defined in Proposition 4.1.4. Item (ii) of Examples 4.3.2 includes G, the inverse relation F^{-1} of the functional relation F and it shows that the inverse relation of a functional need not be functional. But the situation is better with compositions.

Proposition 4.3.5. *If F, G are functional relations, then $G \circ F$ is functional.*

Proof. Suppose $\langle x, y\rangle$ and $\langle x, y'\rangle \in G \circ F$. Then $\langle x, z\rangle \in F$ and $\langle z, y\rangle \in G$ for some object z. Similarly, some object z' satisfies $\langle x, z'\rangle \in F$, $\langle z', y'\rangle \in G$. But $z = z'$ because F is functional and thus $y = y'$ because G is functional. □

The order of the relations in the notation $G \circ F$ is justified by the usual writing of the image of one element: $(G \circ F)(x)$ is precisely $G(F(x))$.

Of course, $G \circ F$ is in general different than $F \circ G$. For instance, consider the functional given by $F(x) = x^2$ with domain $\mathbb{R}$; and the one given by $G(x) = e^x$ with the same domain (again, we accept that these are classes, though we cannot prove this yet. It will later be shown that they are sets, but for now we lack the necessary axioms for their construction as such). Then $G \circ F$ contains all ordered pairs $\langle x, y\rangle$ such that $y = e^{x^2}$, while $F \circ G$ contains the pairs having $y = e^{2x}$, giving different values; for instance, for $x = 1$.

We consider now the question of when is F^{-1} a functional relation for a given functional F. The following concept is the key to this issue.

Definition 4.3.6. Let F be a functional. We say that F is *injective* if the following holds whenever $x, y \in \mathrm{D}(F)$:

$$x \neq y \Rightarrow F(x) \neq F(y)$$

That is, a functional is injective when different elements have always different images. The following form of the property is usually more convenient:

$$\forall x, y \in D(F)\ (F(x) = F(y) \Rightarrow x = y)$$

and both assertions are equivalent by the contrapositive rule in Section 2.4.

Example 4.3.7. Let F be the functional $\{\langle x, y\rangle \in \mathbb{R} \times \mathbb{R} | y = \sin x\}$. It is not injective: for instance, $F(0) = F(\pi)$ but $0 \neq \pi$. However, $G = \{\langle x, y\rangle \in [-\pi/2, \pi/2] \times \mathbb{R} | y = \sin x\}$ is injective, because different values in the interval $[-\pi/2, \pi/2]$ have different images. By the way, also $H = \{\langle x, y\rangle \in [-\pi/2, \pi/2] \times [0, 1] | y = \sin x\}$ is injective: the domain is now $[0, \pi/2]$ and different elements in the domain have different images.

Proposition 4.3.8. *Let F be a functional. F^{-1} is functional if and only if F is injective.*

Proof. Let F be injective and $\langle x, y\rangle, \langle x, y'\rangle \in F^{-1}$. Then $\langle y, x\rangle, \langle y', x\rangle \in F$, so that $F(y) = F(y')$. By the hypothesis, $y = y'$ and F^{-1} is functional.

Conversely, assume that F^{-1} is functional and $F(x) = F(y)$ for $x, y \in \mathrm{D}(F)$. Then $\langle F(x), x\rangle, \langle F(x), y\rangle \in F^{-1}$ so that $x = y$ by Definition 4.3.1. □

It follows from Proposition 4.3.8 that if F is injective, then $F^{-1} \circ F = \mathcal{E}_{D(F)}$ and $F \circ F^{-1} = \mathcal{E}_{cD(F)}$.

4.4 Partitions and Equivalence Relations

The concept we are going to introduce in this section is natural and simple, but some complication may arise because of a particular detail. The idea is to divide a given class into parts; for instance, if $\mathbb{N}$ is a $\mathcal{U}$-class in some universe $\mathcal{U}$, then it can be decomposed in two parts: the odd numbers, say A, and the even numbers, B, so that $\mathbb{N} = A \cup B$. But there are many other ways of making a division of $\mathbb{N}$. For instance, $\{C, D, F, \{0, 1\}\}$ with C consisting of the prime numbers, D the composite even numbers, F the composite odd numbers. Or $\{A_1, A_2, \dots\}$ with A_n containing all natural numbers having n digits in its decimal expansion; and this divides $\mathbb{N}$ into infinitely many parts.

But, as suggested, there is a problem in some cases. Say we divide the universe $\mathcal{U}$ in two parts: $\mathcal{S}_1$, the class of all sets that are not elements of themselves, seen in Proposition 3.1.11; and the class of all the other objects, $\mathcal{U} \setminus \mathcal{S}_1$. The problem is that we cannot speak of $\{\mathcal{S}_1, \mathcal{U} \setminus \mathcal{S}_1\}$ as a class or even as a collection; because $\mathcal{S}_1$ is not an object of the universe, as it was proved in that proposition; and classes consist, of course, of objects. So our definition of a *partition* of a class has to take this into account: the parts must be $\mathcal{U}$-objects, hence sets.

Definition 4.4.1. Let A be a non-empty class. A class P of subsets of A is called a *partition of* A if the following two conditions hold.

$$(1)\ x \in P \Rightarrow x \neq \emptyset; \qquad (2)\ \forall x \in A\ (\exists! u \in P\ (x \in u))$$

We have introduced in this definition a new symbol, $\exists! u \in P$ (read: exists a unique u in P). As with the related notation $\exists x \in P$, this asserts that there is some object x which is an element of P and satisfies the relation which follows, but additionally it asserts that there is **only one** element of P satisfying the property.

According to the definition, the elements of the partition P are non-empty sets included in A. The crucial fact is that each element of A belongs to precisely one of the sets in the partition. If $x \in A$, we write $\overline{x}$ for the unique element u of P such that $x \in u$; that is, the part to which x belongs.

Examples 4.4.2. The idea of dividing some collection into parts appears in all human activities and not only in set theory. If we consider the year as a collection of 365 days, it has been divided into 12 parts, the months; and each month can be seen as a collection of days, hence a subcollection of that of the days of the year. In arithmetic, a natural division of the natural numbers is a partition consisting of three sets: the set of prime numbers, the set of composite numbers, the set $\{0, 1\}$. The set of real numbers can also be seen as the union of three sets: positive real numbers (a set which is usually written $\mathbb{R}^+$), negative numbers ($\mathbb{R}^-$) and the set $\{0\}$. The originality of set theory is that it is able to view partitions as objects of the universe (when they are sets), following Definition 4.4.1.

Another example with real numbers is the partition $\mathbb{R}^-, [0, 1), [1, 2), [2, 3), \ldots,$ which gives a partition consisting of infinitely many elements. In this example, $\overline{\pi} = [3, 4)$, or $\overline{\sqrt{2}} = [1, 2)$.

There is a direct relationship between partitions and a certain type of binary relations.

Proposition 4.4.3. *Let P be a partition of a non-empty class A. We may consider the following relation (it is clearly a class):*

$$R = \{\langle a, b\rangle \in A \times A | \exists p \in P\ (a \in p \wedge b \in p)\}$$

Then R is reflexive, symmetric and transitive in A.

Proof. Employing the more usual notation, aRb means here that both a and b belong to the same element of the partition. The symmetry of R is obvious. Also, condition (2) in Definition 4.4.1 entails that R is reflexive in A. Finally, suppose aRb and bRc, so that there exist $p, q \in P$ with $a \in p, b \in p, b \in q, c \in q$. Condition (2) again implies that $p = q$ whence $a, c \in p$ and aRc, so that R is transitive. □

Definition 4.4.4. Given a class A, a relation R is called an *equivalence relation in A* when R is reflexive, symmetric and transitive in A.

Example 4.4.5. Let R be given thus: $xRy \Leftrightarrow ((x, y \in \mathbb{Z}) \wedge (x|y \text{ and } y|x))$.

R is an equivalence relation in $\mathbb{Z}$, as can be easily seen; one has, for example, $-5R5$. Also, the following relation is an equivalence relation in $\mathbb{R}$: $xRy \Leftrightarrow ((x, y \in \mathbb{R}) \wedge (\exists k \in \mathbb{Z}\ (y = x + 2k\pi)))$. Here $0R6\pi$ and $-\pi R\pi$.

Proposition 4.4.3 shows that partitions yield equivalence relations, with elements a, b related when they belong to the same element of the partition.

Even more interesting is the fact that things work almost equally well in the other direction. To this end, we need a definition and a lemma.

Definition 4.4.6. Let A be a non-empty class, and R a relation. For each $x \in A$ we write

$$[x]_R = \{y \in A | yRx\}$$

When R is an equivalence relation in A, then $[x]_R$ is called the *equivalence class* of x in A (with respect to R).

$[x]_R$ depends not only on R, but also on A, though the notation does not reflect this. Each $[x]_R$ is a subclass of A because R, A are classes, and $[x]_R$ is defined through formulas after these classes. When the class A and the relation R are understood, we write $[x]$ instead of $[x]_R$.

Lemma 4.4.7. *Let A be a non-empty class and R an equivalence relation in A. For any $a, b \in A$, the following holds: $aRb \Leftrightarrow [a] = [b]$.*

Proof. We show first $aRb \Rightarrow [a] \subseteq [b]$ for the equivalence relation R. Suppose aRb and take $x \in [a]$. Then xRa and aRb imply xRb hence $x \in [b]$.

We next complete the proof that $aRb \Rightarrow [a] = [b]$. From aRb, it follows bRa by symmetry, and this implies $[b] \subseteq [a]$ by the first part of this proof, so the principle of extension allows us to conclude that $[a] = [b]$.

Finally, $[a] = [b]$ implies $a \in [b]$ because $a \in [a]$ by reflexivity. Hence aRb. □

We may now complete the connection between equivalences and partitions.

Proposition 4.4.8. *Let R be an equivalence relation in a non-empty class A. If for each $x \in A$, the equivalence class $[x]$ is a set, then the class*

$$P = \{p \subseteq A | \exists x \in A\ (p = [x])\}$$

is a partition of A.

Proof. Again, it is clear that P is a class. Each $[x]$ is a non-empty subset of A because $x \in [x]$ by the reflexive property. Thus P satisfies condition (1) of Definition 4.4.1.

To prove (2), note that for $x \in A$, $[x]$ is a set by hypothesis so $[x] \in P$ and again $x \in [x]$. Now if $x \in [z]$ then xRz and $[z] = [x]$ follows from Lemma 4.4.7. □

If R is an equivalence relation which gives a partition P as in Proposition 4.4.8, we call the partition the *quotient class* of A modulo R, written $\frac{A}{R}$, but for typographical reasons we normally prefer to write this in the form $P = A/R$. If the equivalence R in A satisfies the hypothesis in Proposition 4.4.8 and $P = A/R$, then $\overline{x} = [x]$ for any $x \in A$. This introduces a double notation for one concept, but $\overline{x}$ always denotes a set, while $[x]$ could, in the general situation, denote a proper class.

Example 4.4.9. Consider the relation R given as follows, for $x, y \in \mathbb{Z}$:

$$xRy \Leftrightarrow ((x = y) \text{ or } (x \text{ is odd and } y = x + 1) \text{ or } (x \text{ is even and } x = y + 1))$$

It is easy to check that this is an equivalence relation in $\mathbb{Z}$, though one has to consider each of the possible cases giving xRy in order to prove, for instance, transitivity. One may also show that $[x] = \{x, x+1\}$ if x is odd, and $[x] = \{x, x-1\}$ if x is even. Thus $[x]$ is a set by the axiom of pairs (assuming that each number is an object of $\mathcal{U}$) and this equivalence gives a partition P by Proposition 4.4.8. Here P is the class of all sets of two numbers formed with an odd number and the integer following it: $\{1, 2\}, \{13, 14\}, \{-5, -4\}$, etc.

Proposition 4.4.10. *Let P be a partition of a class A. The relation R with domain A given by $xRp \Leftrightarrow (p \in P$ and $x \in p)$ is functional.*

Proof. This follows immediately from condition (2) of Definition 4.4.1. □

With the notation introduced after Definition 4.3.1, we would write $R(x) = \overline{x}$ for the above functional. If E is an equivalence relation in A which determines the partition P as in Proposition 4.4.8 (hence $P = A/E$), then the functional R of Proposition 4.4.10 is called the *canonical projection of A modulo E*.

When R is an equivalence relation in A and $x \in A$, each element $z \in [x]$ is said to be a *representative* of the equivalence class $[x]$. A subcollection of A that contains exactly one element of each equivalence class is a *complete collection of representatives.* In Example 4.4.9, the collection of all odd numbers is such a complete collection, as well as the collection of all even numbers. Also the collection of all integers that are either positive and even or negative and odd is a complete collection of representatives. However, the collection of all numbers that are positive and odd or negative and even, does not contain a representative of each equivalence class (though there is only one missing equivalence class), hence it is not a complete collection of representatives.

4.5 Exercises

1. In the third paragraph of Examples 4.1.2, two relations R, S are introduced and compared. Justify the assertion in the text that R and S do not correspond.
2. In each of the cases that follow, we admit that C is a class and R is a relation. Justify for each of those relations when it is reflexive, symmetric, anti-symmetric or transitive in C; and give a counterexample for any property that fails to hold.

   (i) $C = \{a, b, c, d, e, f, g\}$ (unspecified objects).

   $R = \{\langle a, d\rangle, \langle b, e\rangle, \langle c, f\rangle, \langle d, g\rangle, \langle e, a\rangle, \langle f, b\rangle, \langle g, c\rangle, \langle c, c\rangle, \langle a, g\rangle\}$.

   (ii) $C = \{a, b, c, d, e\}$.

   $R = \{\langle a, a\rangle, \langle c, c\rangle, \langle e, e\rangle, \langle d, b\rangle, \langle b, c\rangle, \langle d, c\rangle, \langle b, d\rangle, \langle c, b\rangle, \langle c, d\rangle, \langle b, b\rangle\}$.

   (iii) $C = \mathbb{R}^+$ contains all positive real numbers.

   $R = \{\langle x, y\rangle \in \mathbb{R}^+ \times \mathbb{R}^+ | x^2 < y\}$.

   (iv) $C = \mathbb{N}$.

   $R = \{\langle x, y\rangle \in \mathbb{N} \times \mathbb{N} | x + y = 100\}$.

   (v) $C = \mathbb{Z}^+$, consists of all integer numbers which are > 0.

   $R = \{\langle x, y\rangle \in \mathbb{Z}^+ \times \mathbb{Z}^+ | 10x - y < 100\}$.
3. Let R be a relation and A a class. Prove that R is reflexive in A if and only if $\mathcal{E}_A \subseteq R$; and that R is anti-reflexive in A if and only if $\mathcal{E}_A \cap R = \emptyset$ ($\mathcal{E}$ is the equality relation, as in Definition 3.1.2, and $\mathcal{E}_A$ is its restriction to A, introduced in general just before Example 4.1.5).

4. For a relation R and a class A, give the necessary and sufficient conditions for R to be symmetric (respectively, anti-symmetric, transitive) in A, in terms of the operations of inversion and composition of the restricted relation R_A.

5. Let F, G be relations. Show that $\mathrm{cD}(G \circ F) = G[\mathrm{D}(G) \cap \mathrm{cD}(F)]$ and $\mathrm{D}(G \circ F) = F^{-1}[\mathrm{D}(G) \cap \mathrm{cD}(F)]$.

6. For each of the following relations R and the given classes C, say which of them are order relations in C. For those cases, say if the order is a total order, and give the maximum, minimum, maximal elements or minimal elements if they exist.

   (i) $C = \mathbb{R}$, all real numbers.

   R is given through the condition $xRy \Leftrightarrow x^2 \geq y^2$.

   (ii) $C = \mathbb{N}$, all natural numbers.

   R is given through $xRy \Leftrightarrow y|2x$.

   (iii) $C = \mathbb{N}^*$, all the non-zero natural numbers.

   R is given thus: $xRy \Leftrightarrow$ (x, y are both even and $x|y$) or (x, y are both odd and $y|x$).

   (iv) $C = \mathbb{N}^*$ and R is the relation of item (iii).

   S is given thus: $xSy \Leftrightarrow$ ((xRy) or x is odd and y is even).

7. Prove Lemma 4.2.2.

8. Let R be a relation with field $F \subseteq A$. Justify that R satisfies any of the properties 2, 3, 4 or 5 of Definition 4.2.1 in A if and only if R satisfies the corresponding property in F. But that R may be reflexive in F and not reflexive in A.

9. Let A be an ordered class and $C \subseteq A$. Prove that the following conditions are equivalent: (1) C has a minimum element; (2) There exists the infimum of C, $\inf(C)$ and $\inf(C) \in C$.

10. Show an example of an ordered class (under the same assumption made in the text that $\mathbb{N}, \mathbb{R}$, etc. are classes) with no maximum but with exactly two maximal elements.

11. Suppose that there exists a set F whose elements are all the subsets of $\mathbb{N}$ which have finitely many elements – this is the set of all the finite subsets of $\mathbb{N}$. Consider the relation $R = \{\langle X, Y \rangle | X, Y \in F \wedge X \subseteq Y\}$. Show that this is an order relation in F and find (if they exist) maximum, minimum, maximal and minimal elements of R in F, in $F \setminus \{\emptyset\}$ and in $F \setminus \{\emptyset, \mathbb{N}\}$.

12. If R is a relation and C is a non-empty class, we say that R is circular in C when $R \cap (C \times C) \neq \emptyset$ and the following implication holds for any $x, y, z \in C : (xRy \wedge yRz) \Rightarrow zRx$.

Prove that R is an equivalence relation in C if and only if R is circular and reflexive. But there are circular relations in C that are not reflexive in C.

13. For each of the relations that follow, say if it is an equivalence in the given class. In that case, describe the equivalence classes.

    (i) $C = \mathbb{N}$. $xRy \Leftrightarrow \exists n \geq 0\ (x, y < 10^{n+1} \wedge 10^n \leq x, y)$.

    (ii) $C = \mathbb{R}$. $xRy \Leftrightarrow |x - y| \leq 1$.

    (iii) $C = \mathbb{R}$. $xRy \Leftrightarrow xLy \wedge yLx$, where L denotes the relation given in item (i) of Exercise 6.

    (iv) $C = \mathbb{N}$. xRy if and only if, when written in decimal notation, the last digit of x is the same as the last digit of y.

14. Let A be a non-empty class and R a relation. Prove that each $[x]$, for $x \in A$, is a class.

15. In Definition 4.4.1 the symbol $\exists!$ is introduced. Show how to replace this symbol so that the same meaning can be represented through a formula in the sense of the definition of formula in Section 3.2. Find the correct formula for the statement in item (2) of Definition 4.4.1.

16. Prove that item (2) of Definition 4.4.1 can be replaced by the following conditions: $P \subseteq \mathbb{P}(A), A = \cup P$ and $p, q \in P \Rightarrow p \cap q = \emptyset$.

17. In this exercise, we admit without proof the following property of positive natural numbers: if x is a power of natural numbers n, m, then there exists a natural number d such that both n and m are powers of d.

    We consider in $\mathbb{N}^*$ (non-zero natural numbers) the relation R given through $xRy \Leftrightarrow \exists\, n, a, b \geq 1\ (n^a = x \wedge n^b = y)$. Prove that R is an equivalence relation in $\mathbb{N}^*$ and find a complete collection of representatives of the equivalence classes.

18. Let $A = \mathbb{R}^2 \setminus \{(0,0)\}$ – we know this corresponds, through the use of cartesian coordinates, to the collection of all points in the plane except the origin. We consider the relation S in A given thus: $(a, b)S(c, d) \Leftrightarrow ad = bc$.

    Prove that S is an equivalence in A, and identify the equivalence classes. Show that the elements (x, y) with $x^2 + y^2 = 1$ form a complete collection of representatives of the equivalence classes.

19. Let R be a relation and A a class. Show that if R is functional on A, then R_A is a functional, but that R_A may be functional without R being functional on A.

20. In the relations that follow, we accept that every component of each of the ordered pairs in the lists is an object of the universe (also the equation $y = x^2$ is considered to the effect of this exercise an object

of $\mathcal{U}$). Justify whether each of the relations is functional; and when it is, say if it is injective.

(i) $\{\langle 2, -2\rangle, \langle 3.2, e^2\rangle, \langle 0, 0\rangle, \langle \pi, 5\rangle, \langle 2, y = x^2\rangle\}$.

(ii) $\{\langle \pi, 0\rangle, \langle 0, y = x^2\rangle, \langle 3.2, 0\rangle, \langle 2, e^2\rangle, \langle -5, 5\rangle\}$.

(iii) $\{\langle \pi, e^2\rangle, \langle 3.2, e^2\rangle, \langle 2, 5\rangle, \langle -5, 5\rangle\}$.

21. For each of the instances in the previous exercise, replace each pair $\langle a, b\rangle$ with $\langle b, a\rangle$ (i.e., write the relation R^{-1} instead of R) and answer the same questions as above for these new relations.

22. Accepting that the collections described below are relations, justify which of them are functional.

    (i) $\{\langle x, y\rangle \in \mathbb{Z} \times \mathbb{R} | x^2 + y^2 = 1\}$.

    (ii) $\{\langle x, y\rangle \in \mathbb{R} \times \mathbb{R} | 2y + sin(x) = 1\}$.

    (iii) $\{\langle x, y\rangle \in \mathbb{N} \times \mathbb{R} | 2xy = 1\}$.

    (iv) $\{\langle x, y\rangle \in [-1, 1] \times \mathbb{R} | sin(y + 1) = x\}$.

23. Justify that the collections given below are classes, with the help of Corollary 3.2.7.

    All pairs $\langle x, y\rangle$ such that $y = \mathrm{D}(x)$.

    All pairs $\langle x, y\rangle$ such that y is a set and a transitive relation in the set x.

    All triples $\langle x, y\rangle$ such that x is functional on the set y.

5

Maps, Orderings, Equivalences

5.1 The Central Axioms of Set Theory and the Principle of Separation

5.1.1 The axioms

According to the definitions of sets and classes, sets are precisely those classes that are objects of the universe $\mathcal{U}$. We now want to investigate the behavior of sets in relation with the operations we have considered for general collections and for classes: intersections, unions, cartesian products, differences, and so on. And to this end we will need several axioms, the central axioms of the theory. These axioms stipulate precisely that these operations and constructions give sets when applied to sets. The next two axioms are due to Zermelo and the third one, to Fraenkel[1].

Axiom 3 (of unions). If c is a set, its union $\cup c$ is a set.

Remark 5.1.1. If $a, b \in \mathcal{U}$, we know from the axiom of pairs that $\{a, b\}$ is a set. But for three $\mathcal{U}$-objects a, b, c, the axiom of pairs does not imply that $\{a, b, c\}$ is a set. With the axiom of unions, this is possible: $\{a, b\} = x$ and $\{c\} = y$ are sets by the axiom of pairs, and also $\{x, y\} = z$ is a set by the same axiom. Now, $\cup z$ is a set by axiom 3. And $\cup z = \{a, b, c\}$ – see the definition of union in Section 3.2.3. The same argument shows that any finite collection of $\mathcal{U}$-objects is a set.

It also follows from axiom 3 that $a \cup b$ is a set if a, b are sets because $\{a, b\} = c$ is a set by the axiom of pairs, and then $a \cup b = \cup c$ is a set.

Another worthy consequence of the axiom of unions is the next proposition. For any set x, $x \cup \{x\}$ is a set. This set is called the *successor* of x, and written $s(x)$. It contains all the elements of x plus the object x itself.

Proposition 5.1.2. *The collection of all ordered pairs $\langle x, s(x)\rangle$ for x a set, is a functional relation.*

[1]Though it would be more just to say that it is due to Fraenkel and Skolem – see [6, pp. 37–38].

DOI: 10.1201/9781003449911-5

Proof. Note that $s(x) = \{u | u \in x \vee u = x\}$. As usual, $\mathcal{S}$ will denote the class of all sets; and $x \in \mathcal{S}$ is an abbreviation of a formula according to Remark 3.2.9. Then

$$\{y | \exists x, u\ (x, u \in \mathcal{S} \wedge y = \langle x, u \rangle \wedge (\forall v (v \in u \Leftrightarrow (v \in x \vee v = x)))\}$$

is the collection of all ordered pairs $\langle x, s(x) \rangle$ such that x is a set; and it is a class by Corollary 3.2.7. □

Axiom 4 (of the power set). If c is a set, then its power class $\mathbb{P}(c)$ is a set, called the *power set* of x.

Example 5.1.3. The power set of finite sets is easy to describe. If a, b, c are three distinct $\mathcal{U}$-objects, then $x = \{a, b, c\}$ is a set by axiom 3. The elements of $\mathbb{P}(x)$ are all those sets that are included in x, and we may name them all: $\{a\}, \{b\}, \{c\}, \{a, b\}, \{a, c\}, \{b, c\}$ are sets by axiom 2, and they are subsets of x, hence elements of $\mathbb{P}(x)$. Also x itself satisfies $x \subseteq x$, hence $x \in \mathbb{P}(x)$. And $\emptyset \in \mathbb{P}(x)$ is a set by axiom 1. This gives these eight elements of $\mathbb{P}(x)$. There are no more, because no other collection is included in x. In general, $\mathbb{P}(x)$ has 2^k elements when x has k elements.

For infinite sets, the explicit description of the power set is not easy. For instance, if $\mathbb{N}$ is a $\mathcal{U}$-object, hence a set, we may name many subsets of $\mathbb{N}$: $\{1, 2, 3, 6\}, \{x \in \mathbb{N} | x \leq 1000\}$ or $\{x \in \mathbb{N} | x \text{ is even}\}$, etc. But there could also be other infinite subsets of $\mathbb{N}$ which are difficult to describe. In fact, we will see later that this is indeed the case.

Axiom 5 is not as simple as the foregoing two. Trying to make it easier to understand, we present it in three steps. First, we show a weak version of the axiom. That is, we state an assertion that is hopefully easy to grasp and which is a consequence of the axiom; it gives an idea of the content of the axiom, but it is not as powerful. Second, we refine this first version by giving another statement which is still a weak version of the true axiom. Only then do we state the correct form of axiom 5.

The first weak version that we now state is similar to the previous two axioms in that it claims that a certain construction from a set gives a set. In all, the axioms 2–5 allow the mathematician to make operations with sets in the way one is accostumed to: the basic constructions give sets from sets. Here, the functional image of a set is a set.

Assertion 1. If F is a functional class and $\mathrm{D}(F)$ is a set, then $\mathrm{cD}(F)$ is a set.

For instance, if we accept that $F = \{\langle x, e^x \rangle | x \in \mathbb{R}\}$ is a class, then F is a functional class with domain $\mathbb{R}$; if $\mathbb{R}$ is a set, then $\mathbb{R}^+$ (the collection of positive real numbers) is a set by the assertion, because $\mathbb{R}^+ = \mathrm{cD}(F)$. As a consequence, $\{\langle x, e^x \rangle | x \in \mathbb{R}^+\}$ is a class, because it is $F \cap (\mathbb{R}^+ \times \mathcal{U})$; and since it is functional with domain the set $\mathbb{R}^+$, we deduce that its codomain $(1, \infty)$ is a set. By using the logarithmic function we would obtain the converse result: F^{-1} is a class because F is a class (Definition 4.1.4). Since $\mathbb{R}^+$ is its domain, if $\mathbb{R}^+$ is a set, then $\mathbb{R}$ is a set because $\mathbb{R} = \mathrm{cD}(F^{-1})$.

It is important to note the following feature of Assertion 1. We have already remarked that, since classes are not (in general) objects of the universe, any assertion which gives apparently some property of all classes (or of infinitely many classes) has to be understood as a potential infinity of assertions, each of them giving the property for one particular class (read Comment 3.1.13 if you skipped it before). This is also what happens with Assertion 1: as it stands, it does not give a property of the universe, unless we identify one particular class F to which the assertion is applied; for that concrete functional, the assertion gives indeed a property of the universe $\mathcal{U}$. We say that each of the assertions of that potential infinity is an *instance* of the statement. Like this, if F is a concrete class, we could state the F-instance of Assertion 1 as: if F is functional and $\mathrm{D}(F)$ is a set, then $\mathrm{cD}(F)$ is a set.

We take this observation into account to state our second form of the axiom: we give an instance of the assertion for each class C.

C-instance of Assertion 2: For any set s the following holds. If C is functional on s (see Definition 4.3.4), then $C[s]$ is a set – recall that $C[s] = \mathrm{cD}(C|_s)$.

This is quite similar to Assertion 1, but somewhat stronger. It is clear that the F-instance of Assertion 2 implies the F-instance of Assertion 1: if F is functional and $\mathrm{D}(F) = s$ is a set, then F is functional on s, hence $F[s] = \mathrm{cD}(F)$ is a set. It would seem that it is also possible to prove the converse implication, by the following argument. Given the class C and the set s, if C is functional on s then $C|_s$ is functional, and from the $C|_s$-instance of Assertion 1 it follows that $(C|_s)[s] = C[s]$ is a set. But this does not work. First, because we do not know whether the domain of $C|_s$ is a set. More importantly, because to prove the C-instance of Assertion 2 in this way, we need all the $C|_s$-instances of Assertion 1 with different sets s, and these could be infinitely many. And it has been already explained that we cannot use infinitely many classes like this, because we would be accepting that we can actually construct infinitely many classes. This is why we cannot prove instances of Assertion 2 from Assertion 1.

We add an example of an application of Assertion 2. Assume that $C = \{\langle x, y\rangle | x, y \in \mathbb{R} \wedge y = x^2\}$ is a class of pairs of real numbers. The C-instance of Assertion 1 would show us that if $\mathbb{R}$ is a set, then $\{0\} \cup \mathbb{R}^+$ is a set, because $\mathbb{R} = \mathrm{D}(C)$ and C is functional. But there are many more deductions from the C-instance of Assertion 2: if $[0, 2)$ is a set then $[0, 4)$ is a set; if $[0, 3)$ is a set, then $[0, 9)$ is a set, etc. As another example, let $C = \{\langle x, y\rangle | (x \in (-\infty, 0] \wedge x = \tan y) \vee (x \in (0, \infty) \wedge y = \tan x)\}$. C is not functional because, for instance, $\langle 0, 0\rangle, \langle 0, \pi\rangle \in C$. But it is functional, for instance, on any positive interval of $\mathbb{R}$. Thus, for example, if $I = (0, \pi/2)$ is a set, then $C[I] = \mathbb{R}^+$ is a set.

We are now ready to present Axiom 5. As explained, this is not a single axiom, but a collection of axioms or *instances* of the axiom, one for each class.

Globally we speak of the axiom as an *axiom scheme*: an infinite collection of axioms, all with the same content applied to each particular class.

(C-instance of) Axiom 5 (of replacement). Let C be a class that consists of ordered triples. For any $u \in \mathcal{U}$ and $s \in \mathcal{S}$, the following holds: if C is functional on the class $\{u\} \times s$, then $C[\{u\} \times s]$ is a set.

Note that $C[\{u\} \times s] = \mathrm{cD}(C|_{\{u\}\times s})$ in the above statement. Each instance of Assertion 2 is a consequence of one instance of the axiom.

Proposition 5.1.4. *Assume axiom 5 and let C be a given $\mathcal{U}$-class: if $s \in \mathcal{S}$ and C is functional on s, then $C[s]$ is a set.*

Proof. Given C, we consider the class C' consisting of the triples $\langle \emptyset, b, c\rangle$ such that $\langle b, c\rangle \in C$. This is easily shown to be a class by taking $\{\emptyset\} \times C$ and converting its pseudo-triples into triples by Proposition 3.1.14. Then, the C'-instance of the axiom of replacement for $u = \emptyset$ and $s \in \mathcal{S}$ gives that $C[s]$ is a set. □

Remark 5.1.5. Axiom 5 is stronger than Assertion 2 because of the introduction of the "parameter" u. For instance, it is easy to show from the axiom, that $\{u\} \times s$ is a set for any object u and any set s; we just apply the C-instance of the axiom with $C = \mathcal{E} \cap (\mathcal{U} \times \mathcal{U} \times (\mathcal{U} \times \mathcal{U})) = \{\langle t, x, \langle t, x\rangle\rangle\}$: for any u and $s \in \mathcal{S}$, $C|_{\{u\}\times s}$ is functional (because C is functional), and thus its codomain is a set and it is $\{u\} \times s$. With Assertion 2, the same argument is possible, but needs the C_u-instance (the above class C, but with a fixed u) for each $u \in \mathcal{U}$; so this would be only a potential proof; i.e., one could give a proof that $\{u\} \times s$ is a set for each particular u, but not the complete proof.

We may also speak, and we shall eventually do, of the C-instance of axiom 5, even when C is not a class of triples. In that case, the C-instance of the axiom is just the $(C \cap T)$-instance with T being the class of all ordered triples (Proposition 3.1.7).

A 2-universe $\mathcal{U}$ that satisfies axioms $3, 4, 5$ (here we understand of course that $\mathcal{U}$ satisfies each instance of the axiom) will be called a 5-*universe*. For the rest of results in this chapter, **we assume that $\mathcal{U}$ is a 5-universe**, except otherwise stated.

Comment 5.1.6. Recall our comments in Section 3.1.2 about the two levels of our study. Before Axiom 5, the study of classes has no consequences on the level of the universe. It is by this axiom that classes are linked to the study of the universe, and through this link the study of classes will now pay off. Because each instance of the axiom establishes that by constructing a class we may get new objects of the universe from old ones. And so we may obtain properties of the universe $\mathcal{U}$.

The conjunction of the axiom of replacement and the axiom of unions produces a result which may be seen as a strengthening of the axiom of replacement itself.

Proposition 5.1.7. *Let C be any class of ordered triples. For any $u \in \mathcal{U}$ and $s \in \mathcal{S}$, the following holds. If for any $x \in s$ we have that $C[\{\langle u, x\rangle\}]$ is a set, then $C[\{u\} \times s]$ is a set.*

Proof. Given C, let $C_1 = \{\langle u, x, z\rangle | z \in \mathcal{S} \wedge \forall t\, (t \in z \Leftrightarrow (\langle u, x, t\rangle \in C))\}$. By Corollary 3.2.7, C_1 is a class; and by the principle of extension, C_1 is functional.

Take any $u \in \mathcal{U}, s \in \mathcal{S}$ and suppose that for any $x \in s$, $C[\{\langle u, x\rangle\}]$ is a set, call it $C_{u,x}$. Thus $\langle u, x, C_{u,x}\rangle \in C_1$ for any $x \in s$. Since $C_1|_{\{u\}\times s}$ is functional, its codomain is a set W by the C_1-instance of the replacement axiom. Therefore $\cup W$ is a set by axiom 3, but $\cup W$ is just $\mathrm{cD}(C|_{\{u\}\times s}) = C[\{u\} \times s]$. □

This statement shall be called the C-instance of the *strong form of the axiom of replacement*. It is clear that the C-instance of the strong form of replacement implies directly the C-instance of the axiom of replacement: given C, u, s as above so that $C|_{\{u\}\times s}$ is functional, the hypothesis of Proposition 5.1.7 is satisfied (since unitary or empty collections are sets) and hence $\mathrm{cD}(C|_{\{u\}\times s})$ is a set. It will be shown in the exercises that the strong form of replacement implies also the axiom of unions.

Comment 5.1.8. Since any set is a class, there is the X-instance of the axiom of replacement for each set X. But all such instances follow from just one particular instance. Thus all the X-instances for sets X do not count as infinitely many instances of replacement.

Proposition 5.1.9. *Let $\mathcal{U}$ be a 4-universe. There is a class C such that if the C-instance of the axiom of replacement holds in $\mathcal{U}$, then the x-instance of the axiom of replacement holds for every $x \in \mathcal{S}$.*

Proof. Let C be the class of 4-tuples $\langle x, u, y, z\rangle$ such that $x \in \mathcal{S}$ and $\langle u, y, z\rangle \in x$. Suppose that the C-instance of replacement holds (by viewing C as a class of ordered triples); we will see that the x-instance of the axiom holds for any $x \in \mathcal{S}$.

Given x, let $u, s \in \mathcal{U}$ with s a set, and suppose that the class of all triples of x is functional on $\{u\} \times s$. But then C is functional on $\{\langle x, u\rangle\} \times s$, from which it follows by the hypothesis on C that $\{z | \exists t \in s\, (\langle x, u, t, z\rangle \in C)\}$ is a set. And this set is $x[\{u\} \times s]$. □

The infinity of instances of the axiom of replacement may be reduced to those C-instances where C is a neat class, i.e., a class with a class sequence where rule 1 (introduction of arbitrary sets) of Definition 3.1.2 is not applied. The result that follows will be important for Part II of the book, but it is not used in the rest of Part I. So, it may be skipped in a first reading.

Proposition 5.1.10. *Let C be a class in a 4-universe $\mathcal{U}$. We may construct a neat class B such that if the B-instance of the axiom of replacement holds in $\mathcal{U}$, then the C-instance of the axiom holds as well.*

Proof. Let $C_1, \ldots, C_n = C$ be a class sequence leading to C; and let $i_1, \ldots, i_k$ be the lines of the sequence such that $A_{i_j} = t_j$ is introduced by rule 1. Let us write $\underline{x}$ for $\langle x_1, \ldots, x_k \rangle \in \mathcal{S}^k$ – so that each x_i is a set. In particular, $\underline{t} = \langle t_1, \ldots, t_k \rangle$.

The idea of the proof can be explained easily when $k = 1$ (so only one line of the class sequence introduces a set t_1 by rule 1), and assuming that C is already a class of triples. C can be described by a formula in which the particular set t_1 appears; thus a formula like $P(t_1, x, y, z)$. Then B will be the neat class determined by the formula $P(w, x, y, z)$, w being a new variable; and with B considered as a class of triples $\langle \langle w, x \rangle, y, z \rangle$. To prove the C-instance of replacement, we choose some $u \in \mathcal{U}$ and $s \in \mathcal{S}$, and we must see that the elements z with $\langle u, y, z \rangle \in C$ for some $y \in s$, is a set X. These elements are those satisfying $P(t_1, u, y, z)$. So, if we choose the object $\langle t_1, u \rangle$ and the set s, the B-instance of replacement asserts that the same X is a set.

We now give the actual proof. We construct, step by step, a sequence $B_1, \ldots, B_n$ of neat classes with the following properties: (1) Each element of B_i is of the form $\langle \underline{x}, y \rangle$ with $\underline{x} \in \mathcal{S}^k$: (2) For $i = 1, \ldots, n$, $C_i = \{y | \langle \underline{t}, y \rangle \in B_i\}$. What we show is that, if (1)–(2) are valid for the neat classes $B_1, \ldots, B_{i-1}$ $(i \leq n)$, then we may add the term B_i which still satisfies the properties.

If C_i is introduced by rule 1 and $C_i = t_j$, then take $B_i = \{\langle \underline{x}, y \rangle | \underline{x} \in \mathcal{S}^k \wedge y \in x_j\}$; this is a class by Propositions 3.1.10 and 3.2.2; and from the proofs of these propositions, one sees that B_i is a neat class. Moreover, conditions (1) and (2) are obvious.

If rule 2 is applied and $C_i = \mathcal{M}$ or $C_i = \mathcal{E}$, then we set $B_i = \mathcal{S}^k \times \mathcal{M}$ or $\mathcal{S}^k \times \mathcal{E}$ respectively, hence they are neat classes, and again (1) and (2) are clear.

If $C_i = C_r \setminus C_s$ is introduced by rule 3, then choose $B_i = B_r \setminus B_s$, and B_i is a neat class because B_r, B_s are neat classes. Condition (1) is clear; as for (2), $\langle \underline{t}, y \rangle \in B_i$ if and only if $\langle \underline{t}, y \rangle \in B_r \setminus B_s$ and this happens if and only if $y \in C_i$ by the assumption that B_r, B_s satisfy (2).

When $C_i = C_r \times C_s$ is introduced by rule 4, then define $B_i = \{\langle \underline{x}, \langle y, z \rangle \rangle | \langle \underline{x}, y \rangle \in B_r \wedge \langle \underline{x}, z \rangle \in B_s\}$. B_i is obtained from $B_r \times B_s$ intersected with the class of pairs $\langle \langle a, b \rangle, \langle a, c \rangle \rangle$; and then reordered by cancelling the second component of the form $\underline{x}$ and converting to pseudo-triples. This gives B_i as a neat class satisfying condition (1). Then $\langle y, z \rangle \in C_i \Leftrightarrow \langle \underline{t}, y \rangle \in B_r \wedge \langle \underline{t}, z \rangle \in B_s$ by the assumption on B_r, B_s; and in turn this is equivalent to $\langle \underline{t}, \langle y, z \rangle \rangle \in B_i$.

Consider now $C_i = \mathrm{D}(C_r)$ obtained by rule 5. Then $B_i = \{\langle \underline{x}, y \rangle | \exists z \, (\langle \underline{x}, \langle y, z \rangle \rangle \in B_r)\}$. B_i is a neat class since it is obtained from B_r by taking the pseudo-triples of B_r, converting them into triples $\langle \underline{x}, y, z \rangle$, and taking its domain. (1) is obvious and (2) is easily seen: $y \in C_i \Leftrightarrow \exists z \, (\langle y, z \rangle \in C_r \Leftrightarrow \exists z \, (\langle \underline{t}, \langle y, z \rangle \rangle \in B_r) \Leftrightarrow \langle \underline{t}, y \rangle \in B_i$.

Finally, if C_i comes from C_r by some of the transformations of rules 6, 7 of Definition 3.1.2, we apply the corresponding transformation to obtain B_i from B_r. For instance, if rule 6 is applied, then $B_i = \{\langle \underline{x}, \langle a_2, a_1, a_3 \rangle\rangle | \langle \underline{x}, \langle a_1, a_2, a_3 \rangle \in B_r\}$. By using Propositions 3.1.14 and 3.1.15, one sees that B_i is a (neat) class. Again, (1) and (2) are immediate.

In particular $\{y | \langle \underline{t}, y \rangle \in B_n\} = C$. To construct the required class B of triples, we set $B = \{\langle \langle \underline{x}, y_1 \rangle, y_2, y_3 \rangle | \langle \underline{x}, \langle y_1, y_2, y_3 \rangle \rangle \in B_n\}$, a neat class of triples. We assume now that the B-instance of replacement holds and consider $u \in \mathcal{U}, s \in \mathcal{S}$ such that $C|_{\{u\} \times s}$ is functional. Then the restriction of B to the domain $\{\langle \underline{t}, u \rangle\} \times s$ is again functional and its codomain is the same $\text{cD}(C|_{\{u\} \times s})$, which is therefore a set. □

5.1.2 The principle of separation

Another consequence of replacement is one of the original axioms of Zermelo, which is usually referred to as the *separation axiom.* We show it in the form of a scheme, i.e., an infinity of statements. The following is the *C-instance of the principle of separation.*

Proposition 5.1.11. *Let C be a class of ordered pairs. For any $s \in \mathcal{S}$ and $u \in \mathcal{U}$, $\{x \in s | \langle u, x \rangle \in C\}$ is a set.*

Proof. Construct the class C' of all triples $\langle u, x, x \rangle$ with $\langle u, x \rangle \in C$. For each u the induced relation is functional, hence by the C'-instance of axiom 5, $\{x \in s | \langle u, x \rangle \in C\}$ is a set. □

As an application of this principle, we show that $a \cap b$ is a set whenever a, b are sets: we apply Proposition 5.1.11 to the inverse-membership class $\mathcal{M}^{-1}$ with $u = b$ and $s = a$. This gives that $\{x \in a | x \in b\}$ is a set, and obviously it is $a \cap b$. But, more generally, the proposition yields a practical way (indeed, the most practical way) of proving that certain collections are sets. Recall from Corollary 3.2.7 that a formula $P(x_1, \ldots, x_n)$ determines the class $\{\langle x_1, \ldots, x_n \rangle | P(x_1, \ldots, x_n)\} \subseteq \mathcal{U}^n$. For example, the formula $x \in y$ can be seen as $P(x, y, z)$ and, as such, it determines the class $\{\langle x, y, z \rangle | x \in y\}$.

Corollary 5.1.12. *Let $P = P(x_1, \ldots, x_n, y)$ be a given formula where the variable y is free, and let $C \subseteq \mathcal{U}^{n+1}$ be the class determined by P. Then, for any set s and objects $u_1, \ldots, u_n \in \mathcal{U}$, the collection $\{x \in s | P(u_1, \ldots, u_n, x)\}$ is a set.*

Proof. The C-instance of the principle of separation, applied with $u = \langle u_1, \ldots, u_n \rangle$ and the set s shows that $\{x \in s | \langle u_1, \ldots, u_n, x \rangle \in C\}$ is a set. □

As classes, formulas are our constructions. Thus what we are asserting here is that there is a proof of the statement in the corollary for each formula. The case $n = 0$ says that if $P(x)$ is a given formula, then for any set s, $\{x \in s | P(x)\}$ is a set (of course, we know that $P(x)$ is the condition for the object x to belong to the class defined by P). This is basically Zermelo's *Axiom der Aussonderung*, with the somewhat vague notion of a "class statement" (Klassenaussage) replaced with the concept of an effectively constructible "formula". The general form of Corollary 5.1.12 considers formulas with more variables to which arbitrary values, or "parameters" u_i can be given; any election of such values produces a set.

Corollary 5.1.13. *If a, b are sets, then $a \cap b$, $a \cup b$ and $a \setminus b$ are sets.*

Proof. By Corollary 5.1.12, the formulas $x \in u$ and $x \notin u$ give that, for any sets a, b, $\{x \in a | x \in b\}$ and $\{x \in a | x \notin b\}$ are sets. Of course, these are $a \cap b$ and $a \setminus b$. That $a \cup b$ is a set has been seen right before Proposition 5.1.2. □

This property is abbreviated by saying that sets are *closed under* the operations of intersection, union and difference. Generally, a class $\mathcal{C}$ is closed under some operation with objects if the result of applying the operation to objects of the class $\mathcal{C}$ produces objects of the class $\mathcal{C}$.

The form of the principle of separation (Proposition 5.1.11) which eliminates the parameter u, is extremely simple but also useful.

Proposition 5.1.14. *Given a class C, $C \cap s$ is a set for any set s.*

Proof. Apply Proposition 5.1.11 to the class $\{\emptyset\} \times C$ with the set s. □

Corollary 5.1.15. *The class $\mathcal{S}$ of all sets and the universe $\mathcal{U}$ are proper classes.*

Proof. Recall from Proposition 3.1.11 that $\mathcal{S}_1$ (the class of all those sets x such that $x \notin x$) is a proper class and $\mathcal{S}_1 \subseteq \mathcal{S}$; and suppose that $\mathcal{S}$ is a set. Then $\mathcal{S}_1 \cap \mathcal{S} = \mathcal{S}_1$ is a set by Proposition 5.1.14 and this is contradictory. The same proof works for the universe $\mathcal{U}$. □

5.1.3 Some consequences of the axioms

The axioms are useful in studying constructions with sets. In this connection, we may define the intersection of a class C of sets: $\cap C = \{x | \forall u \in C \ (x \in u)\}$.

Proposition 5.1.16. *Let C be a class such that $\emptyset \neq C \subseteq \mathcal{S}$. Then $\cap C$ is a set.*

Proof. Let $u \in C$. Then $\cap C = \{x \in u | (\forall y \ (y \in C \Rightarrow x \in y))\}$ is a set by Corollary 5.1.12. □

Proposition 5.1.17. *If A, B are sets, then the cartesian product $A \times B$ is a set.*

Proof. By Corollary 5.1.13, $A \cup B$ is a set, and $\mathbb{P}(A \cup B), \mathbb{P}(\mathbb{P}(A \cup B))$ are sets by axiom 4. By the definition of ordered pairs, every element of $A \times B$ is a set included in $\mathbb{P}(A \cup B)$, hence it is an element of $\mathbb{P}(\mathbb{P}(A \cup B))$. This shows the inclusion $A \times B \subseteq \mathbb{P}(\mathbb{P}(A \cup B))$.

Now, $A \times B = \{u \in \mathbb{P}(\mathbb{P}(A \cup B)) | \exists x, y\ (x \in A \wedge y \in B \wedge u = \langle x, y\rangle)\}$. This is a set for any sets A, B, by Corollary 5.1.12, with the set $s = \mathbb{P}(\mathbb{P}(A \cup B))$ and the formula with parameters A, B. □

Proposition 5.1.18. *(i) Let R be a set. Then $D(R)$ and $cD(R)$ are sets.*

(ii) Let R be a relation. R is a set if and only if $D(R)$ and $cD(R)$ are sets.

Proof. (i) If $x \in \mathrm{D}(R)$, then $\{x\} \in \cup R$ and hence $x \in \cup(\cup R)$. Thus $\mathrm{D}(R) = \{x \in \cup(\cup R) | \exists y\ (\langle x, y\rangle \in R)\}$ and $\mathrm{D}(R)$ is a set by Corollary 5.1.12. The same argument works with $\mathrm{cD}(R)$.

(ii) $R = \{x \in \mathrm{D}(R) \times \mathrm{cD}(R) | x \in R\}$. So, if $\mathrm{D}(R)$ and $\mathrm{cD}(R)$ are sets, then $\mathrm{D}(R) \times \mathrm{cD}(R)$ is a set by Proposition 5.1.17 and we may apply Corollary 5.1.12. The converse follows by item (i). □

We insist one last time: statement (i) above gives a property of objects of the universe and hence it is proved from finitely many axioms. Statement (ii) is valid for any class R which is a relation, and it has a different proof for each class R – though all these proofs follow the same argument, (ii) may be used in future proofs by applying its property to one or finitely many classes.

Proposition 5.1.19. *If the relations F, G are sets, then F^{-1} and $G \circ F$ are sets too.*

Proof. $F^{-1} = \{x \in \mathrm{cD}(F) \times \mathrm{D}(F) | \exists u, v, y\ (y = \langle u, v\rangle \wedge x = \langle v, u\rangle \wedge y \in F)\}$; F^{-1} is a set by Corollary 5.1.12 since $\mathrm{cD}(F) \times \mathrm{D}(F)$ is a set by Proposition 5.1.18.

Similarly, $G \circ F = \{x \in \mathrm{D}(F) \times \mathrm{cD}(G) | \exists u, v, y\ (x = \langle u, v\rangle \wedge \langle u, y\rangle \in F \wedge \langle y, v\rangle \in G)\}$. □

5.2 Maps

5.2.1 Functions and maps

Definition 5.2.1. A *function* is a functional relation which is a set.

A *map* is an ordered triple $\langle f, A, B\rangle$ such that f is a function, $A = \mathrm{D}(f)$, B is a set and $\mathrm{cD}(f) \subseteq B$. This is called a *map from A to B*.

As with general functionals, the fact that $\langle x, y\rangle \in f$ may be represented by $f(x) = y$. By the functionality, $f(x)$ is unique and there is no ambiguity with this notation. The same fact is also written $x \overset{f}{\mapsto} y$.

The map $\langle f, A, B\rangle$ will also be represented as $f : A \to B$, and as $A \overset{f}{\to} B$. The domain A is called the *initial set* of the map and B is the *final set*. The codomain of f is called the *image* of the map. It is also called the *image* of f, written $\mathrm{Im}(f)$ (thus $\mathrm{Im}(f) = \mathrm{cD}(f)$).

Remarks 5.2.2. The terminology introduced here is unusual. More common in the mathematical literature is to take the words function and map as synonymous. However, since there are two concepts and two words on use, it seems adequate to distinguish the meanings by name. A map consists essentially of a function, but also of its domain and some set including the codomain. Since a map is an ordered triple, two maps are to be considered equal when their three components coincide one by one. Like this, the following three maps are different (assuming that they are indeed maps, which we cannot guarantee for now): (1) $f : \mathbb{R} \to \mathbb{R}$ with $f(x) = sin(x)$; (2) $g : \mathbb{R} \to [-1, 1]$ with $g(x) = sin(x)$; (3) $h : [\pi/2, 3\pi/2] \to [-1, 1]$ with $h(x) = sin(x)$. And this difference has some consequences: for instance, there is an inverse map (a concept to be defined later) for the last one but not for the other two.

When dealing with functions and maps, we will be frequently using this convention: a map is identified by a capital letter and the same lower-case letter identifies its function. For instance, we might denote the three maps that have been just mentioned as F, G, H. They are three different maps and f, g, h are the functions in each of them (but here the functions f, g are the same).

As happens with ordered pairs or general relations, functions and maps give an interpretation of a classical concept of mathematics (function), as an object of the universe of sets. Like this, we view as sets all the traditional functions like polynomial functions, for instance $f(x) = 4x^2-2$ (giving a map $f : \mathbb{R} \to \mathbb{R}$, or from $\mathbb{R}$ to $[-2, \infty)$); trigonometric functions like $g(x) = tan(x - 1)$ (which gives a map $g : \mathbb{R} \setminus \{(1 + \pi/2) + k\pi | k \in \mathbb{Z}\} \to \mathbb{R}$); or other transcendental functions as $h(x) = e^x$ (which can give a map $h : \mathbb{R} \to \mathbb{R}$, but also from $\mathbb{R}$ to $(-\pi, \infty)$ or from $(1, \infty)$ to $\mathbb{R}$).

The difference between map and function is significant also in connection with compositions. From Propositions 5.1.19 and 4.3.5, we know that a composition of functions $g \circ f$ is a function. Its domain and codomain are described after Proposition 4.1.4, but this description is more complicated than the simple definition of composition of maps:

Definition 5.2.3. Let $f : A \to B$ and $g : B \to C$ two maps. The map $g \circ f : A \to C$ is called the *composition map* of the given maps.

Note that this definition is correct (in that it uniquely identifies the defined map) because $\mathrm{cD}(f) \subseteq B = \mathrm{D}(g)$, hence $\mathrm{D}(g \circ f) = A$ and $\mathrm{cD}(g \circ f) \subseteq C$. If we write $F = \langle f, A, B\rangle$ and $G = \langle g, B, C\rangle$, then its composition $\langle g \circ f, A, C\rangle$ will be written $G \circ F$. The composition of maps is associative:

Proposition 5.2.4. *Let $F = \langle f, A, B\rangle$, $G = \langle g, B, C\rangle$ and $H = \langle h, C, D\rangle$ be three maps. Then $H \circ (G \circ F) = (H \circ G) \circ F$.*

Proof. By Definition 5.2.3, both compositions have the same initial and final sets, so we only need to show that $h \circ (g \circ f) = (h \circ g) \circ f$. But it is simple to see that both functions coincide for each $x \in A$, hence they are equal. □

Examples 5.2.5. A basic example of a map is inclusion: when $A \subseteq B$ are sets, then $\langle \mathcal{E}_A, A, B\rangle$ is a map with $\mathcal{E}_A = \mathcal{E} \cap (A \times A)$ which is a set by separation. When $B = A$, this is called the *identity map of* A and written 1_A (in this case, it is usual to write the attached function $\mathcal{E}_A$ as 1_A too), giving the map $1_A : A \to A$. It is also obvious that if $F = \langle f, A, B\rangle$ is a map, then $1_B \circ F = F \circ 1_A = F$.

Another basic example of a map is projection – see Proposition 4.3.3. Let A be a set of ordered pairs; then there is a map $p_1 : A \to \mathrm{D}(A)$ with $p_1(\langle x, y\rangle) = x$; and another map $p_2 : A \to \mathrm{cD}(A)$ with $p_2(\langle x, y\rangle) = y$. To see that p_1 is a function, we simply observe that $p_1 = \{x \in A\times \mathrm{D}(A) | \exists u, v, y\ (y = \langle u, v\rangle \wedge x = \langle y, u\rangle)\}$ and apply Corollary 5.1.12. The case of p_2 is similar.

Remark 5.2.6. It is very common to intentionally confuse not only the map $\langle f, A, B\rangle$ and its corresponding function f, but also the function itself and an equation (when it exists) that gives the explicit values of the images. This is understandable because of the rooted use in calculus of expressions like *the function* $y = e^x$ or *the function* f *with* $f(x) = cos(x)$, and this metonymy is harmless if one can understand which are the initial and the final sets. But it may be confusing sometimes; with reference to the three maps for the sine function given in Remarks 5.2.2, the first two are two different maps with the same function, while the function of the third one is different: the pair $\langle 0, 0\rangle \in f$ but $\langle 0, 0\rangle \notin h$.

Proposition 5.2.7. *The following collections are classes: (1) all sets that are relations; (2) all functions.*

Proof. $\mathbb{P}(\mathcal{U} \times \mathcal{U})$ is the collection of all sets that are relations, and it is a class. If we call it R_0, then the collection of all functions can be described by the condition:

$$x \in R_0 \wedge [\forall u, v, w, y, y'((v = \langle u, y\rangle \wedge w = \langle u, y'\rangle \wedge v \in x \wedge w \in x) \Rightarrow y = y')]$$

Since this is a (abbreviated) formula, the collection is a class by Corollary 3.2.7. □

Proposition 5.2.8. *The collection of all maps is a class. Moreover, given sets A, B, the collection of all maps from A to B is a set.*

Proof. Let $\mathcal{F}$ be the class of all functions (Proposition 5.2.7), and consider the class $\mathcal{F} \times \mathcal{S} \times \mathcal{S}$. The intersection of this class and the class $\{\langle f, y, z\rangle | y = \mathrm{D}(f) \wedge \mathrm{cD}(f) \subseteq z\}$ is then the class of all maps.

If M is the class of all maps, then $\{u \in (\mathbb{P}(A \times B) \times \{A\} \times \{B\}) | u \in M\}$ is the collection of maps from A to B and is a set by Corollary 5.1.12. □

In the above proof, besides using for instance $u \in M$ as an abbreviation of a formula according to Remark 3.2.9, we have also $y = \mathrm{D}(f)$ or $\mathrm{cD}(f) \subseteq z$ which are not properly formulas as we have constructed them. But we may consider these also as abbreviations and some of the exercises will ask for finding formulas that correspond to such "formulas". In the future, we will use simple abbreviations of this kind without further notice.

The meaning of $R[X]$ and $R^{-1}[Y]$ for a relation R was given just after Proposition 4.1.4.

Proposition 5.2.9. *Let $f : A \to B$ be a map, and let $X \subseteq A$ and $Y \subseteq B$ be subsets. Then $f[X]$ and $f^{-1}[Y]$ are sets.*

Proof. Recall that $f[X] = \{f(u) | u \in X\}$ and $f^{-1}[Y] = \{v \in A | f(v) \in Y\}$. Then $f[X] = \{y \in B | \exists x\, (x \in X \wedge \langle x, y\rangle \in f)\}$. This is a set by Corollary 5.1.12. The case of $f^{-1}[Y]$ is similar. □

$f^{-1}[Y]$ is called the *inverse image* of Y (with respect to the given map).

5.2.2 Injective and surjective maps

Definition 5.2.10. A map $F = \langle f, A, B\rangle$ is called *surjective* when $B = \mathrm{cD}(f)$. On the other hand, F is called *injective* when the function f is injective.

Examples 5.2.11. Let us revise the examples of the maps F, G, H of Remarks 5.2.2. The functions $f = g$ are not injective: for instance, $f(0) = 0 = f(\pi)$ so there exist different elements with the same image; and the maps F, G are not injective. However, h is injective and thus H is an injective map. On the other hand, $\mathrm{cD}(f) = \mathrm{cD}(g) = [-1, 1]$ so that F is not surjective and G is a surjective map. H is surjective too.

We may also consider the maps at the end of Remarks 5.2.2. The function f consisting of all pairs $\langle x, 4x^2 - 2\rangle \in \mathbb{R} \times \mathbb{R}$ is not injective; for instance, $f(1) = 2 = f(-1)$. So, any map with that function is not injective, but if the final set of the map is $[-2, \infty)$ then the map is surjective, because this is the codomain of the function. The map g from $\mathbb{R} \setminus \{(1 + \pi/2 + k\pi | k \in \mathbb{Z}\}$ to $\mathbb{R}$ is surjective, but not injective, but its restriction $g|_{(1-\pi/2, 1+\pi/2)}$ is injective, therefore it gives an injective map $(1 - \pi/2, 1 + \pi/2) \to \mathbb{R}$. This map is also surjective.

Note that for a function f, there is a unique map with function f which is surjective, namely $\langle f, \mathrm{D}(f), \mathrm{cD}(f)\rangle$ – see Proposition 5.1.18.

It would seem that, for a map, the properties of being injective or surjective are quite unrelated, as the first one depends on the function while the second depends on the codomain. Nevertheless, there is a correspondence between some properties of these two concepts.

Proposition 5.2.12. *Let $f : A \to B$ and $g : B \to C$ be maps, call them F, G. Then:*

(i) If F, G are injective (respectively, surjective), then $G \circ F$ is injective (resp., surjective).

(ii) If the composition $G \circ F$ is injective (respectively, surjective), then F is injective (resp., G is surjective).

Proof. (i) If G, F are injective, then g, f are injective functions by definition. If $x \neq y$ in A, then $f(x) \neq f(y)$ and $g(f(x)) \neq g(f(y))$, thus $g \circ f$ is an injective functional and $G \circ F$ is injective. Similarly, if G, F are surjective, then $\mathrm{cD}(f) = B$ and the codomain of $g \circ f$ is $g[\mathrm{cD}(f)] = g[B]$, which is C by the hypothesis.

(ii) If $G \circ F$ is injective, then $g \circ f$ is injective by definition. Thus if $x, y \in A$ and $f(x) = f(y)$, then $(g \circ f)(x) = (g \circ f)(y)$ and $x = y$ by the hypothesis, and thus f is injective and F is injective. Next, suppose $G \circ F$ is surjective so that $C = \mathrm{cD}(g \circ f) \subseteq \mathrm{cD}(g)$, hence $C = \mathrm{cD}(g)$ and G is surjective. □

Proposition 5.2.13. *Let $F = \langle f, A, B\rangle$ be injective and surjective. Then $\langle f^{-1}, B, A\rangle$ is a map, which we write F^{-1}, and $F^{-1} \circ F = 1_A$ and $F \circ F^{-1} = 1_B$.*

Conversely, suppose that $f : A \to B, g : B \to A$ are maps (call them F, G) such that $G \circ F = 1_A$ and $F \circ G = 1_B$. Then both F and G are injective and surjective, and $G = F^{-1}$, $F = G^{-1}$.

Proof. Suppose F is injective and surjective. Since f is injective, f^{-1} is a function by Propositions 4.3.8 and 5.1.19. Besides $\mathrm{D}(f^{-1}) = \mathrm{cD}(f) = B$ because f is surjective; and $\mathrm{cD}(f^{-1}) = \mathrm{D}(f) = A$, so $f^{-1} : B \to A$ is a map F^{-1}. The relations $f^{-1} \circ f = 1_A$ and $f \circ f^{-1} = 1_B$ are obvious.

Conversely, suppose the equalities $G \circ F = 1_A$ and $F \circ G = 1_B$. Since $1_A : A \to A$ and $1_B : B \to B$ are clearly injective and surjective, F and G are injective and surjective by Proposition 5.2.12. Moreover, for each $x \in A$, $g(f(x)) = x = f^{-1}(f(x))$, and it follows immediately that $g = f^{-1}$ and $G = F^{-1}$. By symmetry, we have the same property by interchanging G, F. □

In the situation of Proposition 5.2.13, the map F^{-1} is called the *inverse map* of F. By Proposition 5.2.13, F^{-1} is the unique map that satisfies the equalities $F \circ F^{-1} = 1_B$ and $F^{-1} \circ F = 1_A$. A map is *bijective* (or a *bijection*) when it is injective and surjective, equivalently when it has an inverse map. According to Proposition 5.2.12, the composition of two bijections is a bijection; and we give next its inverse map.

Corollary 5.2.14. *If $f : A \to B$ and $g : B \to C$ are bijective maps F, G, then $(G \circ F)^{-1} = F^{-1} \circ G^{-1}$.*

Proof. By Proposition 5.2.4,

$$(F^{-1} \circ G^{-1}) \circ (G \circ F) = F^{-1} \circ 1_B \circ F = F^{-1} \circ F = 1_A$$

and the analogous equality is obtained for $(G \circ F) \circ (F^{-1} \circ G^{-1})$.

□

5.2.3 Relations, partitions, operations

When dealing with functionals we have at our disposal the words "function" and "map" to distinguish between a functional class (a functional), a set that is a functional (a function) and the triple which identifies also the initial and final sets (a map). In the case of general (binary) relations, we are not so fortunate, and we only have the word "relation" to be used in a variety of situations. When A is a set and $R \subseteq A \times A$, then the relation R is also a set by separation (Proposition 5.1.14). This is the situation we want to address.

Definition 5.2.15. If A is a set and R is a relation with $R \subseteq A \times A$, we call the pair $\langle R, A\rangle$ a *relation in A*.

Lacking a different term, we also say that the relation R itself is in such case a *relation in A*. Accordingly, we will say that the relation $\langle R, A\rangle$ is reflexive, symmetric, transitive, anti-reflexive or anti-symmetric when R is, respectively, reflexive, symmetric, transitive, anti-reflexive or anti-symmetric in A – see Definition 4.2.1. In particular, when $\langle R, A\rangle$ is reflexive, transitive and anti-symmetric, then we say that $\langle R, A\rangle$ is an *ordering* (or that R is an ordering of A or that R orders A; or even that A is an ordered set, if we may forget the relation R itself). Similarly, $\langle R, A\rangle$ is called an *equivalence* when it is reflexive, transitive and symmetric.

Of course, if R is a relation in A in this sense, R is also a relation in B for other sets, for instance for any $B \supseteq A$. In such event, R is the same relation, but $\langle R, A\rangle$ and $\langle R, B\rangle$ are different. The ambiguity thus produced is perhaps balanced by the gain in ease of speech, letting the context to decide the actual meaning. Besides, the properties that R be transitive in A, symmetric in A, etc. do not change when A is replaced with B in such situation, by Lemma 4.2.2. Thus the ambiguity will hardly produce any damage.

For an arbitrary relation R and object x we have defined the collection $[x]$ in Definition 4.4.6: $[x] = \{y | yRx\}$.

Proposition 5.2.16. *Let $\langle R, A\rangle$ be a relation in A. If $x \in A$, then $[x]$ is a set.*

Proof. Let p_1, p_2 be the projections determined by R, as presented in Examples 5.2.5. Then $[x] = p_1[p_2^{-1}[\{x\}]]$, so it is a set by Proposition 5.2.9. □

This entails that the relationship between partitions and equivalences is tighter when we deal with sets than it is for classes.

Proposition 5.2.17. *Let $\langle R, C\rangle$ be an equivalence. Then there is an induced partition C/R which is a set.*

Proof. It can be seen as usual from Corollary 5.1.12, that all ordered pairs $\langle R, A\rangle$ that are relations in a set form a class L. Bearing in mind that $[x]$ is a set by Proposition 5.2.16, consider the class $\{\langle T, x, z\rangle | \exists R_1, A_1\ (T \in L \wedge T = \langle R_1, A_1\rangle \wedge x \in \mathrm{cD}(R_1) \wedge z = [x])\}$. Applying the replacement axiom for that class with object $u = \langle R, C\rangle$ and set $s = C$, we obtain that C/R is a set. □

Thus, in the frame of sets, partitions and equivalences are essentially the same concept. The quotient class C/R is then called the *quotient set* of the set C with respect to the equivalence R in C (or quotient set of C modulo R). Recall that the elements of the partition C/R are the sets $\overline{x} = [x]$ for $x \in C$; and we will indistinctly use one or another symbol in this context, for they represent the same set. Also, if $\langle R, C\rangle$ is an equivalence, the *canonical projection* $\{\langle x, \overline{x}\rangle | x \in C\}$ gives a map $p : C \to C/R$.

Example 5.2.18. Consider the following relation R in the set $\mathbb{Z}^+$ of positive integers. $xRy \Leftrightarrow (\exists n \in \mathbb{N}\ (10^n \leq x < 10^{n+1} \wedge 10^n \leq y < 10^{n+1}))$. This is an equivalence in $\mathbb{Z}^+$, and the elements of the associated partition or quotient set are: the sets of 1-digit numbers (i.e., from 1 to 9), the set of 2-digit numbers (from 10 to 99), the set of 3-digit numbers, and so on. These are the "equivalence classes" (see definition in Definition 4.4.6) $C_1, C_2, C_3, \ldots$ in the quotient set $\mathbb{Z}^+/R$.

Let us now define the following map $f : \mathbb{Z}^+ \to \mathbb{Z}$. Given $x \in \mathbb{Z}^+$, $f(x)$ is the smallest number n such that $x \geq 10^n$. Like this, $f(735) = 2, f(9999) = 3$ and $f(10000) = 4$. The function f has the property that it is constant on the elements of each equivalence class for the above equivalence R: every 4-digit number x will be ≥ 1000 but < 10000, so $f(x) = 3$ for any of the members of that class $[1000]$. This is the kind of functions we want to consider now. By the way, the map $g : \mathbb{Z}^+ \to \mathbb{Z}$ with $g(k) = 0$ when $0 < k < 10000$ and $g(k) = 1$ when $k \geq 10000$ has the same property; for example, if $x < 10000$ and $y \in [x]$ then $y < 10000$ too.

Definition 5.2.19. Let $f : D \to C$ be a map and let E be an equivalence in D. We say that f is *compatible with* E when the following holds for any $x, y \in D$:

$$xEy \Rightarrow f(x) = f(y)$$

Proposition 5.2.20. *Let E be an equivalence in the set D and $f : D \to C$ a map which is compatible with E. There exists a unique map $\widehat{f} : D/E \to C$ such that $\widehat{f}(\overline{x}) = f(x)$ for each $x \in D$.*

Proof. Under the given hypotheses, $f[\overline{x}]$ has only one[2] element, $f(x)$. Consequently the set $\{\langle \overline{x}, y\rangle | \exists u\ (y = f(u) \wedge u \in \overline{x})\}$ is functional, hence a function with domain D/E and codomain included in C. Therefore it gives a map $\widehat{f}$ which has the stated property. The uniqueness is obvious. □

Examples 5.2.21. For the equivalence R of Example 5.2.18, the first compatible map f induces the map $\widehat{f} : \mathbb{Z}^+/R \to \mathbb{Z}$ which gives $[x] \mapsto$ number of digits of x, minus one. For instance, $\widehat{f}(\overline{9999}) = f(9999) = f(1005) = 3$. The compatible map g of the same example gives $\widehat{g}(C_1) = \widehat{g}(C_2) = \widehat{g}(C_3) = \widehat{g}(C_4) = 0$; the rest of elements of $\mathbb{Z}^+/R$ have image 1.

As another example, let us consider a function of elementary calculus like $y = tan(x)$. When we identify it as the collection of ordered pairs $\langle x, y\rangle$ with $y = tan(x)$, this is a function in our sense provided that it is a set, which we accept for now and will be justified later. Its domain D consists of all real numbers that are not of the form $\frac{\pi}{2} + k\pi$ for $k \in \mathbb{Z}$, hence it gives a map $f : D \to \mathbb{R}$ (again, assuming that $\mathbb{R}$ is a set). Let us consider in $\mathbb{R}$ the following relation E, which we accept to be a set:

$$E = \{\langle x, y\rangle | x, y \in \mathbb{R},\ \text{and } |x - y| = k\pi \text{ for some integer } k \geq 0\ \}$$

It is easy to see that E is an equivalence, hence the restriction E_D is also an equivalence, according to Lemma 4.2.2; it gives the quotient set D/E. By the properties of $\tan(x)$ (namely, $\tan(x) = \tan(x \pm \pi)$), f is compatible with E. Proposition 5.2.20 shows that there is a map $\widehat{f} : D/E \to \mathbb{R}$ such that $\widehat{f}(\overline{x_0}) = tan(x_0)$.

Remark 5.2.22. Proposition 5.2.20 yields the more practical way for defining maps from a quotient set D/E: the map f is defined in D and proved to be compatible with E; this determines uniquely the map in D/E by the proposition.

The property that a map f is compatible with an equivalence E is usually referred to by saying that f is *independent of representatives* (of the equivalences classes modulo E); or that $\widehat{f}$, defined as above from f, is *well defined.*

We now turn to binary operations.

Definition 5.2.23. Let C be a set. A *binary operation in the set C* is a map $* : C \times C \to C$.

[2] Beware with this notation $f[\overline{x}]$; it is the notation for $R[A]$ introduced just before Example 4.1.5 that we are using here. And $f[\overline{x}]$ could also have been written $f[[x]]$, as $\overline{x} = [x]$; in that case, the symbol $[\dots]$ would have been employed with two different meanings.

Remark 5.2.24. This abstract concept embraces many of the operations we use in Mathematics: sum of natural numbers is an operation in $\mathbb{N}$ if we identify the operation with a set which contains all ordered pairs $\langle\langle n,m\rangle, n+m\rangle$. We have the same for multiplication of real numbers or addition of 2×3 matrices over $\mathbb{R}$, etc. In these cases, it is customary to write, say $x+y$ or $x\cdot y$ to denote the image in the operation of the pair $\langle x,y\rangle$. We will follow this same usage; moreover, similarly to the use of variables $x, y, \ldots$ to stand for elements of the considered collections, we use also symbols for variable operations, for instance $\star, \diamond$; and write then $a\star b$ or $u\diamond v$ to refer to the images of $\langle a,b\rangle$ or $\langle u,v\rangle$ under the operation denoted by the variable symbol. We also use known operation symbols, like $+, \cdot$, to represent abstract operations. When using $+$ or $\cdot$ in this sense, we do not attach to these symbols their usual meaning; instead they represent variable operations.

The analogous to Definition 5.2.19 is:

Definition 5.2.25. Let C be a set and $\cdot$ a binary operation in C. Let E be an equivalence relation in C. We say that the operation $\cdot$ is *compatible with the equivalence* E when the following holds for arbitrary elements $a, b, u, v \in C$:

$$(aEu, bEv) \Rightarrow ((a\cdot b)E(u\cdot v))$$

Corresponding to Proposition 5.2.20, we have:

Proposition 5.2.26. *Let C be a set, E an equivalence in C and $\cdot$ a binary operation in C which is compatible with E. There is a unique operation $\widehat{\cdot}$ in C/E such that, for every $u, v \in C$, the following holds:*

$$\overline{u}\ \widehat{\cdot}\ \overline{v} = \overline{u\cdot v}$$

Proof. By the principle of separation (Corollary 5.1.12) and Proposition 5.2.17, $P = \{\langle x,y,z\rangle \in (C/E)^3 | \exists u, v\,(u\in x \wedge v\in y \wedge u\cdot v\in z)\}$ is a set. Moreover, the compatibility hypothesis, together with condition (2) in the definition of partition (Definition 4.4.1) entails that it is a function: $u'\in x\wedge v'\in y \Rightarrow u'Eu \wedge v'Ev \Rightarrow u'\cdot v'\in z$. Thus P is a binary operation in C/E which we denote as $\widehat{\cdot}$. The same reason proves the equation in the statement, in view of Proposition 4.4.8. Uniqueness is obvious. □

When the operation $\cdot$ is compatible with the equivalence E, we usually say that $\cdot$ is *independent of representatives*; or that the new operation $\widehat{\cdot}$ is *well defined*. Again, this is the more common way for defining an operation on a quotient set C/E.

5.3 Exercises

1. Prove that if $\mathcal{U}$ is a 3-universe and C is a collection of n objects of $\mathcal{U}$ (for n a natural number), then C is a set.

2. Let $\mathcal{U}$ be the universe consisting only of the objects $\emptyset, \{\emptyset\}, \{\emptyset, \{\emptyset\}\}$. Prove that $\mathcal{U}$ satisfies the axiom of unions, but does not satisfy the axiom of the power set.

3. Find a formula which can be abbreviated as $y = \mathrm{D}(f)$. Same question for "y is a function with domain included in x".

4. Consider $\emptyset$ as the function of a map $\emptyset : \emptyset \to A$. When is $\emptyset$ an injective or a surjective map?

5. We admit that there is a function f whose domain is $\mathbb{R}^2$ and given through $f(\langle x, y\rangle) = \langle x + y, xy\rangle$. Is it injective? Identify $\mathrm{cD}(F)$, $F[(\mathbb{R}^+)^2]$ and $F^{-1}[(\mathbb{R}^+)^2]$.

6. Assuming that $\mathbb{R}$ is a set, let us consider two functions with domain $\mathbb{R}$ given by $f(x) = \sqrt{x^2+2}$ and $g(x) = 2x + 7$. Determine the composed functions $g \circ f$ and $f \circ g$. Say if any of the two functions f, g has an inverse function and compute that inverse in such case.

7. The integer part $[x]$ of a real number x is the unique integer that satisfies $[x] \leq x < 1 + [x]$. Let $\sim$ be a relation in $\mathbb{R}$ given by $x \sim y \Leftrightarrow [x] = [y]$. We assume that $\mathbb{R}$ is a set and $\sim$ is an equivalence in $\mathbb{R}$. Call $\mathbb{R}'$ the associated partition (i.e., the quotient set) and determine for each of the following equations whether it gives a well-defined function with domain $\mathbb{R}'$:

   (i) $F(\overline{x}) = [x - 2]$.

   (ii) $F(\overline{x}) = [x^2]$.

8. In a 4-universe where the principle of separation holds, let F be a relation and assume that for every set s such that $s = \mathrm{D}(F|_s)$, $F[s]$ is a set. Prove that the F-instance of the simple form of the axiom of replacement (Proposition 5.1.4) holds (Hint: consider the class $\{\langle \emptyset, x\rangle | \exists y\, (\langle x, y\rangle \in F)\}$).

9. State the property corresponding to that of the preceding exercise, but for the normal F-instance of replacement, now with F a class of triples. Prove it if it holds or show that it does not hold if this is the case.

10. Let s be a set, and $s' = s \cap \mathcal{S}$ (the class of the elements of s that are sets). Prove that $\bigcup s = \bigcup s'$.

11. Let s be a non-empty set whose elements are subsets of a set A, and f a function with domain A. Then $\bigcup s \subseteq A$ and $\bigcap s \subseteq A$. Prove that $f[\bigcup s] = \bigcup\{f[u] | u \in s\}$ and that $f[\bigcap s] \subseteq \bigcap\{f[u] | u \in s\}$. Is $f[\bigcap s] = \bigcap\{f[u] | u \in s\}$? Prove or give a counterexample.

12. Find a class C such that the C-instance of the strong form of the axiom of replacement implies the axiom of unions (Hint: consider the class of triples $\langle u, x, y\rangle$ such that $x \in u$ and $y \in x$).

13. Let us admit that $\mathbb{N}, \mathbb{Z}, \mathbb{Q}, \mathbb{R}$ are sets of the universe $\mathcal{U}$. For each of the maps described below (we assume that f is in each case a class, hence a set), identify its image. Also, say which of those maps is injective, which is surjective, which is bijective.

    a) $f : \mathbb{Z} \to \mathbb{Z}, \quad f(x) = -x^3$.

    b) $f : \mathbb{Z} \to \mathbb{Z}, \quad f(x) = x^2 + 5$.

    c) $f : \mathbb{N} \to \mathbb{Q}, \quad f(x) = \frac{x^2}{x+1}$.

    d) $f : \mathbb{R} \to \mathbb{R}^+ \cup \{0\}, \quad f(x) = |x|$.

14. Let $f : A \to B$ be a map. Prove that f is injective if and only if the following holds for any subset $X \subseteq A$: $f[X^c] \subseteq f[X]^c$ (complements are taken relative to A and B respectively).

15. Let F be a map $f : A \to B$; X, Y subsets of A, B respectively. Prove:

    (i) $f[f^{-1}[Y]] \subseteq Y$.

    (ii) $X \subseteq f^{-1}[f[X]]$.

    (iii) The inclusion of (i) (respectively, of (ii)) is an equality when F is surjective (resp., injective).

    The notation follows Proposition 5.2.9.

16. Show that for given sets A, B the collection of all injective maps or the collection of all surjective maps from A to B are sets.

17. A relation R in the set A is called a *pre-order* in A when R is reflexive and transitive in A. Let E be the relation in A defined by $xEy \Leftrightarrow (xRy \wedge yRx)$. Prove that E is an equivalence relation in A and that R is compatible with E (this means that if xRy and $x'Ex, y'Ey$ then $x'Ry'$). Prove that it is possible to define a relation $\widehat{R}$ in the quotient A/E such that $[x]\widehat{R}[y] \Leftrightarrow xRy$ for every $x, y \in A$. Show furthermore that $\widehat{R}$ is an order in A/R.

18. In the same set F of Exercise 11 of Chapter 4, define the relation $X \; R \; Y \Leftrightarrow$ there exists an injective map $f : X \to Y$. Show that R is a pre-order in F, and describe the equivalence E and the relation $\widehat{R}$ mentioned in the previous exercise.

19. Let $F = f : A \to B$ be a map. Prove that f is injective if and only if the following holds for every set X: if G, H are maps from X to A such that $F \circ G = F \circ H$, then $G = H$.

20. Let f be a function whose domain is the set A. Let us define a relation $\sim$ in A by setting: $a \sim b \Leftrightarrow f(a) = f(b)$. Prove that $\sim$ is an equivalence relation in A. Consider the quotient set $A/\sim$ and use Proposition 5.2.20 to show that the function $\widehat{f}$ with domain A/R and given by $\widehat{f}(\overline{x}) = f(x)$ is well-defined (and thus $f = \widehat{f} \circ p$, where

p is the canonical projection $A \to A/R$). Prove moreover that $\widehat{f}$ is injective.

21. Let $F_1 = \langle f, A, B \rangle$ be a map. Show that there is a unique subset $C \subseteq B$ such that $F_2 = \langle f, A, C \rangle$ is a surjective map. Call $j : C \to B$ to the inclusion map. Prove that $F_1 = J \circ F_2$ and that F_2 is injective if and only if F_1 is injective.

22. Use the previous exercises to prove the following result: if $F = f : A \to B$ is a map, then there are maps R, S, T such that $F = R \circ S \circ T$, R is injective, S is bijective and T is surjective.

23. Give a different proof of Proposition 5.2.26, by following the steps: (1) Define an equivalence E' in the set $C \times C$ by extending E; (2) From the natural map $\varphi : C \times C \to (C/E) \times (C/E)$ and applying Proposition 5.2.20, obtain a bijective map $\psi : (C \times C)/E' \to (C/E) \times (C/E)$; (3) Consider the composition $C \times C \dot{\to} C \xrightarrow{p} C/E$ with p the canonical projection, and use again Proposition 5.2.20 to get $h : (C \times C)/E' \to C/E$; (4) Take the composition $h \circ (\psi)^{-1}$ and show that this is the required map.

6

Numbers and Infinity

One of the main aims of set theory is to provide the soil on which all mathematical objects (numbers of any kind, real or complex functions, polynomials, algebraic structures, etc.) may grow. In this chapter, we focus on how set theory fulfills this task, starting with the most basic objects, natural numbers. A new axiom, the axiom of infinity, will be necessary to accomplish this objective.

6.1 The Axiom of Infinity

6.1.1 Dedekind-Peano sets

If set theory has to be a foundation for mathematics, surely natural numbers and the collection of all natural numbers have to be objects of the universe of set theory. Elucidating what are the natural numbers and how we can define them is not a simple task and, in fact, it has been considered by many first-level mathematicians, partly because this has been the battlefield between different views on the nature of mathematical knowledge. The approach of set theory to this issue is somewhat peculiar: instead of trying to capture the essence of natural numbers, it simply constructs objects of the universe whose properties are precisely the properties that the natural numbers have. The fundamental properties that the natural numbers are supposed to have are the *Peano's postulates*. They were formulated by Peano with the intention of characterizing the series of the natural numbers. In our approach, Peano's postulates become a definition: we shall call a *Dedekind-Peano set* to any set which satisfies these postulates. Although our definition of these sets follows closely Peano's views, it is perhaps fit to include also Dedekind's name: in 1888, a year before Peano, Dedekind had given another, roughly similar, characterization for natural numbers. And by naming these sets as Dedekind-Peano (alphabetical order), the contribution of both great figures is acknowledged.

In this section, we introduce a new axiom and find a particular set (essentially unique in each universe) satisfying the conditions in the definition of Dedekind-Peano sets. If not a copy of the collection of ordinary natural numbers, it behaves as it is expected that natural numbers behave.

DOI: 10.1201/9781003449911-6

Definition 6.1.1. A *Dedekind-Peano set* is a triple $\langle N, e, S\rangle$ where N is a set, $e \in N$ and $S : N \to N$ is a map, such that the following properties hold (Note that we are using the same letter S to denote the map and the function associated to it; this does not follow our notational conventions but is necessary to avoid confusion with the functional s which gives the successor $s(x)$ of a set x).

(i) $\mathrm{cD}(S) = N \setminus \{e\}$.
(ii) The map S is injective.
(iii) Let $A \subseteq N$ be a subset of N with $e \in A$ and suppose that the implication $x \in A \Rightarrow S(x) \in A$ holds. Then $A = N$.

One can easily recognize in these conditions the basic properties which describe the natural numbers, if 0 is the interpretation of e and S is seen as associating to each natural number the next one or *successor* (note that, in principle, this sense of successor in the series of natural numbers is not the same as the successors of sets defined before Proposition 5.1.2): each number has a successor, which is the central idea in the construction of numbers; and each natural number except 0 is the successor of some number; finally condition (iii) is a form of the familiar *principle of induction* for natural numbers. Our purpose now is to assure that some member of the universe is a Dedekind-Peano set, thus identifying the collection of all natural numbers (or a collection that has the same properties of the natural numbers) as an object of the universe.

6.1.2 Inductive sets

We assume that the universe $\mathcal{U}$ is a 5-universe. Recall that $\mathcal{S}$ is the class of all sets and for each $x \in \mathcal{S}$, $s(x)$ denotes the successor of x, i.e., $s(x) = x \cup \{x\}$.

Definition 6.1.2. A set c is *inductive* if it satisfies the following two conditions:

(1) $\emptyset \in c$.
(2) $x \in c \cap \mathcal{S} \Rightarrow s(x) \in c$.

Proposition 6.1.3. *The collection of all inductive sets is a class.*

Proof. By Proposition 5.1.2, the successor functional s is a class. The collection of all sets c that are inductive can be identified by the following formula:

$$\emptyset \in c \wedge (\forall x\ ((x \in c \wedge x \in \mathcal{S}) \Rightarrow (\exists y, z\ (z = \langle x, y\rangle \wedge z \in s \wedge y \in c))))$$

hence it is a class by Corollary 3.2.7. $\square$

Proposition 6.1.4. *Suppose that the universe $\mathcal{U}$ contains some inductive set. Then the intersection of the class of all inductive sets is an inductive set, all of whose elements are sets.*

Proof. Let D be the class of inductive sets, which is not empty by hypothesis. Its intersection is a set ω by Proposition 5.1.16 and we show that ω is inductive: $\emptyset \in \omega$ obviously; and if $x \in \omega \cap \mathcal{S}$ then, for every $u \in D$ we have $x \in u$ and $s(x) \in u$, whence $s(x) \in \omega$, showing that ω is inductive.

We see now that $\omega \cap \mathcal{S}$ is an inductive set: first, it is a set by the $\mathcal{S}$-instance of separation, and also $\emptyset \in \mathcal{S}$; and if $x \in \omega$ is a set, then $s(x) \in \omega \cap \mathcal{S}$. Now, since ω is the smallest[1] inductive set and $\omega \cap \mathcal{S}$ is inductive, $\omega = \omega \cap \mathcal{S}$ so that every element of ω is a set. □

The hypothesis of Proposition 6.1.4, that is, the assertion that the universe $\mathcal{U}$ contains some inductive set is essentially one of the original axioms of Zermelo, the *Axiom des Unendlichen*; though Zermelo gave the axiom in a slightly different form, with the condition $x \in c \Rightarrow \{x\} \in c$ substituted for (2) of Definition 6.1.2.

Axiom 6 (of infinity). The universe $\mathcal{U}$ contains some inductive set.

If a 5-universe $\mathcal{U}$ satisfies the axiom of infinity, we say that $\mathcal{U}$ is a 6-universe; a 6-universe $\mathcal{U}$ contains an inductive set which is the intersection of all inductive sets. It is called *the smallest inductive set* and written ω.

We have denoted as s the successor class, formed with all pairs $\langle x, s(x)\rangle \subseteq \mathcal{S} \times \mathcal{S}$. Thus $s \cap (\omega \times \omega)$ is a functional class and a set by separation. It is the function of a map $s : \omega \to \omega$, by the definition of inductive sets. We use now the same letter s to denote this map and its function, besides its general use as the name of the successor functional in $\mathcal{S}$. We hope that the context will clarify the possible ambiguities, and this abuse will not generate confusion.

We see now that ω satisfies Peano's postulates.

Proposition 6.1.5. *Assume that $\mathcal{U}$ is a 6-universe, and let ω be the smallest inductive set. Then the triple $\langle \omega, \emptyset, s\rangle$ is a Dedekind-Peano set.*

[1]The word "smallest" has already been used here and it seems necessary to precise its meaning, since it has not been defined as a technical term. "Smallest" refers to some order relation in a set or class, and means "minimum". For instance, in Example 5.2.18 we read: $f(x)$ is the smallest number n such that $x > 10^n$; the relation there is the usual order of the natural numbers, and "smallest" and "minimum" are synonymous. When the relation to which "smallest" refers is not explicit and not clear from the context, it is normally the inclusion relation. This is the case here: ω is the smallest inductive set means that it is minimum with respect to inclusion in the class of all inductive sets. That is: ω is an inductive set; and it is included in any other inductive set.

Proof. First, $\emptyset \in \omega$ because ω is inductive; and $s : \omega \to \omega$ is a map. We show that the properties of Definition 6.1.1 are satisfied:

(i) It is clear that $\emptyset \neq s(x)$ for any set x, so $\emptyset \notin \mathrm{cD}(s)$. Moreover, $\mathrm{cD}(s) \cup \{\emptyset\}$ is plainly an inductive subset of ω, hence $\mathrm{cD}(s) \cup \{\emptyset\} = \omega$ and $\mathrm{cD}(s) = \omega \setminus \{\emptyset\}$.

ii) Take $A = \{x \in \omega | \forall u \in x, \forall y \in \omega\ (s(u) = s(y) \Rightarrow u = y)\} \subseteq \omega$. A is a set by Corollary 5.1.12. We check that A is inductive and thus we may deduce $A = \omega$. This yields injectivity of $s : \omega \to \omega$, because if $s(x) = s(y)$, it follows from the properties $s(x) \in A$ and $x \in s(x)$ that $x = y$.

To prove that A is inductive: $\emptyset \in A$ is vacuously true. Then let $x \in A$, and suppose that $s(x) \notin A$. Since every $u \in x$ satisfies the property in the definition of A, this property must fail for x, hence there is some $y \in \omega$ with $x \neq y$ and $s(x) = s(y)$. But $y \in s(y) = s(x)$ and $y \neq x$, hence $y \in x$, but this is a contradiction with the property $x \in A$.

(iii) Suppose $B \subseteq \omega$ is as in the hypothesis of (iii) of Definition 6.1.1. Then B is inductive and included in ω, therefore $B = \omega$. □

6.2 A Digression on the Axiom of Infinity

6.2.1 The set ω and natural numbers

A natural question to ask ourselves now is: is the set ω, whose existence in 6-universes is guaranteed by the axiom of infinity, a perfect copy of our intuitive natural numbers? To study this problem, let us introduce the following notation, for any set $x \in \mathcal{S}$ and a natural number (in the ordinary intuitive sense) n: $s^n(x) = s(s(s(\ldots(s(x))\ldots)))$, where n is the number of repetitions of the symbol "s" in the right side of the equation. We consider here $s^0(x) = x$ and thus $s^1(x) = s(x)$ or $s^4(x) = s(s(s(s(x))))$. These objects of $\mathcal{U}$ are perfectly identified following our assumptions on natural numbers in Section 2.6: we start with $x \in \mathcal{S}$ and thus $s(x)$ is a set. We understand from those assumptions that by adding a symbol s at a time, we may obtain $s^2(x), s^3(x)$ as sets; and that we may eventually arrive to the set $s^n(x)$ in finitely many steps.

Starting with $x = \emptyset$, the series of these elements $\emptyset, s(\emptyset), s^2(\emptyset), \ldots$ is completely parallel to the usual series of the natural numbers $0, 1, 2, 3, \ldots$. This series of ordinary natural numbers is commonly described as "$0, 1, 2, 3, 4$, and so on". But it is really hard to give a clear logical explanation for the expression "and so on" here. And yet, the intuitive idea of this infinite sequence is shared without differences by everybody. It is probably because we have been taught that series from our first childhood, but certainly the idea of adding one number at a time and following indefinitely in this way, always in the same form and never ending, is a concept that has become understood with astonishing homogeneity everywhere. It is this intuition, the intuition of potential infinity, that we are accepting here as common knowledge.

Proposition 6.2.1. *For each ordinary natural number n, $s^n(\emptyset)$ belongs to any inductive set; hence, in any 6-universe, $s^n(\emptyset) \in \omega$.*

Proof. Use RAA and suppose that there is an inductive set X and a particular natural number n such that $s^n(\emptyset) \notin X$. This element is $s(s(s(\dots s(\emptyset)\dots)))$ (with n occurrences of s), and it follows from Definition 6.1.2 that, for any set t, $s(t) \notin X$ implies $t \notin X$. Thus, by deleting one s at a time, we will finally deduce (by our assumption of Section 2.6 that we may obtain n after finitely many steps from 0) that $s(\emptyset) \notin X$ and $\emptyset \notin X$, which contradicts the definition of inductive set. □

Thus, in a 6-universe, the set ω contains all the objects $s^n(\emptyset)$; and these objects may be considered as faithful copies of the ordinary natural numbers, each n being represented by $s^n(\emptyset)$. Can we conclude that, in a 6-universe, the set ω consists precisely of all the objects $s^n(\emptyset)$?

What we first observe in this connection is that when set theory is treated as a formal symbolic theory, the question cannot even arise. Because referring to all these objects $s(s(s(\dots(s(\emptyset))\dots)))$ by mentioning them one by one is impossible; and an abstract general description of them is also unavailable without the intuition of potential infinity, which is the idea that may link them all. So, the question is meaningless unless we accept some intuitions on the natural numbers, like those we stated in Section 2.6. Consequently, the option followed in many set theory books is to **define** the natural numbers as the elements of ω, since the issue about the relationship between ω and ordinary numbers cannot be completely solved.

In our presentation of set theory, the question may be posed. And we may consider the following statement, which might be seen as an alternative to the axiom of infinity.

Assertion NNS (for "Natural Numbers form a Set"). There is a set $u \in \mathcal{U}$ which contains $s^n(\emptyset)$ for each natural number n; and contains no other element.

Proposition 6.2.2. *Let $\mathcal{U}$ be a 5-universe where assertion NNS holds, and let u be the set whose existence is asserted in NNS. Then $\mathcal{U}$ is a 6-universe and $u = \omega$.*

Proof. By definition, $\emptyset \in u$. And for any $x \in u$, $x = s^n(\emptyset)$ hence $s(x) = s(s^n(\emptyset)) = s^{n+1}(x) \in u$. Thus u is inductive and $\mathcal{U}$ is a 6-universe. Also $u \subseteq \omega$ by Proposition 6.2.1 and hence $u = \omega$. □

Thus the question we asked above about ω can be now repeated in this form: does any 6-universe satisfy NNS? We know that in a 6-universe, ω contains all the objects $s^n(\emptyset)$. If those objects form a set (which is what assertion NNS says), then we have seen that this set would equal ω. But if we do not assume NNS, then we do not know whether the objects $s^n(\emptyset)$ form a set; and if they do not (and nothing in the axioms so far forces us to believe that there is a set containing them and nothing more), then ω cannot contain just those objects, since ω is a set. Therefore, ω would in such case contain other objects. In general, ω may be seen as an abstract set-theoretic completion for all the elements $s^n(\emptyset)$, but we cannot be sure about which are all the elements of ω because we cannot be sure about which are the objects of the universe, as they are known to us only through the axioms.

Briefly, in 6-universes, either assertion NNS holds and ω consists of the objects $s^n(\emptyset)$; or it does not hold and the objects $s^n(\emptyset)$ do not form a set, hence ω contains other objects.

6.2.2 Axiom 6 and property NNS

The fundamental difference between the axiom of infinity and assertion NNS has already been mentioned: with NNS, we postulate the existence in the universe of a faithful copy of the collection of the usual natural numbers; the axiom of infinity constructs a set with the properties of the natural numbers (the Peano's postulates), but cannot guarantee that this is a perfect copy.

Nevertheless, many mathematicians (as many mathematicians in the past) would prefer to avoid using an assertion which mentions ordinary natural numbers, like NNS, as an axiom of set theory. First, when set theory is viewed as a formal symbolic theory, which is quite usual, there is no choice: NNS cannot be stated. Even if one is willing to follow an intuitive description of the theory, it is possible to object the use of intuitive assumptions about numbers as a preliminary requisite for the theory. For instance, it is possible to have doubts as to the clarity of the idea of the potentially infinite series of the numbers. Also, the intuitive idea of finite numbers as corresponding with some process that can effectively be carried out if we have enough time, is mixing some subjective, perhaps psychological element into the objective realm of numbers; and one may think that this spoils the pure nature of mathematics.

But probably the main reason for rejecting the use of intuitions on natural numbers in set theory, is the belief that set theory should be a sound foundation for all of mathematics, including the arithmetic of natural numbers. The idea that the essence and properties of the natural numbers are derivable from the laws of logic goes back at least to Frege, and was further developed by the logicist school in the first years of the 20th century. Certainly, this idea has not been universally accepted and, for example, Hilbert said that *Mathematics [...] can never be grounded solely on logic. Consequently, Frege's and Dedekind's attempts to so ground it were doomed to failure.* But it has had a considerable impact in the way mathematicians see the relative role of arithmetic and set theory. And if one considers set theory as more abstract and clear than arithmetic, the desire to make the first one the basis for all of mathematics and in particular for arithmetic is understandable.

Let us now focus on the coincidences between NNS and the axiom of infinity. Both in general 6-universes or in 6-universes where NNS holds, ω is the smallest inductive set. If we see $\emptyset$ as a representation of 0 and we accept to identify both, i.e., $0 = \emptyset$, then $s(\emptyset) = \{\emptyset\} = \{0\}$ could be a representation of 1. Calling $1 = \{\emptyset\}$, $s^2(\emptyset) = s(\{\emptyset\}) = s(1) = \{\emptyset, \{\emptyset\}\}$ can be called 2; and $s^3(\emptyset) = \{\emptyset, \{\emptyset\}, \{\emptyset, \{\emptyset\}\}\} = s(2) = \{0, 1, 2\}$ under the above representations. Thus we may identify each number n and $s^n(\emptyset)$, by making n to be seen as the set of all the numbers preceding n, $n = \{0, 1, \ldots, n-1\}$, so n is a set with n elements. Accordingly, we accept to write these elements $s^n(\emptyset)$ of ω as ordinary natural numbers. By the same token, the set ω itself is written as $\mathbb{N}$ in contexts related to arithmetic. And, following the identification, the elements of ω will be called *natural numbers*. When $\mathcal{U}$ satisfies NNS, ω contains only the objects $s^n(\emptyset)$; this cannot be assured for arbitrary 6-universes.

For the rest of Part I in this book, we assume that $\mathcal{U}$ is a 6-universe, but do not assume property NNS; and we construct in this setting the set theoretic version of the different kinds of numbers and traditional objects of mathematics, as well as the theory of cardinals and ordinals. But accepting assertion NNS is closer to our intuition about finiteness, and in the second part of the book we will work in 6-universes where NNS holds. But even then, the only properties of ω that we shall use are those obtained in the first part, and thus everything will be valid in general 6-universes too. The difference appears only with the intuitive interpretation of the concepts and results: without NNS, ω might contain other elements; and these elements contain infinitely many (in the ordinary sense) objects.

Comment 6.2.3. About the finiteness question that has just arisen, let us mention a couple of definitions of finite sets which have been proposed in the literature. The most popular is probably Dedekind's definition: a set S is infinite when there exists a subset $T \subsetneq X$ (we use the notation $T \subsetneq S$ to mean that $T \subseteq X$ and $T \neq X$), and a bijective map $f : T \to S$. Then S is finite if it is not infinite.

A second definition was proposed by Russell and Whitehead: the set S is finite when it belongs to any inductive family (= set) of subsets of S; and a set F of subsets of S is called inductive when it satisfies: (i) $\emptyset \in F$; and (ii) $A \in F, x \in S \Rightarrow A \cup \{x\} \in F$. A third one is Tarski's definition: the set S is finite when every non-empty family of subsets of S contains a minimal element with respect to inclusion.

These three definitions of finite sets are equivalent to the basic definition: a set S is finite when there is a bijective map from some element of ω to S. While this definition matches well our intuitive idea of finite when assertion NNS holds, in general 6-universes it is not necessarily equivalent to that intuitive idea. Let us examine just one example of a fallacious argument that could try to prove the contrary, working in a 6-universe where NNS is not assumed to hold, but where we accept the intuitions about ordinary natural numbers.

Suppose that ω contains some element c such that, for every usual natural number n, $c \neq s^n(\emptyset)$. Now, since $c \neq \emptyset$, there is $p(c) \in \omega$ such that $c = s(p(c))$ – $p(c)$ is the *predecessor* of c. For the same reason, we may obtain infinitely many (in our usual sense of infinity) elements $p(p(c)), p(p(p(c)))$, etc; and none of them can be of the form $s^n(\emptyset)$. We might then construct a descending chain of subsets of c, $c \supsetneq p(c) \supsetneq p^2(c)$ etc., which has no minimal element. Therefore c is infinite in Tarki's sense, but it is proved that every element of ω is finite in Tarski's sense, so this gives a contradiction from the assumption that such element c exists.

The fallacy lies here in that the objects $p^n(c)$ need not form a set, by the same reason that the $s^n(\emptyset)$ may fail to form a set. So, they escape Tarski's condition, as the family in the statement has to be a set. A similar fallacy occurs in a possible argument based on Dedekind's characterization of finite sets. One could think that there is a bijective map between $c \setminus \{\emptyset\}$ and c, obtained by simply taking $s^n(\emptyset) \mapsto s^{n+1}(\emptyset)$ and being the identity elsewhere. As above, this would entail that c is not a finite set. But the function of a map must be a set by definition; and the definition of this function by cases uses the identity function on some "subset", consisting of all the elements other than the $s^n(\emptyset)$, a collection which is not known to be a set.

6.3 The Principle of Induction

For the rest of this chapter, $\mathcal{U}$ will be a 6-universe and ω is the smallest inductive set.

6.3.1 Natural induction

The fact that ω satisfies condition (iii) of Definition 6.1.1 is known as the *principle of natural induction.* Note that a weaker version of this property for ordinary natural numbers is a consequence of our assumptions on numbers in Section 2.6; because if we know that $x \in A \Rightarrow s(x) \in A$ and $0 \in A$, then we may prove $k \in A$ for each number k, by starting with $0 \in A$ and repeatedly using the implication to successively obtain $1 \in A, 2 \in A$, etc., until we prove $k \in A$. But what intuition promises is only that under the hypothesis of condition (iii), we could give a proof that $k \in A$ for any given ordinary natural number k; condition (iii) of Definition 6.1.1 is an audacious extension: the hypothesis implies in just one step that every $x \in \omega$ belongs to A (even those elements of ω, if they exist, that would be infinite sets for the intuitive meaning of "infinite"). This principle is the basis for many proofs, which are called *proofs by (natural) induction.* We use this method of proof from the next result.

We define a set A to be *transitive* if the following is satisfied: $(x \in A \text{ and } y \in x) \Rightarrow y \in A$. Equivalently, $x \in A \cap \mathcal{S} \Rightarrow x \subseteq A$. It is easily seen that transitive sets form a class.

Proposition 6.3.1. *The set ω is transitive. Also, if $x \in \omega$ then x is transitive.*

Proof. We prove that ω is transitive by induction, and this proof will serve us to present some common features of such proofs. In this case, we want to show that every $x \in \omega$ satisfies the property $x \subseteq \omega$ (x is a set by Proposition 6.1.4). Usually, proofs by natural induction seek precisely to prove that every $x \in \omega$ satisfies a certain property $P(x)$, which in this case is $x \subseteq \omega$.

The typical proof consists of three steps. First, to check that the elements $x \in \omega$ that satisfy $P(x)$ form a set $A \subseteq \omega$; when $P(x)$ is a formula with the free variable x (and possibly some parameters), this is a consequence of Corollary 5.1.12 and if it is so, we normally will not even mention this step. Second, to check that $\emptyset \in A$, that is, $P(\emptyset)$ holds – this is frequently an almost trivial step. Third, to prove that $x \in A \Rightarrow s(x) \in A$, which is the same as showing that $P(x) \Rightarrow P(s(x))$. This third step is usually referred to as the *inductive step.* In the conditional assertion $P(x) \Rightarrow P(s(x))$ the condition $P(x)$ is called the *induction hypothesis.* By the principle of induction, this third step will now imply that $A = \omega$ by (iii) of Definition 6.1.1 and Proposition 6.1.5; that is, $P(x)$ holds for every $x \in \omega$. There are several variations on this scheme that we will consider in due time.

In our case, $P(x)$ is $x \subseteq \omega$ which is a formula, so we are done with the first step. The second step is to show that $\emptyset \subseteq \omega$, and this is trivial.

The induction hypothesis is that $x \subseteq \omega$ (for some arbitrary $x \in \omega$), and we must show from this the conclusion $s(x) \subseteq \omega$. But $s(x) = x \cup \{x\} \subseteq \omega$ follows directly from the inductive hypothesis and the fact that $x \in \omega$. This proves that ω is transitive.

To show that every element of ω is transitive, we use induction again, by considering now the set $A = \{x \in \omega | x \text{ is transitive } \}$. This is a set because the collection of all transitive sets is a class. $\emptyset$ is transitive because $x \in \emptyset \Rightarrow x \subseteq \emptyset$ is vacuously true.

To prove the inductive step by RAA, suppose x is transitive but $s(x)$ is not. This would imply that there are $y \in s(x)$ and $z \in y$ such that $z \notin s(x)$. But if $y \in x$, then $z \in x \subseteq s(x)$ by the transitivity of x; otherwise, $y = x$ hence $z \in x \subseteq s(x)$ again, giving a contradiction. □

Recall that $\mathcal{S}_1$ is the (proper) class of all sets x such that $x \notin x$. It seems to pertain to the very idea of a set that any set belongs to $\mathcal{S}_1$; because the existence of a set x such that $x \in x$ contradicts the natural idea that a set x is formed with certain well-determined objects whose existence is a previous condition for the existence of x (see, for instance, [12, p.4]). However, one cannot prove $x \notin x$ from the axioms for 6-universes. But for elements of ω this does hold:

Lemma 6.3.2. *If $x \in \omega$, then $x \notin x$.*

Proof. Let $A = \{x \in \omega | x \notin x\}$, and let us prove that $A = \omega$ by induction. Clearly, $\emptyset \in A$. Suppose $x \in A$ and $s(x) \notin A$ so that $s(x) \in s(x)$. Either $s(x) \in x$ or $s(x) = x$. Bearing in mind the transitivity of x (Proposition 6.3.1), either case implies $x \in x$, which contradicts $x \in A$. □

Proposition 6.3.3. *Let $x, y \in \omega$. Then $x \subsetneq y \Leftrightarrow x \in y$.*

Proof. The implication $x \in y \Rightarrow x \subseteq y$ follows from Proposition 6.3.1. In turn, this implies $x \subsetneqq y$ by Lemma 6.3.2.

For the converse implication, let $A = \{y \in \omega | \forall x\ ((x \in \omega \wedge x \subsetneqq y) \Rightarrow x \in y)\}$, which is a set. We use now induction to show that $A = \omega$ and the result will be proved.

First, the property has to be true for $y = \emptyset$. But this means $x \subsetneqq \emptyset \Rightarrow x \in \emptyset$, which is vacuously true. To see the inductive step, suppose $y \in A, x \in \omega$ and $x \subsetneqq s(y) = y \cup \{y\}$. If $x \subsetneqq y$, then $x \in y$ by the induction hypothesis and thus $x \in s(y)$. If $x = y$, then obviously $x \in s(y)$. Otherwise $x \not\subseteq y$ but $x \subseteq s(y)$, so $y \in x$ and also $y \subseteq x$ by transitivity. Therefore $s(y) \subseteq x$ and $x = s(y)$ which contradicts the assumption $x \subsetneqq s(y)$. □

We say that this proof is made by *induction on y*. This explains that we deal here with an assertion $Q(x, y)$ which depends on two variables; that is, we must prove that some subset of $\omega \times \omega$ is the total $\omega \times \omega$. To use induction, we must reduce the assertion to one variable, $P(y)$ being the hypothesis that $Q(x, y)$ holds for all $x \in \omega$; and this is what we have done in this case. An alternative method, called *double induction* is shown in [21] and will be presented in the exercises.

6.3.2 Well-orderings

Definition 6.3.4. Let $\langle \preceq, A\rangle$ be an ordering in a set A. We say that $\preceq$ *well orders A* (or that $\langle \preceq, A\rangle$ is a *well-ordering*) if every non-empty subset u of A has a minimum element.

Remark 6.3.5. Definition 6.3.4 can be applied without change to the strict order relation $\prec$ which is coupled with the ordering $\preceq$, as seen in Proposition 4.2.7; so that $\langle \preceq, A\rangle$ is a well-ordering if and only if $\langle \prec, A\rangle$ is a well-ordering. Note also that a well-ordering is always a total ordering: given $x, y \in A$, the set $\{x, y\}$ has a minimum, hence either $x \preceq y$ or $y \preceq x$.

It is easy to give examples of orders $\langle \preceq, A\rangle$ that are not well-orderings. $\langle \leq, \mathbb{R}\rangle$, $\langle \leq, \mathbb{Z}\rangle$ are, if they are relations in our universe $\mathcal{U}$, total orderings but not well-orderings: for example, the points in the open interval $(0, 1) \subseteq \mathbb{R}$ or the set of integers < -100 have no minimum. On the other hand, it is easy to observe that if $\langle \preceq, A\rangle$ is a well-ordering and $S \subseteq A$, then $\langle \preceq_S, S\rangle$ is also a well-ordering of S – for the meaning of the restriction $\preceq_S$ see the observations before Example 4.1.5.

As a trivial example, $\emptyset$ with the empty relation is well-ordered, since the property of the definition is vacuously true in this case. The archetypical example of a well-ordered set is ω or any of its elements $x \in \omega$, as will be seen in the next proposition. For a relation $\langle \preceq, A\rangle$ with $A \neq \emptyset$, to be a well-ordering entails that A contains a minimum element, a_0; and if $A \setminus \{a_0\} \neq \emptyset$, then there is some a_1 which is a minimum element of that set; similarly, there is a_2, the minimum of $A \setminus \{a_0, a_1\}$, and so on; this gives an order of the type $a_0 \prec a_1 \prec a_2 \prec a_3 \prec \dots$, with no elements in between. So, a well-ordering is basically like the usual ordering of the natural numbers; except that the definition can be applied to sets much, much bigger than the natural numbers, and than ω.

Proposition 6.3.6. $\mathcal{M} \cap (\omega^2) = \{\langle x, y\rangle | x, y \in \omega \wedge x \in y\}$ *is a relation in* ω *and is a strict ordering of* ω*. The corresponding partial order (see Proposition 4.2.7) is given by* $x \subseteq y$*.*

Proof. $\mathcal{M} \cap (\omega^2)$ is clearly a set, hence a relation in ω. It is anti-reflexive by Lemma 6.3.2 and it is transitive by Proposition 6.3.1, because $x \in y \in z$ imply $x \in z$. Thus membership for elements of ω is a strict ordering. The last assertion follows from Proposition 6.3.3. □

Proposition 6.3.7. *The ordering in* ω *given by* $x \subseteq y$ *well orders* ω*.*

Proof. Let us assume that there is a subset $A \subseteq \omega$ which has no minimum, and consider the set of lower bounds of A, $B = \{x \in \omega | y \in A \Rightarrow x \subseteq y\}$. B is a set by Corollary 5.1.12 and $B \cap A = \emptyset$ because if $x \in B \cap A$, then x is the minimum of A, which contradicts the assumption. Thus it will suffice to show that $B = \omega$ to get that $A = \emptyset$. We use induction; first, $\emptyset \in B$ is obvious. For the induction step, let $x \in B$ and $y \in A$. Then $x \subseteq y$ by the definition of B; and $x \subsetneqq y$ because $B \cap A = \emptyset$. Hence $x \in y$ by Proposition 6.3.3 and $s(x) \subseteq y$, showing that $s(x) \in B$. □

Note that, since $s^n(\emptyset) = \{s^k(\emptyset) | k < n\}$, any possible element $c \in \omega$ which is not of the form $s^n(\emptyset)$ (for an ordinary natural number n) contains all the elements of that form, hence it has infinitely many elements. We shall write $x \leq y$ (respectively, $x < y$) to mean $x \subseteq y$ (resp., $x \in y$) for elements of ω; that is, we use $\leq$ (resp., $<$) to denote the ordering of ω of Proposition 6.3.7 (resp., the strict ordering of Proposition 6.3.6).

6.3.3 Other forms of induction

The general strategy for using the principle of induction has been explained above and seen in particular instances. On the other hand, some other forms of the principle of induction are sometimes useful. The following is the *strong principle of induction.*

Proposition 6.3.8. *Let A be a subset of ω and assume that the following implication holds for each $x \in \omega$: $x \subseteq A \Rightarrow x \in A$.*
Then $A = \omega$.

Proof. We use RAA and suppose that $A \neq \omega$ satisfies the implication in the statement of the proposition. By Proposition 6.3.7 there is a minimum element $x \in \omega \setminus A$ with respect to the ordering $x \subseteq y$. Therefore, every $y \in x$ belongs to A, whence $x \subseteq A$. But then $x \in A$ by the condition, and we have reached a contradiction. □

Remark 6.3.9. The improvement over the normal principle of induction offered by the strong principle can be readily appreciated: the objective in both cases is to show that $A = \omega$. Using the first principle, we need to prove the implication $x \in A \Rightarrow s(x) \in A$; with the strong principle, we just need to show $(\forall u \leq x\ (u \in A)) \Rightarrow s(x) \in A$. This second implication is easier to prove because its hypothesis is stronger; in general, a conclusion is easier to reach if you have a strong hypothesis than if your hypothesis is weak.

Another form of induction is the following.

Proposition 6.3.10. *Let $A \subseteq \omega$, and assume that x_0 is the minimum element of A with respect to the ordering of ω. If the following holds:*

$$x \in A \Rightarrow s(x) \in A$$

then $A = \{y \in \omega | x_0 \leq y\}$.

Proof. Check that $A \cup x_0$ is inductive, hence $A \cup x_0 = \omega$. The property follows. □

It has been mentioned in Section 6.2.2 that, if we assume property NNS, the elements of ω correspond with the ordinary natural numbers; thus $n \in \omega$ consists of the numbers $< n$ and it is a finite set, with the intuitive meaning of "finite". Consequently, if we define that S is a finite set whenever there is a bijective map between S and some natural number, this concept translates faithfully the intuitive idea of a finite set. In a general 6-universe, we still use this same property to define finite sets – but beware that possibly this definition of finiteness does not match the intuitive idea.

Definition 6.3.11. A set A is said to be *finite* when there exist $x \in \omega$ and a bijective map $f : x \to A$. A set is *infinite* when it is not finite.

With this definition, all the elements of ω are obviously finite. If $x \in \omega$ and $f : s(x) \to A$ is a map (not necessarily bijective), we say that f is a *finite sequence* of elements of A, and $s(x)$ is the *length* (or number of terms) of the sequence. A usual notation for sequences employs subscripts. Like this, a sequence $u : s(x) \to A$ would be written as $u_0, u_1, \ldots, u_x$, where $u_y = u(y) \in A$. A map $f : \omega \to A$ is called an *infinite sequence.*

6.4 The Recursion Theorems

As natural induction, natural recursion is also an extremely useful tool. Just as natural induction is an ambitious extension of intuitive finite induction, natural recursion extends what is an intuitive way of defining a sequence of objects. Indeed, if we choose some object c_0, and then we describe a way (by means of some rule, which may be given by a functional class) for assigning an object $c_{s(n)}$ to any value c_n, then we might effectively determine any particular c_k we are asked for. This is an intuitive justification of the possibility of constructing finite sequences (in the intuitive sense) by using some rule for getting new elements from old ones. Natural recursion asserts, under the corresponding hypotheses, the existence of an infinite sequence, that is, a function with domain ω which takes values according to the chosen rule – thus extended to elements that cannot be reached from 0 in finitely many steps, if ω contains such "infinite" elements.

The following lemma will be used in the proof of the recursion theorem, but we state and prove it in greater generality, because it will also be useful in other situations.

Lemma 6.4.1. *Let C be a class whose elements are functions. If C is totally ordered by the inclusion relation, then $\cup C$ is a functional, whose domain is the union of the domains of the elements of C.*

Proof. Each element of $\cup C$ belongs to some function $f \in C$, hence these elements are ordered pairs and $\cup C$ is a relation. To see that $\cup C$ is functional, let $\langle x, y\rangle, \langle x, y'\rangle \in \cup C$. Thus $\langle x, y\rangle \in f$ and $\langle x, y'\rangle \in g$ for some functions $f, g \in C$. Since inclusion is a total order in C, we may assume w.l.o.g. that $f \subseteq g$. Then $\langle x, y\rangle \in g$, and $y = y'$ because g is a function. The identification of $\mathrm{D}(\cup C)$ is clear. □

We show two versions of recursion, which roughly correspond to the principle of induction and the strong principle. Before stating the first one, let us explain its content in an informal manner. What we want is to show that an infinite sequence of objects can be constructed from the following data: the initial term and a rule that gives the $s(x)$-th term from both the x-th term and the value x itself. In the statement, the objects are supposed to belong to some class C, so in particular the first term $c_0 \in C$. The rule must be a functional, depending on two values, one corresponding to the x-th term of the sequence to be constructed (hence an element of C) and the other being x itself (hence an element of ω). Therefore the functional will have domain $C \times \omega$.

Theorem 6.4.2. *Let C be a class, $c_0 \in C$ and F a functional class with $D(F) = C \times \omega$ and $cD(F) \subseteq C$. Then there is a unique function g with $D(g) = \omega$ and $cD(g) \subseteq C$ such that the following properties are fulfilled:*
(i) $g(0) = c_0$.
(ii) $\forall x \in \omega\ [g(s(x)) = F(\langle g(x), x\rangle)]$.

Proof. We first prove the following claim, under the assumption of the existence of C, c_0 and F.

For each $x \in \omega$, there is a unique function f such that $\mathrm{D}(f) = s(x)$, $\mathrm{cD}(f) \subseteq C$, $f(0) = c_0$ and $f(s(t)) = F(\langle f(t), t\rangle)$ for every $t \in x$.

We use induction to prove the claim. For $x = 0$, f has the only value $f(0) = c_0$ and it is obvious that f satisfies the properties in the claim and is unique. Now, let $y = s(x)$. By the inductive hypothesis, there exists a unique f_1 with domain $s(x) = y$ and the supposed properties. Then $f = f_1 \cup \{\langle y, F(\langle f_1(x), x\rangle)\rangle\}$ has domain $s(y)$ and satisfies the required properties, since we have just added $f(y)$ in the required form, to the previous function f_1. As for uniqueness, suppose f_2, f_3 fulfill the conditions in the claim and have domain $s(y)$. The restrictions of f_2, f_3 to the domain $y = s(x)$ do coincide by the inductive hypothesis, as they again fulfill those conditions for the domain $s(x)$. In particular, $f_2(x) = f_3(x)$, and therefore $f_2(y) = F(\langle f_2(x), x\rangle) = F(\langle f_3(x), x\rangle) = f_3(y)$, hence $f_2 = f_3$.

Let us call f_x to the unique function with domain $s(x)$ that satisfies the conditions in the claim. By Corollary 5.1.12, there is a set $S = \{f_x | x \in \omega\}$. We now check that S is totally ordered by inclusion so that we can apply Lemma 6.4.1.

Take f_x, f_z and assume w.l.o.g. that $x \subseteq z$. Then $h = f_z|_{s(x)}$ is a function satisfying $h(0) = c_0 = f_x(0)$; and if we had $h(s(t)) \neq f_x(s(t))$ for some other element $s(t)$, there would exist, by Proposition 6.3.7, a smallest $s(t)$ for which that inequality holds. Then $h(t) = f_x(t) = f_z(t)$ whence $h(s(t)) = f_z(s(t)) = F(f_z(t), t) = F(f_x(t), t) = f_x(s(t))$, a contradiction. The conclusion is that $f_z|_{s(x)} = f_x$ so $f_x \subseteq f_z$. This means that inclusion is a total ordering in S, so $\cup S$ is a function with domain ω, by axiom 3 and the lemma.

Finally, $g = \cup S$ satisfies condition (ii) of the statement as can be seen by induction, since $g|_{s(x)} = f_x$. Uniqueness follows by the same argument that gives the uniqueness for each f_x. □

Remark 6.4.3. The claim at the beginning of the above proof shows that recursion may be used also to obtain functions with domain $s(x)$ for any $x \in \omega$; and for this we only need that the functional F be defined on $C \times x$.

Comment 6.4.4. We have used Lemma 6.4.1 in the above proof and the lemma gives a property of classes, hence we may apply it, in principle, only for a particular class. But in this case S is a set, and the version of the lemma for sets holds in complete generality by Proposition 5.1.9.

Example 6.4.5. Just to illustrate the idea of recursion, we present some examples where ω is identified to the ordinary natural numbers. Let us show first an application of the preceding theorem to form the sequence giving the number of diagonals in a polygon of n sides. If we denote that number as $p(n)$, the first values are $p(0) = p(1) = p(2) = p(3) = 0$, as there are no polygons with less than three sides, and a triangle has no diagonals. Then, an easy geometric argument shows that for $n \geq 3$, $p(n+1) = p(n) + n - 1$ (consider three consecutives vertices A, B, C; draw AC and consider the n-sided polygon obtained by deleting the sides AB, BC and adding AC as a side. Then one may find the relation between $p(n+1)$ and $p(n)$). Thus we identify the class C of the statement as ω, the first values $c_0 = c_1 = c_2 = c_3 = 0$, and the functional F has $F(m,n) = m + n - 1$, when $n \geq 3$ – for the other cases, $F(m,n) = 0$. We accept that this is indeed a (functional) class.

The theorem implies that there is a function $g : \omega \to \omega$ with $g(0) = g(1) = g(2) = g(3) = 0$; and $g(n+1) = F(g(n), n) = g(n) + n - 1$. That is, $g(n) = p(n)$ gives the number of diagonals of polygons of n sides. For example, $g(4) = 0 + 3 - 1 = 2$, $g(5) = 2 + 4 - 1 = 5$, $g(6) = 5 + 5 - 1 = 9$, etc.

In the second theorem, the $s(x)$-th term of the sequence is obtained (by some functional F as in the first theorem) not from the preceding x-th term, but from the set of all the previous elements. For that reason, the functional F has to be defined on $\mathbb{P}(C) \times \omega$ instead of $C \times \omega$.

Theorem 6.4.6. *Let C be a class and F a functional class with $D(F) = \mathbb{P}(C) \times \omega$ and $cD(F) \subseteq C$. Then there exists a unique function g with $D(g) = \omega$ and $cD(g) \subseteq C$ such that for each $x \in \omega$, $g(x) = F(\langle g[x], x \rangle)$.*

Proof. The proof is very similar to that of Theorem 6.4.2 and will be left as an exercise. □

Example 6.4.7. A simple rule for obtaining a sequence of natural numbers which are pairwise coprime (that is, any two numbers in the sequence are coprime) is the following. We start with $c_0 = 1$; and each of the c_n with $n \geq 1$ is 1+ the product of all the previous numbers in the sequence. That is, $c_1 = 2, c_2 = 2 + 1 = 3, c_3 = 6 + 1 = 7, c_4 = 42 + 1 = 43$, etc. By the above recursion theorem, we can obtain this sequence as an object of $\mathcal{U}$, a function with domain ω. In this case, $C = \omega$, $c_0 = 1$, and the functional F can be defined as $F(\langle X, n\rangle) = 1 + \prod_{m \in X} m$.

The theorem asserts then that there exists $g : \omega \to \omega$ such that $g(0) = 1$ and $g(n) = F(\langle g[n], n\rangle)$; like this, $g(n) = 1$ plus the product of the elements of $g[n]$, that is, 1 plus the product of all the previous values in the sequence.

Remark 6.4.8. These recursion theorems will be frequently used in a simplified form: instead of asking that $\mathrm{D}(F) = C \times \omega$, we may take $\mathrm{D}(F) = C$ (or $\mathbb{P}(C)$ for the second version). In that case, the rule for obtaining the image of an element $s(x)$ depends only on the previous value, and not on x itself. But this gives the same conclusion of the proposition, the existence of a unique function g with the same properties as in Theorems 6.4.2 or 6.4.6. The proof that this works is easy because the assumed functional F may be transformed into a functional F' with domain $C \times \omega$ by simply letting $F'(\langle c, x\rangle) = F(c)$ and apply the general result.

Note that Theorems 6.4.2 and 6.4.6 have one proof for each particular pair of classes C, F. But when C, F are sets, the recursion theorem is just one theorem, according to Proposition 5.1.9.

Examples 6.4.9. Let us apply the simplified form of the recursion theorem to construct a very popular sequence, the *Fibonacci sequence*. This is a sequence of natural numbers starting with $0, 1$ as the first two terms; and formed under the rule: each term is the sum of the two terms immediately preceding it. That is, the first terms of the sequence are: $0, 1, 1, 2, 3, 5, 8, 13, 21, 34, 55$.

By the recursion theorem, this defines an infinite sequence. The class C of the statement will be $\omega \times \omega$, the starting element $c_0 = \langle 0, 1\rangle$ and the functional class F is defined with $F(\langle n, m\rangle) = \langle m, n + m\rangle$ – it can be justified from later results that this is indeed a class. The simplified form of recursion shows that these conditions define a function $g : \omega \to \omega \times \omega$ satisfying: $g(0) = \langle 0, 1\rangle$, and $g(s(n)) = F(g(n)) = \langle p_2(g(n)), p_1(g(n)) + p_2(g(n))\rangle$, with p_1, p_2 denoting the projection functions from $\omega \times \omega$ to ω. Thus $g(1) = \langle 1, 1\rangle$, $g(2) = \langle 1, 2\rangle$, $g(3) = \langle 2, 3\rangle$, $g(4) = \langle 3, 5\rangle$, etc. The Fibonacci sequence is given by the first element of each pair.

We may now see the relation between ω and any other Dedekind-Peano set.

Proposition 6.4.10. *Let $\langle N, e, S\rangle$ be a Dedekind-Peano set. There is a bijective map $f : \omega \to N$ with $f(0) = e$ and which commutes the functions s and S (i.e., $f(s(x)) = S(f(x))$).*

Proof. We use the simplified form of Theorem 6.4.2, as explained in Remark 6.4.8. Here the class C of the statement will be our set N; the chosen element $c_0 = e$; and the functional F will be that of the map $S : N \to N$. By applying the theorem, we find a map $f : \omega \to N$ such that $f(0) = e$ and $f(s(x)) = S(f(x))$. Thus $e \in \mathrm{cD}(f)$; and if $u \in \mathrm{cD}(f)$ then $u = f(x)$ and $S(u) = f(s(x))$ whence $S(u) \in \mathrm{cD}(f)$. By (iii) of Definition 6.1.1, $\mathrm{cD}(f) = N$ and the map is is surjective.

It remains to see that f is injective. Suppose not; by Proposition 6.3.7, there is a smallest $x \in \omega$ such that there exists $y \in \omega$ with $x \neq y$ and $f(x) = f(y)$. Clearly $x \neq 0$ because if $x = 0$ then $y = s(z)$ and $e = f(y) = f(s(z)) = S(f(z))$ so that $e \in \mathrm{cD}(S)$, contrary to (i) of Definition 6.1.1. Symmetrically, $y \neq 0$ and thus $x = s(u)$ and $y = s(v)$. Then $f(x) = f(s(u)) = S(f(u)) = f(y) = S(f(v))$. By (ii) of Definition 6.1.1, $f(u) = f(v)$; and $u \neq v$ because $u = v \Rightarrow x = y$. This contradicts the assumption on x since $u < x$. □

The relation here between ω and N is that of a perfect translation from one language to another. We may imagine that ω is some kind of language, where the elements of ω are the basic bricks (the words, we could say) and the successor function is the only grammatical rule, connecting some elements to others. N would be a language of the same kind, with its elements and the function S connecting them. The function f is the translator: it gives a word x' of N for each $x \in \omega$; and if we have a relation in ω of the form $y = s(x)$, then the corresponding relation $y' = S(x')$ appears in the language of N, by the property in the statement that those functions commute with f.

By this proposition, ω is the unique set in the universe $\mathcal{U}$ that satisfies the Peano's postulates.

Here, unique has to be understood in the precise sense of the proposition: other sets could satisfy the same properties, but there is a perfect correspondence between any of such sets and ω; for instance, we could take some $x \in \omega$, consider $N = \omega \setminus s(x)$ and define $e = s(x)$ and $S(y) = s(y)$. In this form, N contains all the elements of ω above x. The new triple $\langle N, e, S\rangle$ satisfies the conditions for a Dedekind-Peano set which is not strictly equal to ω. But it is essentially the same set as ω with a change in the names of the elements.

6.5 The Operations in ω

Addition and multiplication in ω are defined by using natural recursion (Theorem 6.4.2). We prove the main statements but skip the detailed proof of the properties of these operations in many cases. Instead of the correct notation $f(\langle x, y\rangle)$ for functions defined on $\omega \times \omega$, we shall normally employ $f(x, y)$ for simplicity.

Proposition 6.5.1. *There is a unique binary operation in ω, $a : \omega \times \omega \to \omega$ such that the following conditions hold.*

$$a(x, 0) = x, \quad a(x, s(y)) = s(a(x, y)$$

for any $x, y \in \omega$.

Proof. Let F be the set of all maps $\omega \to \omega$ (Proposition 5.2.8). Let $\phi : F \to F$ be the map defined by $\phi(f) = s \circ f$. By Theorem 6.4.2 (simplified form) there exists a unique map $A : \omega \to F$ such that $A(0) = 1_\omega$ and $A(s(x)) = \phi(A(x)) = s \circ A(x)$. Then we define $a : \omega \times \omega \to \omega$ by $a(x, y) = A(y)(x)$ and one can immediately check that a satisfies the properties of the operation given in the statement of the proposition.

As for uniqueness, suppose that $a' : \omega \times \omega \to \omega$ is a function satisfying the same properties. From a' we get $A' : \omega \to F$ with $A'(x)(y) = a'(y, x)$. Hence $A'(0) = 1_\omega$ and $A'(s(x))(y) = a'(y, s(x)) = s(a'(y, x)) = s(A'(x)(y))$ and thus $A'(s(x)) = s \circ A'(x) = \phi(A'(x))$. From the uniqueness of A, we deduce that $A' = A$ and $a' = a$. □

From now on we turn to usual notation: $A(\langle x, y\rangle)$ will be written as $x + y$. Thus the properties of Proposition 6.5.1 are $x + 0 = x$ and $x + s(y) = s(x + y)$. The basic properties of the addition are easily obtained by induction and will be considered in the exercises: associativity, commutativity, existence of a neutral element 0 for this operation; the neutral element is also called identity element in some operations. The same applies to the cancellation property, which states that $x + y = x + z \Rightarrow y = z$.

In the previous section, we have seen that the relation $x \subseteq y$ is a well-ordering of ω; and it is also written $x \leq y$. This ordering is related to the addition, as known from elementary arithmetic.

Proposition 6.5.2. *Given $x, y \in \omega$ we have $x \leq y \Leftrightarrow \exists t \in \omega\ (t + x = y)$. This t is unique and $x < y$ if and only if $t \neq 0$.*

Proof. We prove implication from left to right by induction on x. For $x = 0 = \emptyset$, $x \subseteq y$ is always true and the conclusion is also true by taking $t = y$. Suppose now the implication is valid for some $x \in \omega$ and let $s(x) \subseteq y$. Then $x \subseteq y$ which entails by the induction hypothesis $t + x = y$ for some t; and also $x \neq y$, hence $t \neq 0$ so $t = s(u)$ for some $u \in \omega$. Thus $y = s(u) + x = x + s(u) = s(x + u) = s(u + x) = u + s(x)$. This completes the induction.

For the converse, it is enough to see that $x \leq t + x$ for any t. This can be shown easily by induction on x: if $x \subseteq t + x$, then $s(x) \subseteq s(t + x) = t + s(x)$.

Uniqueness follows by the cancellation property. This also implies that $x = y \Leftrightarrow t = 0$. □

The second basic operation in ω is multiplication.

Proposition 6.5.3. *There is a unique binary operation in ω, $m : \omega \times \omega \to \omega$ such that the following conditions hold.*

$$m(x, 0) = 0, \quad m(x, s(y)) = m(x, y) + x$$

for any $x, y \in \omega$.

Proof. This is similar to the proof of Proposition 6.5.1, with the map $\psi : F \to F$ given by $\psi(f)(x) = f(x) + x$ substituted for the map ϕ of that proof. Again

Theorem 6.4.2 gives $M : \omega \to F$ with $M(0) = 0$ (i.e., the constant function 0) and $M(s(x)) = \psi(M(x))$, and we proceed exactly like in the above proof. □

The usual notation gives $x \cdot 0 = 0$ and $x \cdot s(y) = (x \cdot y) + x$; and the properties of associativity, commutativity, neutral (or identity) element and distributivity (that is, $x \cdot (y+z) = x \cdot y + x \cdot z$) can also be proven by induction. In a similar form, one can also define the operation of exponentiation in ω: $x^0 = 1$ and $x^{s(y)} = x^y \cdot x$. The well-known properties of this operation (for instance, $x^{n+m} = x^n \cdot x^m$) are easily proven by induction.

6.6 Countable Sets

Lemma 6.6.1. *Let $x \leq y$ be elements of ω. There are bijective maps $\omega \to \omega \setminus \{x\}$ and $y \to s(y) \setminus \{x\}$.*

Proof. In both cases, the bijection is formed with the pairs $\langle u, u \rangle$ for $u < x$; and $\langle u, s(u) \rangle$ for $u \geq x$. □

Proposition 6.6.2. *If $x, y \in \omega$ and there are bijective maps $f : x \to A$ and $g : y \to A$ for some set A, then $x = y$.*

Proof. The composition $g^{-1} \circ f : x \to y$ is a bijection by Propositions 5.2.12 and 5.2.13. We now employ RAA to show that for numbers $u \neq v$, there is no bijection $u \to v$. Thus, we assume the contrary and take u the smallest number such that there is a bijective map $h : u \to v$ with $u \neq v$. Obviously, $u, v \neq 0$ so $u = s(r)$ and $v = s(t)$, from which we have $r \neq t$. Let $h(r) = k \in v$, so that by deleting the pair $\langle r, k \rangle$ in h we get a bijection $r \to s(t) \setminus \{k\}$. By composing with the bijection $s(t) \setminus \{k\} \to t$ of Lemma 6.6.1, we obtain a bijection $r \to t$ and this contradicts the hypothesis on u because $r < u$ and $r \neq t$. □

When $f : x \to A$ is a bijection with $x \in \omega$, x will be called the *number of elements of the set* A and the preceding result shows that it is unique. We now turn to sets which are not finite – in the sense of finite for 6-universes, given in Definition 6.3.11.

Definition 6.6.3. A set A is called *denumerable* when there exists a bijective map $\omega \to A$. A set is called *countable* if it is finite or denumerable.

Proposition 6.6.4. *Every denumerable set is infinite.*

Proof. It is enough to see that ω is infinite. We prove this by induction. Obviously, there is no bijective map $\emptyset \to \omega$. Assume that $x \in \omega$ is such that there is no bijection $x \to \omega$, and suppose that $h : s(x) \to \omega$ is a bijection. If $h(x) = y$, there is a bijection between x and $\omega \setminus \{y\}$, hence a bijection $x \to \omega$ by Lemma 6.6.1, contradicting the inductive hypothesis. □

We have seen that for finite sets, the existence of a bijection is equivalent to the property that the sets have the same number of elements, i.e., they have the same size. For infinite sets, things seem to be different: according to Lemma 6.6.1, there is a bijection between ω and a proper subset of ω. Indeed, there are many other examples of bijections between an infinite set and proper subsets; for instance, a bijection between ω and the subset of even numbers is formed with all pairs $\langle x, 2x\rangle$. Another example will be useful very soon and thus we present it here: a bijection $\omega \to \omega \setminus \{0\}$ contains all pairs $\langle 2x, 2x+2\rangle$ and the pairs $\langle 2x+1, 2x+1\rangle$; that is, each odd number corresponds to itself; and each even number u is taken to $u+2$.

Theorem 6.6.5. *Let $A \subseteq B$ be sets. If there exists an injective map $f : B \to A$, then there is a bijective map $B \to A$.*

Proof. If $A = B$ or f is bijective the property is obvious, so we may assume $A \neq B$ and $f[B] \neq A$. This entails that $B_0 = B \setminus A$ and $B_1 = A \setminus f[B]$ are not empty and, of course, disjoint subsets. We now use simplified recursion to define subsets $B_t \subseteq B$ for any $t \geq 2$ in ω. For this we call $B_t = f[B_{t-2}]$.

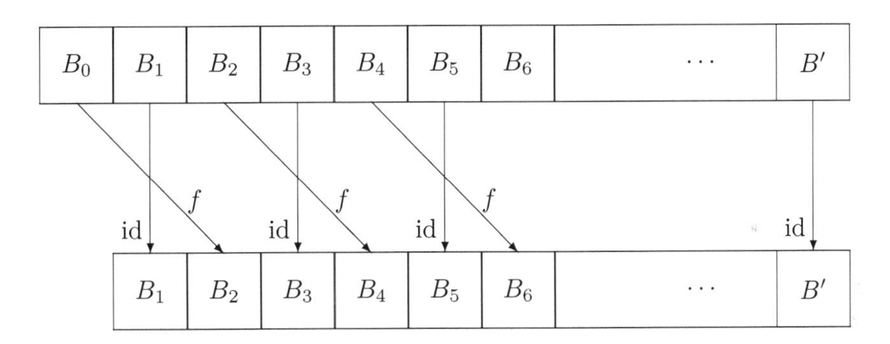

Figure 6.1
The CSB-theorem.

To see that all the sets B_t are pairwise disjoint (i.e., any two of them are disjoint) we use natural induction and assume that the sets B_x with $x < t+2$ (and $t \geq 0$) are pairwise disjoint. Suppose that $u \in B_{t+2}$ belongs to some B_x with $x < t+2$, and obviously, $x \geq 2$. Then $u = f(v) = f(y)$ with $v \in B_t$ and $y \in B_{x-2}$ with $x - 2 < t$. Since f is injective, $v = y$ and this is impossible by the inductive hypothesis. If we set $B' = B \setminus (\cup\{B_t | t \in \omega\})$, then B' and the sets B_t form a partition of B; and consequently, B' and the sets B_t (with $t \geq 1$) form a partition of A.

We may now construct the bijection $h : B \to A$, inspired by the last example above of a bijection between ω and $\omega \setminus \{0\}$ (see the figure with the partitions of B and A): we define $h(x) = f(x)$ if $x \in B_{2u}$ for some u; and

$h(x) = x$ if $x \in B'$ or $x \in B_{2u+1}$ for some u. This takes bijectively B_0 to B_2, B_2 to B_4, etc.; and B_1 to B_1, B_3 to B_3, etc., so it is a bijection $h : B \to A$. □

A crucial result in the theory is the following corollary of the above theorem. It is known as the CSB-theorem[2].

Corollary 6.6.6. *(The CSB-theorem) If A, B are sets, and there exist injective maps $f : A \to B$ and $g : B \to A$, then there exists a bijection $A \to B$.*

Proof. $f \circ g$ induces an injective map $B \to f[A]$ and $f[A] \subseteq B$. By Theorem 6.6.5 there is a bijection $h : f[A] \to B$; since $f : A \to f[A]$ is a bijective map, the composition $h \circ f : A \to B$ is a bijection. □

In what follows we use the CSB-theorem to obtain that certain sets are countable. But its importance largely exceeds these applications, since it can be used for arbitrarily big sets.

Proposition 6.6.7. *Any subset C of ω is countable (i.e., either finite or denumerable).*

Proof. Suppose C has an upper bound u in ω, that is, $C \subsetneq s(u)$. We use induction to prove that the set $\{u \in \omega | (u \text{ is an upper bound of } C \subseteq \omega) \to (C \text{ is finite})\}$ is all of ω; that is, we prove by induction that if u is an upper bound of C, then C is finite. When $u = 0$, then $C = \emptyset$ or $C = \{0\}$, and finite. If $u = s(v)$ is an upper bound for C and $u \notin C$, then $C \subseteq u = s(v)$ and C is finite by the inductive hypothesis. If $u = s(v)$ is an upper bound and $u \in C$, then v is an upper bound for $C' = C \setminus \{u\}$ and C' is finite. By the inductive hypothesis, there is a bijection $C' \to t$ for some $t \in \omega$; and this bijection can be extended to a bijection $C \to s(t)$, so C is finite.

Suppose now that C has not an upper bound, and consider the well-ordering of C restriction of the natural ordering of ω. We define by natural recursion a map $\omega \to C$ by setting $f(0)$ the minimum element of C; and $f(s(u))$ will be the minimum of the set $\{x \in C | x > f(u)\}$, which is not empty by the hypothesis. This function satisfies the property $x < y \Rightarrow f(x) < f(y)$, hence it is injective. Since the inclusion $C \to \omega$ is an injective function, there is a bijection $\omega \to C$ by Corollary 6.6.6 and C is denumerable, hence countable. □

[2]From Cantor, Schröder and Bernstein. This has been also referred to as Cantor-Bernstein and Schröder-Bernstein theorem. According to experts (for instance, [4, p. 239], and [14]), Cantor conjectured the theorem, Schröder gave a flawed proof and Bernstein gave the first correct published proof. But Dedekind had also proven it as early as 1887.

Proposition 6.6.8. *The set $\omega \times \omega$ is denumerable.*

Proof. There exist injective maps $\omega \to \omega \times \omega$; for instance, we may take $x \mapsto \langle x, 0\rangle$. So, by Corollary 6.6.6 it will be enough to show that there is an injective map $g : \omega \times \omega \to \omega$. To this end, we define $g(x, y) = 2^x \cdot 3^y$ and prove that it is injective. Suppose $g(x, y) = g(u, v)$ and, w.l.o.g., that $u \leq x$ so that $x = u + t$ by Proposition 6.5.2. Then $2^u \cdot 3^v = 2^x \cdot 3^y = 2^u \cdot 2^t \cdot 3^y$ hence $3^v = 2^t \cdot 3^y$, so that $t = 0$ because 3^v is odd[3]. Therefore $x = u$ and $3^y = 3^v$, from which $y = v$ follows by induction. □

A very useful property states that the union of countably many countable sets is countable. But this property needs another axiom and we can only prove it here under an additional assumption.

Proposition 6.6.9. *Let X be a denumerable set whose elements are sets. Assume that there is a function φ with $D(\varphi) = X$ and such that for each $z \in X$, $\varphi(z) : z \to \omega$ is an injective map. Then $\cup X$ is a countable set.*

Proof. By hypothesis, there is a bijective map $f : \omega \to X$. Consider the map $\alpha : \cup X \to \omega$ given by $\alpha(x) = u \Leftrightarrow u$ is the smallest element such that $x \in f(u)$. Take then the set of pseudo-triples $\{[x, u, t] \in \cup X \times \omega^2 | (u = \alpha(x) \wedge (z = f(u) \Rightarrow \varphi(z)(x) = t)\}$. It is plain that this set is the function of a map $\cup X \to \omega^2$; it is moreover injective: if $[x, u, t], [y, u, t]$ belong to the defined set of pseudo-triples, then $x, y \in f(u) = z$ and $\varphi(z)(x) = t = \varphi(z)(y)$ whence $x = y$ because $\varphi(z)$ is injective. By composing with a bijection $\omega^2 \to \omega$ obtained from Proposition 6.6.8, we obtain an injective map from $\cup X$ to ω which shows that $\cup X$ is countable, according to Proposition 6.6.7. □

Remark 6.6.10. In the hypotheses of Proposition 6.6.9, every element of X is countable by Proposition 6.6.7. Thus the property gives a condition under which the denumerable union of countable sets is countable. When X is finite instead of denumerable, no additional assumption is needed to obtain that $\cup X$ is countable, provided each $u \in X$ is a countable set. This will be seen in the exercises.

After seeing that natural constructions lead from countable sets to countable sets, the next result shows a different side of the story. Though Cantor contributed to set theory with many basic results, when one hears the term "Cantor's theorem" it is highly probable that it is this one which is referred to.

---

[3]It could be objected that we are using here properties of ordinary natural numbers, which have not been proved as properties of ω. But it is easy to show that even and odd elements give a partition of ω, that the product of two odd numbers is odd, and, using induction, that a power of an odd number is odd.

Theorem 6.6.11. *(Cantor's theorem) If A is any set, then there is no injective map $\mathbb{P}(A) \to A$.*

Proof. We use RAA and assume that there is an injective map $h : \mathbb{P}(A) \to A$. Then there exists a subset $S \subseteq A$ defined as follows: $S = \{u \in A | \exists X \in \mathbb{P}(A)\ (h(X) = u \wedge u \notin X)\}$. By the definition of S and the injectivity of h, any subset $X \subseteq A$ satisfies the following equivalence: $h(X) \in S \Leftrightarrow h(X) \notin X$. Since S is a subset of A, we infer that $h(S) \in S \Leftrightarrow h(S) \notin S$. This is a contradiction and the theorem is proved. □

Applying this result to ω, we see that there is no injective map $\mathbb{P}(\omega) \to \omega$; hence $\mathbb{P}(\omega)$ is not countable.

6.7 Integers and Rationals

Up to now, a set which may represent the natural numbers (that is, ω), and corresponding substitutes for the inequality relation and elementary arithmetical operations have been constructed as objects of a 6-universe $\mathcal{U}$. From this, we may show that also some substitutes for the systems of numbers[4] which are basic to mathematics: integers, rationals, real or complex, are objects of $\mathcal{U}$, and so are the main operations and relations in those sets, and the usual functions with them. A complete account of these constructions would be long and boring, with not much to be learned from it. We shall content ourselves with presenting the fundamental constructions for these sets and operations, and justify some simple properties, leaving some other as exercises. The interested reader can find more detailed explanations in the literature[5].

In this section we present the constructions of sets which mimic the behavior of integer and rational numbers, along with their operations. Since we will be dealing with quotient sets, a word about the compatibility of operations and equivalence relations mentioned in Chapter 5, is in order.

Proposition 6.7.1. *Let C be a set and E an equivalence relation in C. Let $+, \cdot$ be two binary operations in C. Suppose that both operations are compatible*

[4] But beware that, as happens with natural numbers, these substitutes are not faithful copies of our familiar systems of numbers unless property NNS is assumed; for instance, it might happen that they contain numbers which are infinite in the sense that they are bigger than any ordinary number.

[5] For instance, [18], [19], [22], and also [17, Appendix A].

with E and let $\hat{+}, \hat{\cdot}$ the operations induced on the quotient set C/E according to Proposition 5.2.26. Then:

(i) If any of the operations in C is associative or commutative, so is the corresponding operation in C/E.

(ii) If $\cdot$ is distributive with respect to $+$, then $\hat{\cdot}$ is distributive with respect to $\hat{+}$.

(iii) If e is an identity element with respect to $+$ or $\cdot$, then $[e]$ is an identity element of C/E with respect to $\hat{+}$ or $\hat{\cdot}$ respectively.

Proof. All the proofs are simple checks. We see for instance the case of (ii):

$$([x] \hat{\cdot} [y]) \hat{+} ([x] \hat{\cdot} [z]) = [x \cdot y] \hat{+} [x \cdot z] = [(x \cdot y) + (x \cdot z)]$$

and

$$[x] \hat{\cdot} ([y] \hat{+} [z]) = [x] \hat{\cdot} [y + z] = [x \cdot (y + z)]$$

and the distributivity in C completes the proof. □

6.7.1 The ring $\mathbb{Z}$ of integers

The elementary vision of integers is that they include the natural numbers $0, 1, 2, \ldots$ along with the corresponding negatives $-1, -2, \ldots$; and sum and product obey the well-known rules. Inside set theory, we may obtain a set corresponding to the integers starting with the set ω which corresponds to the set of natural numbers. The idea is that ordinary integers are differences of ordinary natural numbers; by extending this relation, our set-theoretic integers will be identified through ordered pairs of elements of ω.

Proposition 6.7.2. *Consider the following binary relation in the set $\omega \times \omega$:*

$$\langle a, b \rangle \ E \ \langle c, d \rangle \Leftrightarrow a + d = b + c$$

Then E is an equivalence.

Proof. E is clearly a set, hence a relation in $\omega \times \omega$. Proving that it is an equivalence is almost trivial, only transitivity deserves some comment: suppose $\langle a, b \rangle E \langle c, d \rangle$ and $\langle c, d \rangle E \langle f, g \rangle$. Then $a + d = b + c$ and $c + g = d + f$. By the properties of the operations in ω, $a + c + g = a + d + f = b + c + f$, whence $a + g = b + f$ by cancellation and $\langle a, b \rangle \ E \ \langle f, g \rangle$. □

Let us define the following operations in $\omega \times \omega$:

$$\langle a, b \rangle + \langle c, d \rangle = \langle a + c, b + d \rangle$$

and

$$\langle a, b \rangle \cdot \langle c, d \rangle = \langle ac + bd, ad + bc \rangle$$

It is straightforward to prove that:

$+$ and $\cdot$ are commutative and associative, $\cdot$ is distributive with respect to $+$, and both have identity elements: $\langle 0, 0 \rangle$ for $+$ and $\langle 1, 0 \rangle$ for $\cdot$.

Proposition 6.7.3. *The operations + and · defined on $\omega \times \omega$ are compatible with the equivalence relation E.*

Proof. We prove only the property for the product, the one for the sum being similar. In view of the commutativity of the product in $\omega \times \omega$ and the transitivity of E, it is enough to prove that, for $x, y \in \omega$,

$$\langle a, b\rangle \; E \; \langle c, d\rangle \Rightarrow (\langle a, b\rangle \cdot \langle x, y\rangle) \; E \; (\langle c, d\rangle \cdot \langle x, y\rangle)$$

That is,

$$a + d = b + c \Rightarrow ax + by + cy + dx = ay + bx + cx + dy$$

which is clear because $(a + d)x + (b + c)y = (c + b)x + (a + d)y$. □

By Proposition 6.7.1, this entails that we have corresponding operations of addition and multiplication, which we now denote simply as + and · respectively, in the quotient set $(\omega \times \omega)/E$:

$$[\langle a, b\rangle] + [\langle c, d\rangle] = [\langle a + c, b + d\rangle], \quad [\langle a, b\rangle] \cdot [\langle c, d\rangle] = [\langle ac + bd, ad + bc\rangle]$$

By Proposition 6.7.1, these operations are commutative and associative, multiplication is distributive with respect to the sum, and both have identity elements: $[\langle 0, 0\rangle]$ and $[\langle 1, 0\rangle]$. Moreover, each element has a symmetric element in the sum:

$$[\langle a, b\rangle] + [\langle b, a\rangle] = [\langle a + b, a + b\rangle] = [\langle 0, 0\rangle]$$

Proposition 6.7.4. *Every element of the quotient set $(\omega \times \omega)/E$ is of one and only one of the following three types: either it is $[\langle k, 0\rangle]$ or $[\langle 0, k\rangle]$ (for some $k \in \omega$ and $k \neq 0$); or else it is $[\langle 0, 0\rangle]$.*

Proof. If $a > b$, then $a = k + b$ by Proposition 6.5.2 and $\langle a, b\rangle \; E \; \langle k, 0\rangle$. If $a < b$, then $k + a = b$ and $\langle a, b\rangle \; E \; \langle 0, k\rangle$. Finally, $\langle a, a\rangle \; E \; \langle 0, 0\rangle$. It is clear that these three cases are pairwise incompatible. □

In this way, we see that the elements of $(\omega \times \omega)/E$ behave as our familiar integer numbers: $[\langle a, 0\rangle]$ is identified with the positive a, while $[\langle 0, b\rangle]$ corresponds to the negative $-b$. Of course, $[\langle 0, 0\rangle]$ is the number 0, the identity for the sum. Like this, this quotient set is written $\mathbb{Z}$ and its elements are called the integers – of set theory. We must remind that if ω contains some element c which is not equal to any ordinary natural number, then c and $-c$ will also be integers. But all the ordinary integers can be seen as elements of $\mathbb{Z}$. Addition and multiplication as defined above correspond to usual operations with integers. There is also the obvious ordering, which is a total ordering:

$$[\langle a, b\rangle] \leq [\langle c, d\rangle] \Leftrightarrow a + d \leq b + c$$

a definition which is independent of the representatives chosen, as can be easily checked. Note that, with this definition, the positive integers are precisely those > 0. In all, this constructs a kind of copy of the integers within set theory.

The above properties show that $\langle \mathbb{Z}, +, \cdot \rangle$ is a *commutative ring* with identity element: this means that the sum is associative, commutative, has an identity or neutral element and each element has a symmetric element; the product is distributive with respect to the sum and is also associative and commutative with identity element. Moreover, $\mathbb{Z}$ is a domain: i.e., $(xy = 0 \Rightarrow (x = 0 \vee y = 0)$.

6.7.2 The field of rational numbers

The construction of the set $\mathbb{Q}$ of rational numbers follows very much the lines of the preceding construction of $\mathbb{Z}$, and hence we will only give the main definitions and some comments. In what follows, $\mathbb{Z}^*$ will denote the set $\mathbb{Z} \setminus \{0\}$.

Proposition 6.7.5. *The following relation in the set* $\mathbb{Z} \times \mathbb{Z}^*$ *is an equivalence:*

$$\langle a, b \rangle \sim \langle c, d \rangle \Leftrightarrow ad = bc$$

As with integers, we define operations in $\mathbb{Z} \times \mathbb{Z}^*$:

$$\langle a, b \rangle + \langle c, d \rangle = \langle ad + bc, bd \rangle$$

where $bd \in \mathbb{Z}^*$ because $\mathbb{Z}$ is a domain.

$$\langle a, b \rangle \cdot \langle c, d \rangle = \langle ac, bd \rangle$$

Again, it is simple to check that these operations are commutative and associative. $\langle 0, 1 \rangle$ is an identity element for the sum, and $\langle 1, 1 \rangle$ is an identity element for the product.

Proposition 6.7.6. *The operations* $+$ *and* $\cdot$ *in* $\mathbb{Z} \times \mathbb{Z}^*$ *are compatible with the equivalence* $\sim$.

Proof. Straightforward. □

By Proposition 6.7.1, the quotient set has operations (which we write again as $+$ and $\cdot$) corresponding to the above which are commutative and associative and with identity elements. It is moreover easy to check that the product is distributive with respect to the sum, that each element has a symmetric with respect to the sum, and that each nonzero element (which is any element $[\langle a, b \rangle]$ with $a \neq 0$) has a symmetric with respect to the product, i.e., an inverse element.

The identification of the elements of $(\mathbb{Z} \times \mathbb{Z}^*)/\sim$ having ordinary integers as components, and our familiar rational numbers is immediate: in fact, $[\langle a, b \rangle]$ is the rational number which is given by the fraction $\frac{a}{b}$. This can be extended so that the rational numbers of set theory can be represented as fractions. The set $(\mathbb{Z} \times \mathbb{Z}^*)/\sim$ is thus denoted as $\mathbb{Q}$, and the above-mentioned properties of the operations $+$ and $\cdot$ make $\mathbb{Q}$ a *field*.

Definition 6.7.7. For $q \in \mathbb{Q}$, we say that q is *positive* when some pair $\langle a, b\rangle \in q$ satisfies the relation $(ab > 0)$. If $q \neq 0$ is not positive, then we say that q is *negative*. The set of positive elements of $\mathbb{Q}$ is written $\mathbb{Q}^+$. The inequality relation is defined thus: $q_1 < q_2$ if and only if $q_2 - q_1 \in \mathbb{Q}^+$.

Note that q is positive if and only if $\forall\langle a, b\rangle \in q\ (ab > 0)$. Also, q is negative if and only if $-q$ is positive.

All the usual properties of rational numbers are shown to hold in this set $\mathbb{Q}$ and thus $\mathbb{Q}$ is the set-theoretic version of the ordinary rational numbers. For instance, it is possible to prove the *density* property of the rationals: if $q < q'$ belong to $\mathbb{Q}$, then there exists $q'' \in \mathbb{Q}$ such that $q < q'' < q'$.

A field F is called an *ordered field* when it contains a subset P such that (i) $0 \notin P$; (ii) For each $x \in F$, either $x \in P$ or $-x \in P$ or $x = 0$; (iii) The set P is closed for addition and multiplication, i.e., if $x, y \in P$, then $x+y, xy \in P$. Thus, we have just seen that $\mathbb{Q}$ is an ordered field with the set $\mathbb{Q}^+$ of positive elements as the set P.

6.8 The Field of Real Numbers

Definition 6.8.1. A subset $C \subseteq \mathbb{Q}$ is called a *cut* of rationals when the following conditions are satisfied.

(i) $C \neq \emptyset$ and $C \neq \mathbb{Q}$.
(ii) If $x, y \in \mathbb{Q}$ and $x < y$, then $x \in C \Rightarrow y \in C$.

The collection of all cuts of rationals is a class and therefore a set by separation. A cut of rational numbers is a *Dedekind cut* (or, shortly, a *D-cut*) when it has no minimum element. We define, for any cut C, another cut $\hat{C} = C \setminus \{q \in C | q \text{ is the minimum of C}\}$. When C is a cut with a minimum m, then $\hat{C} = C \setminus \{m\}$ is a D-cut, by the density of $\mathbb{Q}$; when C has no minimum, then $\hat{C} = C$. Therefore the codomain of the function $C \mapsto \hat{C}$, which is a set by Proposition 5.1.18, is the set of all D-cuts.

The set of all Dedekind cuts will be denoted as $\mathbb{R}$, and we identify it as the set-theoretic version of the real numbers.

The ordering in $\mathbb{R}$ is given by the relation: $C_1 \leq C_2$ if and only if $C_2 \subseteq C_1$ (and the corresponding strict ordering is $C_1 < C_2 \Leftrightarrow C_2 \subsetneqq C_1$), and it is easily seen to be a total ordering. If $q \in \mathbb{Q}$, then $C(q) = \{x \in \mathbb{Q} | q < x\}$ is a D-cut (by the density property of $\mathbb{Q}$) hence a real number. This embeds $\mathbb{Q}$ into $\mathbb{R}$ and this inclusion clearly preserves the orderings. This identification of rationals as real numbers gives $0 = \mathbb{Q}^+$, and we say that the real number C is *positive* when $C \subsetneqq \mathbb{Q}^+$, and *negative* when it is neither positive nor 0. The set of positive real numbers is denoted $\mathbb{R}^+$. Also, the density property of $\mathbb{Q}$ in $\mathbb{R}$ (i.e., given $C_1 < C_2 \in \mathbb{R}$, there exists $q \in \mathbb{Q}$ with $C_1 < C(q) < C_2$) can be proved from these definitions.

Proposition 6.8.2. *Let $C_1, C_2 \in \mathbb{R}$. The set $C_1 + C_2 = \{q_1 + q_2 | q_i \in C_i\}$ is a D-cut, hence a real number. This operation in $\mathbb{R}$ is commutative and associative and $0 = \mathbb{Q}^+$ is the zero element. For $C \in \mathbb{R}$, $C' = \{q \in \mathbb{Q} | q + c > 0 \ (\forall c \in C)\}$ is a cut, and $\hat{C}' + C = 0$. Therefore $\langle \mathbb{R}, + \rangle$ is an abelian group.*

We leave the proof as an exercise. We may then define for any real number r its absolute value $|r|$ as usual: $|r| = r$ when $r \geq 0$ and $|r| = -r$ when r is negative.

Proposition 6.8.3. *Let C_1, C_2 be non-negative real numbers (thus $C_1, C_2 \subseteq \mathbb{Q}^+$). The set $C_1C_2 = \{q_1q_2 | q_i \in C_i\}$ is a D-cut, hence a non-negative real number. This operation in the set of non-negative reals is commutative and associative, and 1 is the identity element. This product is distributive with respect to the sum in the same set of real numbers ≥ 0. Moreover, for $C > 0$ and $C' = \{q \in \mathbb{Q}^+ | qc > 1 \ (\forall c \in C)\}$, we obtain that $\hat{C}'$ is a non-negative real number with $C\hat{C}' = 1$.*

Proof. Exercise. □

We may now define a product $r \cdot r'$ for arbitrary real numbers r, r': it suffices to apply the signs rule and take $|r \cdot r'| = |r| \cdot |r'|$. From Proposition 6.8.3 it is easy to see that $\mathbb{R}$ is a field with these operations $+$ and $\cdot$. Moreover, it is an ordered field, $\mathbb{R}^+$ being the set of positive elements.

Proposition 6.8.4. *If Σ is a non-empty subset of $\mathbb{R}$ which has an upper bound, then the supremum sup(Σ) exists.*

Proof. Consider the set $\Gamma = \{r \in \mathbb{R} | \forall c \in \Sigma \ (c \leq r)\}$; that is, Γ is the set of all upper bounds of Σ, which is not empty by hypothesis. By definition, if $r \in \Gamma$ and $r' \in \Sigma$, then $r \subseteq r'$. Take $C = \cup\Gamma \subseteq \mathbb{Q}$: $C \neq \emptyset$ because Γ contains non-empty sets, and $C \neq \mathbb{Q}$ because $r' \in \Sigma, q \notin r' \Rightarrow q \notin C$. C is a cut of rationals: if $q > q'$ for $q' \in C$ then $q' \in r$ for some $r \in \Gamma$ and thus $q \in r$ and $q \in C$. Since $C \subseteq r'$ for every $r' \in \Sigma$, $\hat{C}$ is an upper bound of Σ and it follows that $\hat{C}$ is the supremum of Σ: indeed, if $r \in \Gamma$ and $r \not\subseteq \hat{C}$ then $\hat{C} \neq C$, C has a minimum, $r = C$ so r is not a real number, a contradiction. □

An ordered field is *complete* when every non-empty subset which is upper bounded has a supremum – equivalently, every non-empty subset which is lower bounded has an infimum. Proposition 6.8.4 says that $\mathbb{R}$ is a complete ordered field. These conditions identify the set of real numbers in the same way as the Peano's axioms identify the set ω of the (set-theoretic) natural numbers.

Proposition 6.8.5. *Let $\langle F, +, \cdot, P \rangle$ be a 4-tuple where F is a set, $+$ and $\cdot$ are binary operations on F and P is a subset of F. If F (with $+, \cdot, P$) is a complete ordered field, then there is a bijective map $g : \mathbb{R} \to F$ which preserves the operations and the positive elements.*

We leave the details of the proof of this result to the exercises. But note that the uniqueness of $\mathbb{R}$ (in this sense) is relative to the 6-universe $\mathcal{U}$ we consider. Different universes might have different objects, possibly resulting in different collections of real numbers. On the other hand, for universes where property NNS is satisfied, the sets $\omega, \mathbb{Z}, \mathbb{Q}$ would be basically the same in every such universe. Still, $\mathbb{R}$ could be different in different universes, as it depends on which are the subsets of $\mathbb{Q}$.

This chapter has provided us with set-theoretic versions of the basic sets of numbers, since also complex numbers can be seen as sets (as well as $\mathbb{C}$ itself), by considering them as ordered pairs of real numbers. This is only the beginning of the trend that leads to the conclusion that most of mathematics can be developed inside a universe of sets. But we will not go farther on this way. Suffice it to say that the usual real or complex functions may be seen as objects of the universe, starting with polynomial functions (which are based on the operations of sum and product that have already been shown to be sets) and using the normal definitions for limits or sums of series, etc. From this, the classical mathematical concepts find their place inside set theory.

6.9 Exercises

1. Prove that a triple $\langle N, e, S\rangle$ that satisfies conditions (ii) and (iii) of Definition 6.1.1, and (i') $e \notin \mathrm{cD}(S)$ (instead of (i)), is a Dedekind-Peano set.

2. Let $\mathcal{U}$ be a 5-universe and define the concept of a Z-inductive set in $\mathcal{U}$ as follows. $A \in \mathcal{S}$ is Z-inductive when (1) $\emptyset \in A$; and (2) $x \in A \cap \mathcal{S} \Rightarrow \{x\} \in A$. Let us say that $\mathcal{U}$ is a Z-6-universe when $\mathcal{U}$ contains some Z-inductive set.

   Prove that $\mathcal{U}$ is a Z-6-universe if and only if $\mathcal{U}$ is a 6-universe.

3. Prove the double induction method: let $A \subseteq \omega \times \omega$ which satisfies the following two conditions: (i) $\langle x, 0\rangle \in A$ for any $x \in \omega$; (ii) given $x, y \in \omega$, $((\langle x, y\rangle \in A) \wedge (\langle y, x\rangle \in A)) \Rightarrow (\langle x, s(y)\rangle \in A)$. Then $A = \omega \times \omega$.

4. Give a complete proof of the strong form of recursion (Theorem 6.4.6) (Hint: Follow the idea of the proof of Theorem 6.4.2, and prove the claim to obtain a function for each $x \in \omega$ with the adequate properties but with domain x instead of $s(x)$.).

5. Prove the simplified form (stated just after Theorem 6.4.6) of the principle of recursion of Theorem 6.4.2 as a consequence of that theorem.

6. Give a detailed proof of the properties of the sum in ω: associative, commutative, neutral element, cancellative. It may be useful to start proving that $0 + n = n$ and $1 + n = n + 1$.

7. We consider the following operation in $\omega \times \omega$:

$$\langle a, b \rangle * \langle c, d \rangle = \langle a + d, b + c \rangle$$

Study if this operation is compatible with the equivalence relation of Proposition 6.7.2. If it is, identify the operation that it determines in the quotient set $\mathbb{Z}$.

8. In the game of the towers of Hanoi we have $n \geq 2$ disks with a central hole which are set on a cylindrical peg. From bottom to top, the disks are progressively of smaller size. There are two other similar pegs, which are initially empty. The game consists in transferring disks from one peg to another, subject to the following rules: (i) each movement transfers just one disk from one peg to another; (ii) it is not allowed to place a disk on top of another disk of smaller size. The goal of the game is to move all n disks from the initial peg to some of the other two. Prove that it is possible to achieve this in a total of $2^n - 1$ movements.

9. We are given n coins all having the same weight, except for one of them. To discover which is the wrong coin we have a pair of scales with which we may compare whether two sets of coins have the same weight or not. Prove that for $n = \frac{3^{k+1}-1}{2} = 3^k + 3^{k-1} + \cdots + 3 + 1$ coins, $k + 1$ comparisons are enough to determine which is the wrong coin and whether it is heavier or lighter than the rest. Important: we dispose at the start of one other coin which we know is the correct weight (Hint: It could be useful to compare with the following setting: there are two groups of coins and as before just one coin is wrong; say that group A has $3^k + 3^{k-1} + \cdots + 3 + 1$ coins, and group B has $3^k + 3^{k-1} + \cdots + 3 + 2$ coins. But we know in addition that if the wrong coing belongs to group A then it is heavier and if it belongs to group B then it is lighter than the rest. Again $k + 1$ comparisons are enough to discover the wrong coin in this case).

10. Prove that if there are bijective maps $A_1 \to B_1$, $A_2 \to B_2$, then there are also bijections $A_1 \times A_2 \to B_1 \times B_2$ and $\mathbb{P}(A_1) \to \mathbb{P}(B_1)$.

11. Prove that the set of integers is countable.

12. Prove that the set of rationals is countable.

13. Prove the following variation of Proposition 6.6.9: if X is a finite set and every element of X is a countable set, then $\cup X$ is countable. Beware! Here X is finite in the sense of Definition 6.3.11, not in the intuitive sense.

14. Develop the following sketch of a different proof of Proposition 6.6.8 for the case that the universe $\mathcal{U}$ satisfies the NNS property. Let $f : \omega \times \omega \to \omega$ be given by $f(\langle x, y \rangle) = x + \frac{(x+y)(x+y+1)}{2}$. Use the NNS to prove that f is bijective by looking at its geometric meaning: it orders the elements of ω^2 by starting with $(0,0)$ and then going downwards successively along the lines $(0,1)$ to $(1,0)$, then $(0,2), (1,1)$ to $(2,0)$, $(0,3), (1,2), (2,1)$ to $(3,0)$, etc., as shown in the figure below. Then, consider the general case of a 6-universe and give a non-geometrical, direct proof of the bijectivity of f.

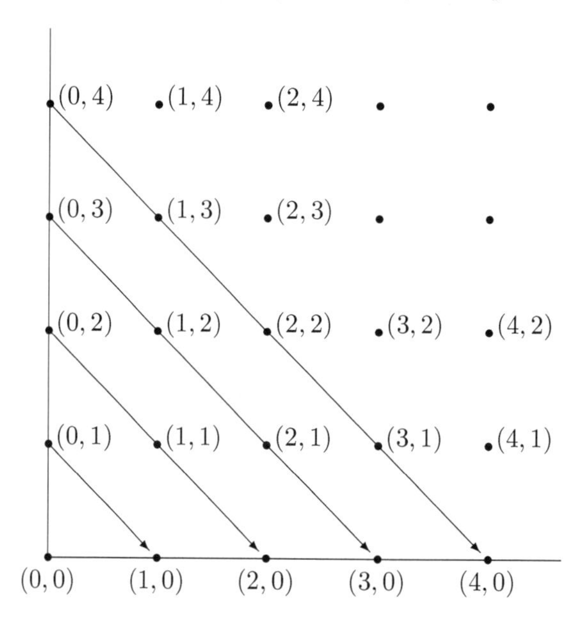

Figure 6.2
$\omega \times \omega$ is countable.

15. Prove that there is a bijection between the set of real numbers and $\mathbb{P}(\omega)$ (Hint: To construct an injective map from $\mathbb{P}(\omega)$ to $\mathbb{R}$, consider for each $S \subseteq \omega$, the sequence $a_1, a_2, \ldots$ where $a_x = 1$ if $x \in S$ and $a_x = 0$ if $a_x \notin S$; then consider the sequence[6] of rational numbers $q_1, q_2, \ldots$ with $q_n = 0.a_1a_2 \ldots a_x$ (adapting the usual values for rational numbers in decimal notation); and observe that the set of rational numbers x such that $x > q_x$ for every x, is a cut of rationals and thus defines a Dedekind cut, i.e., a real number.).

16. For each of the following subsets S of $\mathbb{Z}$, determine whether the restriction of the usual inequality ordering is a well-ordering of S: (i) odd positive integers; (ii) even negative integers; (iii) numbers > -7;

[6]This sequence has to be defined by recursion.

(iv) odd integers > 249. Is there a general rule which identifies which are the subsets of $\mathbb{Z}$ that are well-ordered?

17. In the proof of Proposition 6.6.7, natural recursion is employed to obtain the map f. Make explicit this application of recursion, by identifying which is the functional F of Theorem 6.4.2 and the element c_0 which are used in this application.

18. Prove items (i) and (iii) of Proposition 6.7.1.

19. Following the examples of the definitions of compatibility of functions and of operations, define the corresponding concept of when a relation R in a set A is compatible with an equivalence E in A. Show then that the relation of the set $\omega \times \omega$, as (implicitly) defined after Proposition 6.7.4, is compatible with the equivalence E, so it defines an ordering of $\mathbb{Z}$.

20. Prove the compatibility of the operations in Proposition 6.7.6 with respect to the equivalence defined in Proposition 6.7.5.

21. Let $\langle <, A\rangle$ and $\langle \prec, B\rangle$ be two (strict) well-orderings (of the sets A, B respectively). Consider the following order in the set $A \times B$: $\langle x, y\rangle < \langle u, v\rangle \Leftrightarrow (x < u) \vee ((x = u) \wedge (y \prec v))$ (this is called the *lexicographic order* for a cartesian product of ordered sets). Prove that $\langle <, A \times B\rangle$ is a well-ordering.

22. Let S be the set of all finite sequences of elements of ω (i.e., all maps $f : x \to \omega$ with $x \in \omega$). Consider the following order relation in S: $f \prec g \Leftrightarrow [\exists z \in \mathrm{D}(g)\ ((x < z \Rightarrow f(x) = g(x)) \wedge (z \notin \mathrm{D}(f) \vee f(z) < g(z)))]$. Show that this relation well-orders S.

23. S is again the set of all finite sequences of elements of ω. Let T be the set of all non-empty maps $f : S \to \omega$, and choose an element $t_0 \in T$ with the property that $t_0(\sigma) = \sigma(0)$, for $\sigma \in S$. Choose also a bijective map $h : \omega^2 \to \omega$ and consider a function K with domain $T \times \omega$ and codomain included in T, giving $K(t, x)(\sigma) = h(t(\sigma|_x), \sigma(x)))$. Use the recursion theorem to define a map $g : \omega \to T$ and prove that $g(x) : S \to \omega$ is injective on the subset of all sequences of length x.

    Apply Proposition 6.6.9 to deduce that the set S is countable.

24. Prove that the set F of all maps $\omega \to \omega$ is not denumerable (Hint: using RAA, assume the existence of a bijective map $h : \omega \to F$ and find some $g \in F$ such that $g \neq h(x)$ for every $x \in \omega$).

25. Prove that the rational numbers form a field with the sum and product.

26. Use the definition of positive rational numbers (Definition 6.7.7) to prove that the set $\mathbb{Q}^+$ of positive rational numbers is closed under sums and products. Prove first that the set of positive integers has those same properties.

27. Prove that the inequality in $\mathbb{Q}$ is a total ordering. Moreover, show the following two properties of this ordering of $\mathbb{Q}$: (i) if $x < y$ are elements of $\mathbb{Q}$, then there exists $z \in \mathbb{Q}$ with $x < z < y$; (ii) if $x, y \in \mathbb{Q}$ and $x > 0$, then there exists $u \in \omega$ such that $ux > y$ (these are, respectively, the density and the archimedean properties).

28. Let $\mathbb{Q}^+$ be the set of positive rational numbers. In what follows, we shall write $\frac{a}{b} \in \mathbb{Q}^+$ meaning that $\frac{a}{b} = [\langle a, b\rangle]$ in $\mathbb{Q}$ and that $a, b > 0$.

    Prove that the condition $\frac{a}{b} \preceq \frac{c}{d} \Leftrightarrow ad|bc$ for $\frac{a}{b}, \frac{c}{d} \in \mathbb{Q}^+$, defines a binary relation in $\mathbb{Q}^+$. Prove also that it is an ordering and that

$$\frac{a}{b} \preceq \frac{c}{d} \Rightarrow \frac{a}{b} \leq \frac{c}{d}$$

29. Prove Proposition 6.8.2.

30. Prove Proposition 6.8.3.

31. Prove the archimedean property of $\mathbb{R}$: if $x, y \in \mathbb{R}$ and $x > 0$, then there exists $u \in \omega$ such that $ux > y$.

32. A sequence $(a_x)_{x\in\omega}$ of rational numbers is a *Cauchy sequence* if for every rational $q > 0$ there exists $x \in \omega$ such that $x < y \Rightarrow |a_x - a_y| < q$. Prove that the natural definitions of sum and product of two Cauchy sequences give Cauchy sequences.

33. Let $s = (a_x)$ be a Cauchy sequence of rational numbers, and let $C(s) = \{q \in \mathbb{Q} | \{x \in \omega | q < a_x\}$ is finite $\}$. Prove that $C(s)$ is a cut of rationals. By taking the corresponding Dedekind cut, show that this defines a map $g : S \to \mathbb{R}$ where S is the set of all Cauchy sequences.

34. Determine the subset of the set S of all Cauchy sequences consisting of those $s \in S$ such that $g(s) = 0$.

35. Show that if F is a complete ordered field, then the elements $0, 1, 2, \ldots$ form a subset N (again, this set must be defined by recursion) which is a Dedekind-Peano set. Prove that this set is not upper bounded.

    Deduce that there is also a subset $Q \subseteq F$ which is a kind of copy of the set of rational numbers. Similarly, repeat the process of the construction of $\mathbb{R}$ to obtain a subset $R \subseteq F$ which serves as a copy of $\mathbb{R}$.

7

Pure Sets

7.1 Transitivity

7.1.1 Transitive closure of a set

In this chapter, $\mathcal{U}$ is assumed to be a 6-universe. Just before Proposition 6.3.1 the notion of a transitive set was introduced. We repeat it here and extend it to arbitrary classes.

Definition 7.1.1. Let C be a class. We say that C is *transitive* when the implication $(x \in C \wedge u \in x) \Rightarrow u \in C$ holds.

Equivalently, if x is a set and $x \in C$ then $x \subseteq C$. The set ω (as well as any element of ω) is an example of a transitive set, by Proposition 6.3.1. Our first objective is to show that for any set a there is a smallest transitive set including a.

Proposition 7.1.2. *The collection* $\{\langle x, \cup x\rangle | x \in \mathcal{S}\}$ *is a functional class,* $\mathcal{S}$ *being the class of all sets.*

Proof. It can be described as the collection of ordered pairs $\langle x, y\rangle$ such that

$$x, y \in \mathcal{S} \wedge \forall z(z \in y \Leftrightarrow \exists u(u \in x \wedge z \in u))$$

It is a class by Corollary 3.2.7, and it is functional by the principle of extension. □

Let us call f to the functional class of Proposition 7.1.2 whose domain is $\mathcal{S}$. Applying natural recursion (Theorem 6.4.2, simplified form), we see that for any set a there is a function g with domain ω such that $g(0) = a$ and $g(s(x)) = f(g(x)) = \cup g(x)$. By Proposition 5.1.18, $\mathrm{cD}(g)$ is a set, and its union is a set by Axiom 3.

Proposition 7.1.3. *For any set* a*, let* g *be the function just defined. Then* $\cup cD(g)$ *is a transitive set.*

DOI: 10.1201/9781003449911-7

Proof. Let $u \in \bigcup \mathrm{cD}(g)$, so there is $x \in \omega$ such that $u \in g(x)$. If $z \in u$ then $z \in \bigcup g(x) = g(s(x))$, hence $z \in \bigcup \mathrm{cD}(g)$. $\square$

$\bigcup \mathrm{cD}(g)$ is called the *transitive closure* of the set a and is written $\mathrm{tc}(a)$. Observe that $\mathrm{tc}(a)$ could be described as an infinite union $\bigcup\{\bigcup^x(a)|x \in \omega\}$, where $\bigcup^x(a)$ is defined by recursion with $\bigcup^{s(x)}(a) = \bigcup(\bigcup^x(a))$, and $\bigcup^0(a) = a$. Another description of $\mathrm{tc}(a)$ is the following.

Proposition 7.1.4. *Let a be a set, C a transitive class and $a \subseteq C$. Then $\mathrm{tc}(a) \subseteq C$. Thus $\mathrm{tc}(a)$ is the smallest transitive set including a.*

Proof. Since $\mathrm{tc}(a) = \bigcup_{x\in\omega}(\bigcup^x(a))$, it will be enough to prove, by natural induction, that $\bigcup^x(a) \subseteq C \Rightarrow \bigcup^{s(x)}(a) \subseteq C$. But this is immediate, because for C transitive, $b \subseteq C \Rightarrow \bigcup b \subseteq C$. $\square$

7.1.2 Pure sets

Definition 7.1.5. Let u be a set. We say that u is a *pure set* if all the elements of $\mathrm{tc}(u)$ are sets.

Since all the elements of ω are sets, and $\mathrm{tc}(\omega) = \omega$, we see that ω is a pure set. Moreover, each element of ω is transitive and a pure set.

Proposition 7.1.6. *The collection of all pure sets of the universe $\mathcal{U}$ is a class.*

Proof. Note first that the collection, say T, of all transitive sets is a class: it is

$$\{x \in \mathcal{S}|\forall y, z((y \in z \wedge z \in x) \Rightarrow y \in x)\}$$

Then x is a pure set if and only if $x \in \mathcal{S}$ and $\exists y\ (y \in T \wedge x \subseteq y \wedge \forall z(z \in y \Rightarrow z \in \mathcal{S}))$. Thus pure sets form a class by Corollary 3.2.7. $\square$

We have shown above that ω and the elements of ω are pure sets. Also, the operations in ω, which are the functions of maps $\omega \times \omega \to \omega$ are pure sets, since their elements are formed with pure sets. The following is useful in dealing with pure sets.

Proposition 7.1.7. *Let s be a set. s is a pure set if and only if every element of s is a pure set. Consequently, any subset of a pure set is a pure set.*

Proof. If s is a pure set and $x \in s$, then x is a set and $x \subseteq \mathrm{tc}(s)$, hence $\mathrm{tc}(x) \subseteq \mathrm{tc}(s)$. Consequently $tc(x)$ consists of sets, and x is a pure set.

Conversely, let every element of s be a pure set. If we take $a = s$ in Proposition 7.1.3, we see that every element of $g(0)$ is a pure set. We show inductively that every element of $g(x)$ is a pure set: indeed, suppose x satisfies this property and consider $g(s(x)) = \cup g(x)$. Every element of $\cup g(x)$ belongs to a pure set and thus it is a pure set by the first part of this proof. Therefore every element of $\mathrm{tc}(s)$ is a set and s is a pure set. □

We have also seen how rational or real numbers and operations and functions with them can be constructed from ω, by taking cartesian products, equivalences, partitions and so on. It turns out that all these constructions from ω lead to sets that are still pure sets. So, the class of pure sets of $\mathcal{U}$ contains (copies of or replacements for) all the basic constructions of mathematics.

7.1.3 Classes as universes

We start again with a 6-universe $\mathcal{U}$ and call U to the class of all pure sets of $\mathcal{U}$. Since the class U of pure sets contains all the basic objects of mathematics, it is natural to ask if we could not view U as a universe; and in particular, to see whether U is then a 6-universe on its own. It will be rewarding to study this problem in a more general setting and consider classes in a 6-universe in order to view these classes as universes by themselves.

There is a problem, however, when one considers an arbitrary class C as a universe on its own. We may consider constructions (power set of a set, unions, etc.) or relations (membership, inclusion) or properties (like being an inductive set), and these constructions and concepts which are defined in the same way for $\mathcal{U}$ or for C may show different behavior when considered in one or the other universe; e.g., it could happen that for a certain property (P), an object of C has property (P) as an object of the universe C while it does not satisfy property (P) in the universe $\mathcal{U}$; or possibly the other way round. Note that the situation is not really unfamiliar: many notions, in mathematics or in other disciplines, are *relative* to an ambient reality. For instance, the polynomial $x^2 + 1$ has the property of being irreducible (i.e., it is not the product of two polynomials of lesser degree) when it is considered in $\mathbb{R}[x]$, the ring of polynomials with real coefficients, but it is not irreducible as a complex polynomial; there $x^2 + 1 = (x - i)(x + i)$. Or the number 7 has an inverse element (1/7) when it is an element of the rational field $\mathbb{Q}$, but has no inverse when we consider it as an element of $\mathbb{Z}$. Likewise, if we admit $\mathcal{U}$ as the universe and $C \subseteq \mathcal{U}$ is a class, a set say $\langle x, y\rangle \in C$ (with $x, y \in C$) could be such that y is a well-ordering of x when viewed in the universe C but is not a well-ordering of x in $\mathcal{U}$. This may happen because subsets of x belonging to $\mathcal{U}$ might fail to belong to C, hence they would not be C-subsets of x; and if such a subset has not a minimum, then the well-ordering property fails in $\mathcal{U}$, but this is not visible in the universe C.

This problem may even affect the most basic notions. Imagine that in the universe $\mathcal{U}_\infty$ of item (2) of Example 2.7.2, the collection of all sets forms a class $\mathcal{S}$. Then a certain line AB is a set in the universe $\mathcal{U}_\infty$; but, viewed in the "universe" $\mathcal{S}$ it has no elements because its points do not belong to $\mathcal{S}$. So, it cannot be considered a set in that universe $\mathcal{S}$, because it is not the empty collection neither. In general, for a given $\mathcal{U}$-class C, the sets of the universe C are collections of C-objects, hence of $\mathcal{U}$-objects; and thus they are also sets in $\mathcal{U}$. But the converse may fail, as we have just seen. Using terms which are frequently employed, let us call the elements of C the *visible* (or C-visible) elements: when we look as if C is the universe, the objects of $\mathcal{U}$ which do not belong to C are *invisible* to us. So the only sets of $\mathcal{U}$ which are sets of C are those $x \in C$ all of whose elements are visible. In this sense, C may be a weird universe: objects of C which are sets in $\mathcal{U}$ could be non-sets in C. There is nevertheless a particular situation where the solution to this problem is optimal.

Proposition 7.1.8. *Let C be a class of the 6-universe $\mathcal{U}$ such that $\emptyset \in C$. If C is transitive and $x \in C$, then x is a set of C if and only if x is a set of $\mathcal{U}$.*

Proof. If $x \in C$ and x is a non-empty set of $\mathcal{U}$, then every element of x is C-visible by the transitivity of C. Therefore x is a collection of C-objects and thus it is a set of C. The converse is obviously true for any class C. □

Transitive classes C are well suited for studying the relationship between properties in $\mathcal{U}$ and properties in C. For example, given $x, y \in C$, we have that the relation $x \subseteq y$ holds in C if and only if it holds in $\mathcal{U}$: because all elements of x or y are C-visible, the condition $\forall z\ (z \in x \Rightarrow z \in y)$ is exactly the same in C or in $\mathcal{U}$. Other properties follow now.

Proposition 7.1.9. *Let C be a transitive class of the 6-universe $\mathcal{U}$, and assume that C is a 2-universe.*

(1) C satisfies the axiom of unions if and only if for every set $x \in C$, $\bigcup x \in C$.

(2) C satisfies the axiom of the power set if and only if for every set $x \in C$, $\mathbb{P}(x) \cap C \in C$.

(In the above statement, $\bigcup x$ and $\mathbb{P}(x)$ denote, respectively, the union of x and the power set of x **in the universe** $\mathcal{U}$).

Proof. We will show that the union of x in the universe C coincides with the union of x in $\mathcal{U}$; while the class of all subsets of x in C coincides with $\mathbb{P}(x) \cap C$. From this, the results follow immediately because the axioms ask for these classes to be elements of the universe C.

For $\bigcup x$, the union of x in the universe C is the collection of all $y \in C$ such that there exists $z \in C$ such that $z \in x$ and $y \in z$. But every $z \in x$ is C-visible (i.e., $z \in C$) by the transitivity of C; and every $y \in z$ is also C-visible for the same reason, hence that union is $\{y | \exists z \in x\ (y \in z)\} = \bigcup x$.

Next, the power class of x in the universe C contains all $y \in C$ which are subsets of x, when viewed in the universe C. But, as explained right before the proposition, this happens if and only if $y \in \mathbb{P}(x)$ (in $\mathcal{U}$). Therefore, the power class of x in C is $\mathbb{P}(x) \cap C$. □

Remark 7.1.10. We have used in this proposition the concept of a C-class, which is legal because C is a 2-universe and class sequences were defined for 2-universes (Definition 3.1.2). C-classes are collections of C-objects, and a reasonable notation for identifying, for example, the C-class which is the power class in C of some set $x \in C$ is $\mathbb{P}^C(x)$. With this notation, the conditions given in the proposition for C to satisfy axioms 3 and 4 should be written: $\cup^C x = \cup x$ and $\mathbb{P}^C(x) = \mathbb{P}(x) \cap C$.

Proposition 7.1.11. *Let $\mathcal{U}$ be a 6-universe and $C \subseteq \mathcal{U}$ a transitive class which is a 3-universe. If $x \in C$, then x is an inductive set as an element of $\mathcal{U}$ if and only if it is an inductive set as an element of C.*

Proof. Let $x \in C$. Either hypothesis implies that $\emptyset \in x$, so condition (1) of Definition 6.1.2 is fulfilled in either case. Let now suppose that x is inductive in C, and let $u \in x$ be a set; $u \in C$ by transitivity, hence it is a set of C by Proposition 7.1.8. In both universes, $s(u) = u \cup \{u\}$ is the same object by Proposition 7.1.9. So $s(u) \in x$ because x is inductive in C, hence condition (2) of Definition 6.1.2 holds also in $\mathcal{U}$. Conversely, if $x \in C$ is inductive in $\mathcal{U}$ and $u \in x$ is a set of C, the same argument shows that $s(u) \in x$ and $s(u) \in C$. □

Proposition 7.1.12. *Let C be a transitive class of the universe $\mathcal{U}$ and assume that C is a 2-universe. Any class of the universe C is a class of $\mathcal{U}$.*

Proof. Let us refer to classes of C and classes of $\mathcal{U}$ as C-classes and $\mathcal{U}$-classes, respectively. The hypotheses on C entail directly that the membership and equality classes of C are $\mathcal{M} \cap C$ and $\mathcal{E} \cap C$, respectively; hence they are $\mathcal{U}$-classes by Proposition 3.1.5. Since sets of C are sets of $\mathcal{U}$, every basic C-class is a $\mathcal{U}$-class. We now give the proof by finite induction on the number of terms of a C-class sequence. Suppose that the last term of a C-class sequence of length $n+1$ is A_{n+1} and the previous terms are $\mathcal{U}$-classes. For example, $A_{n+1} = \mathrm{D}(A_n)$. The hypotheses on C imply that if $\langle x, y \rangle \in A_n \subseteq C$, then $x \in C$, and hence $\mathrm{D}(A_n)$ is the same in C or in $\mathcal{U}$ and thus A_{n+1} is also a $\mathcal{U}$-class. Similar simple arguments show that A_{n+1} is again a $\mathcal{U}$-class when it is obtained through any of the rules 3 to 7. □

7.2 ZF-Universes

7.2.1 The universe of pure sets

Let $\mathcal{U}$ be a 6-universe and let U be the class of pure sets of $\mathcal{U}$.

Proposition 7.2.1. *The class U of pure sets is a transitive class which is a 2-universe.*

Proof. Obviously, $\emptyset \in U$. Also, if $x, y \in U$, then $\{x, y\}$ is a pure set by Proposition 7.1.7 hence it belongs to U, so the axiom of pairs holds. Finally Proposition 7.1.7 shows that U is transitive. □

Proposition 7.2.2. *The transitive class U of pure sets is a 4-universe.*

Proof. By Proposition 7.2.1 U is a 2-universe. Now, if $x \in U$ then every element of $\cup x$ is a pure set by Proposition 7.1.7. Thus $\cup x \in U$ by the same proposition, and U satisfies axiom 3 by Proposition 7.1.9.

Proposition 7.1.7 implies also that any subset of a pure set is a pure set. This entails that if $x \in U$, then $\mathbb{P}(x) \cap U = \mathbb{P}(x)$, hence axiom 4 holds in U by Proposition 7.1.9. □

Proposition 7.2.3. *The class U of pure sets of $\mathcal{U}$ satisfies axiom 5.*

Proof. Let A be a U-class of triples. We know from Proposition 7.1.12 that A is a $\mathcal{U}$-class. So let $u, s \in U$ and suppose that the restriction of A to $\{u\} \times s$ is functional (in $\mathcal{U}$ or in U; these are equivalent by the transitivity). Then the codomain of that restriction is a set $y \in \mathcal{U}$ by the corresponding instance of the axiom of replacement in $\mathcal{U}$. But then $y \in U$ by Proposition 7.1.7, and the A-instance of replacement holds in U. □

Remark 7.2.4. As all statements which refer to all classes, this one means that there is a proof of the A-instance of the axiom of replacement for each U-class A.

Proposition 7.2.5. *If $\mathcal{U}$ is a 6-universe, the class U of pure sets of $\mathcal{U}$ is a 6-universe.*

Proof. It only remains to show that axiom 6 holds. But $\omega = \mathrm{tc}(\omega)$ by Proposition 6.3.1 and all elements of ω are sets by Proposition 6.1.4, hence $\omega \in U$ and is inductive by Proposition 7.1.11. □

7.2.2 The axiom of extension

It makes sense to start now with the 6-universe U and construct the class of pure sets of the universe U. But the result is really simple.

Proposition 7.2.6. *If U is the class of pure sets of the 6-universe $\mathcal{U}$ and $x \in U$, then x is a pure set of the universe U.*

Proof. Let $x \in U$. Since U is transitive, the elements of x are the same in U or in $\mathcal{U}$; by the description of the transitive closure of a set x and the property that the sets $\bigcup x$ and ω are the same in $\mathcal{U}$ or U, we infer that $\mathrm{tc}(x)$ is also the same for $\mathcal{U}$ or for U. Since every element of $\mathrm{tc}(x)$ is a set in $\mathcal{U}$, it is a set in U by Proposition 7.1.8, hence x is a pure set of U. □

Corollary 7.2.7. *Let $\mathcal{U}$ be a 6-universe. The following conditions are equivalent.*

(i) $\mathcal{U} = U$, the class of pure sets.

(ii) Every object of $\mathcal{U}$ is a set.

(iii) For any $a, b \in \mathcal{U}$, $a = b \Leftrightarrow (\forall x(x \in a \Leftrightarrow x \in b))$.

Proof. (i) $\Rightarrow$ (ii): From (i), every $x \in \mathcal{U}$ is a pure set, hence a set.

(ii) $\Rightarrow$ (iii): Since a, b are sets, the property (iii) holds by the principle of extension.

(iii) $\Rightarrow$ (i): Suppose $\mathcal{U} \neq U$, so there is some object of $\mathcal{U}$, say a, which is not a set. Then $x \in a \Leftrightarrow x \in \emptyset$ holds vacuously and thus $a = \emptyset$ by condition (iii), a contradiction. □

Despite the fact that in the beginning of our exposition we have suggested that the objective of the theory is to identify the universe of all mathematical objects, we have progressively come to accept more modestly that it could be enough to present and study a collection of objects that is suitable to serve as a basis for ordinary mathematics. In this sense, the universe U serves this purpose quite satisfactorily: it includes versions of all sorts of usual numbers, relations and functions between them. Though in the original universe $\mathcal{U}$ there could exist objects which are not sets, we observe now that, from the above point of view, such objects are not needed. Thus nothing important seems to be lost if we restrict to a universe where sets and objects are the same thing. This is made possible by Corollary 7.2.7 and suggests adding the next axiom to our list.

Axiom 7 (of extension). $(\forall x(x \in a \Leftrightarrow x \in b)) \Rightarrow a = b$, for any $a, b \in \mathcal{U}$.

Or, equivalently, every object of the universe is a set (and a pure set). The principle of extension (Proposition 2.2.6) shares its name with that of the axiom, because the principle assigns to collections the same property that the axiom assigns to objects.

We say that a universe that satisfies all axioms from 1 to 7 is a *ZF^--universe*. It may seem surprising that all the objects of a ZF^--universe are sets, and sets are formed from just the empty set. But, as we have justified, this is how set theory works.

7.3 Relativization of Classes

7.3.1 Extensional apt classes

For the rest of this chapter, we work in a ZF^--universe U.

In Section 7.1.3, we presented the problem of viewing a class C in a universe U as a universe itself, and we mentioned that C can be a weird universe, though things improve when C is transitive. But the fact is that we will need to consider the case of C not being transitive. Even so, by setting some mild restrictions on the class C, we are able to simplify the study while still letting C to be general enough to cover the future needs of the theory. The next definition states the first of these restrictions.

Definition 7.3.1. Let U be a ZF^--universe, C a U-class. We say that C is *extensional* when for any $a, b \in C$ the following holds: $(\forall x \in C\ (x \in a \Leftrightarrow x \in b)) \Rightarrow (a = b)$.

Let us suppose that C is an extensional U-class and $x \in C$. We write $[x] = \{z \in C | z \in x\} = x \cap C$, a notation that was used in Definition 4.4.6 for an arbitrary relation. Thus $[x] \in U$ by separation, and $[x]$ is a collection of objects of C. The advantage of having C extensional is that each $x \in C$ is uniquely determined by $[x]$, because $[x] = [y] \Rightarrow x = y$ for $x, y \in C$, and we may work with the objects of C through the corresponding collections. For instance, for $x, y \in C$, the relation $x \subseteq y$ viewed in C means, of course, $\forall z \in C\ (z \in x \Rightarrow z \in y)$; and that is $[x] \subseteq [y]$, the C-visible elements of x belong to y.

The following observations have immediate proofs.

Proposition 7.3.2. *If C is an extensional class of a ZF^--universe, then C satisfies the axiom of extension.*

Proposition 7.3.3. *If C is a transitive class of a ZF^--universe, then C is extensional and satisfies the axiom of extension.*

We turn back to the problem of studying the relationship between properties of objects in a ZF^--universe U and in a given U-class C. Since "properties" correspond to classes (that is, to the class of all the objects that satisfy the property), this study will be that of the relation between classes of the universe U and classes of the universe C. Recalling that classes have been defined for 2-universes, the second of the restrictions we place on the class-universe C assures this property and something more.

Definition 7.3.4. Let C be a class in the ZF$^-$-universe U. We call C an *apt class* (or an apt set when C is a set of U) if the following conditions are satisfied:

(1) $\emptyset \in C$.

(2) For $a, b \in U$, $\{a, b\} \in C$ if and only if $a, b \in C$.

Note that condition (2) implies that for any $a, b \in U$, $\langle a, b\rangle \in C \Leftrightarrow a, b \in C$. We will say that the class C is *semi-transitive* when $\{a, b\} \in C \Rightarrow a, b \in C$. Like this, condition (2) means that C is semi-transitive and closed under the formation of pairs.

It must be remarked that this definition is not a standard one; the concept of an apt class has not been introduced before and we cannot say it is unavoidable, but it makes life easier in the present context.

Proposition 7.3.5. *If C is a transitive class in a ZF$^-$-universe and is a 2-universe, then C is an apt class. On the other hand, an apt class is a 2-universe.*

Proof. Since C is transitive, it is obviously semi-transitive; and it is closed under pairs because it satisfies the axiom of pairs and the pair of $\{a, b\}$ of C cannot have non-visible elements by transitivity. The second part is also immediate. □

Since an apt class C is a 2-universe, class sequences may be defined in C with the same Definition 3.1.2. We must, however, be a bit careful when interpreting the application of rule 1 of Definition 3.1.2. The terms of a class sequence are, by definition, collections of objects of the universe; therefore when an object x of C is introduced as a class, it is the collection $[x]$ of the visible elements of x that is considered to be the class. For the rest of rules when C is an extensional and apt class, the membership and equality C-classes are respectively $\mathcal{M} \cap C$ and $\mathcal{E} \cap C$ as the ordered pairs of elements of C are those ordered pairs of U whose components belong to C. By the same reason, the cartesian product in C of two collections A, B of objects of C is exactly $(A \times B) \cap C = A \times B$, the same collection in U or in C.

Proposition 7.3.6. *Let C be an apt and extensional class. Every C-class is a U-class.*

Proof. We use simple finite induction on the length n of the C-class sequence $A_1, \ldots, A_n$, so we admit that $A_1, \ldots, A_r$ are U-classes and want to show that A_{r+1} is also a U-class. According to the rule applied, we have:

(1) If A_{r+1} is obtained from rule 1 applied to $x \in C$, then its elements are those of $[x] = x \cap C$, which is a set of U by separation applied to C, hence a U-class.

(2) If rule 2 is applied, then A_{r+1} is either $\mathcal{M} \cap C$ or $\mathcal{E} \cap C$, by the hypothesis on C. Then A_{r+1} is a U-class by Proposition 3.1.5.

(3) If we have applied rule 3, $A_{r+1} = B_1 \setminus B_2$ with B_1, B_2 classes of U, hence A_{r+1} is a class of U.

(4) The case of rule 4 is analogous, we have $A_{r+1} = B_1 \times B_2$ for U-classes B_1, B_2, again because C is an apt class. Thus A_{r+1} is a U-class.

(5) In rule 5, $A_{r+1} = \mathrm{D}(B)$ where $B \subseteq C$ is supposed to be a class of U, and thus so is A_{r+1} by condition (2) of Definition 7.3.4.

(6) If rule 6 is applied to get A_{r+1} as the collection of triples obtained from the triples of some class B which is a U-class by the hypothesis, then the result is also a U-class by the apt hypothesis. The same holds for rule 7. □

7.3.2 Relativization

Let U be a ZF^--universe and C an apt and extensional U-class, and say we want to compare the class of transitive objects of U and the class of transitive objects of C. These classes can be constructed in U and in C through class sequences, and, since the concept is the same in both universes, these class sequences have exactly the same labels, so that both classes are constructed in the same way from the same elements. In the universe U this process will give the class, say T, of transitive objects of U; and we shall denote as T^C the class obtained through that same process in the universe C. It is the classes T^C and T that we want to compare in order to assess the behavior of the given property (in this example, transitivity) in one universe and the other. We now make explicit this description of the relation between a U-class like T and the C-class T^C obtained from class sequences with the same labels.

Definition 7.3.7. Let $C \subseteq U$ be an apt and extensional class in the ZF^--universe U. A U-class sequence $A_1, \ldots, A_n$ will be called a *class sequence over* C if for each term A_i of the sequence introduced by rule 1, we have $A_i \in C$. Any term of such a sequence is a *class over* C.

Definition 7.3.8. Let C be an apt and extensional class in the ZF^--universe U and let $A_1, A_2, \ldots, A_n$ a U-class sequence over C. The *C-relativization* of this sequence is the sequence $A_1^C, \ldots, A_n^C$ constructed as follows for each step $i = 1, \ldots, n$, according to the label $(e(i), u(i))$ of the step.

(1) If $e(i) = 1$, then $A_i^C = [A_i] = A_i \cap C$.

(2) If $e(i) = 2$, then $A_i^C = A_i \cap C$.

(3) If $e(i) = 3$ with $A_i = A_k \setminus A_j$, then $A_i^C = A_k^C \setminus A_j^C$.

(4) If $e(i) = 4$ with $A_i = A_k \times A_j$, then $A_i^C = A_k^C \times A_j^C$.

(5) If $e(i) = 5$ with $A_i = \mathrm{D}(A_j)$, then $A_i^C = \mathrm{D}(A_j^C)$.

(6) If $e(i) = 6$ or 7 with $A_i = \sigma[A_j]$ for the adequate transformation σ of the triples of A_j, then $A_i^C = \sigma[A_j^C]$.

Remarks 7.3.9. (1) It is important to observe that the C-relativization of a sequence over C is a class sequence of the universe C. In particular, $\mathcal{M}^C$

or $\mathcal{E}^C$ are precisely the membership or the equality relation in C. Conversely, a C-class sequence is the relativization of the $\mathcal{U}$-sequence over C which has exactly the same labels as the given C-sequence – and the same objects $x \in C$ introduced by rule 1.

(2) Each term A_i^C in the C-relativization of a class sequence over C is called the *relativization of the term* A_i of the original sequence. This notation hides an ambiguity with which we must be careful. The same class A may be A_i in a certain class sequence and be also equal to some D_j in a different class sequence. Thus A_i^C and D_j^C could in principle be different, while we identify both as A^C. Thus it will be necessary to interpret always the notation A^C as depending not only on the class A but on the concrete class sequence arriving to A.

7.3.3 C-absolute classes

Definition 7.3.10. Let C be an extensional apt class of a ZF^--universe U, and let $\mathcal{A} = A_1, A_2, \ldots, A_n$ be a U-class sequence over C. For each $i \in \{1, \ldots, n\}$, the class A_i is said to be *C-absolute relative to $\mathcal{A}$* when $A_i^C = A_i \cap C$. The sequence itself will be called *C-absolute* when every term A_i is C-absolute.

The fact that a class A is C-absolute (relative to a class sequence leading to A) means that, if (P) is the property that defines the class A, then any object $x \in C$ satisfies the property (P) when considered as an object of U if and only if x satisfies the property (P) as an object of the universe C. In this situation, the elements of A^C are precisely the C-visible elements of A.

In spite of the fact that the C-absolute condition for the class A depends on the class sequence used for defining A, we frequently relax this limitation by speaking of *C-absolute classes* without mentioning the sequence; this is without risk when we can understand from the context which is the definition of the class A we want to use, or when there is just only one natural definition for that class.

Examples 7.3.11. As an example, consider the class $\mathcal{I}$ of ordered pairs $\langle x, y\rangle$ such that $x \subseteq y$ – with the usual meaning of this relation, i.e., every object of x is an object of y. Thus $x \subseteq y$ in the class C holds when $[x] \subseteq [y]$. But it might happen that $[x] \subseteq [y]$ while there are elements of x which are not visible and not in y, so $x \nsubseteq y$ (in U). In that case $\mathcal{I}^C \nsubseteq \mathcal{I}$; thus the class $\mathcal{I}$ is not C-absolute, in general. Of course, if C is a transitive class, then every element of $x \in C$ is visible and hence $\mathcal{I}^C = \mathcal{I} \cap C$, so inclusion is C-absolute for a transitive class C.

Let us discuss another example. Suppose again that C is an extensional apt class and that $u \in C$ is such that $\mathbb{P}(u) \in C$ (here, $\mathbb{P}(u)$ denotes the power set of u in the universe U). If we obtain $\mathbb{P}(u)$ as the only term in a class sequence where rule 1 has been applied, then this is a sequence over C and, according to item (1) of Definition 7.3.8, $\mathbb{P}(u)^C = \mathbb{P}(u) \cap C$ so $\mathbb{P}(u)$ is absolute relative to this sequence. On the other hand, let us construct $\mathbb{P}(u)$ as a class through the general definition of subsets of a set, which would give $\mathbb{P}(u)$ as a term in another class sequence $\mathcal{A}$. As such, $\mathbb{P}(u)^C = \{y \in C | [y] \subseteq [u]\}$. But $\mathbb{P}(u) \cap C = \{y \in C | y \subseteq u\}$ and now the inclusion means inclusion in U. Since it is possible that $[y] \subseteq [u]$ while $y \nsubseteq u$ in U, $\mathbb{P}(u)$ may fail to be C-absolute relative to $\mathcal{A}$.

There is a related concept that can be applied to classes independently of the sequence in which the class appears. We give the definition and then state the relationship between both concepts.

Definition 7.3.12. Let A be a U-class, C an apt and extensional U-class. We say that A is *C-reliable* when $\mathrm{D}(A) \cap C = \mathrm{D}(A \cap C)$.

The idea of the next result comes from a classical result, known as the *Tarski-Vaught theorem.*

Proposition 7.3.13. *Let C be an extensional apt class and $A_1, \ldots, A_n$ a class sequence over C. Let $D(A_{i_1}), \ldots, D(A_{i_k})$ be the terms of the sequence which have been introduced by rule 5 of Definition 3.1.2. Then the class sequence $A_1, \ldots, A_n$ is C-absolute if and only if the classes $A_{i_1}, \ldots, A_{i_k}$ are C-reliable.*

Proof. First, assuming that the sequence is C-absolute we prove that each A_i with $A_j = \mathrm{D}(A_i)$ obtained in the sequence by rule 5, is C-reliable. By hypothesis, $A_i^C = A_i \cap C$ and $A_j^C = A_j \cap C$. But $\mathrm{D}(A_i) \cap C = A_j \cap C = A_j^C = \mathrm{D}(A_i^C) = \mathrm{D}(A_i \cap C)$ according to Definition 7.3.8.

We prove the converse by finite induction and using Definition 7.3.8. Thus we must show that $A_i^C = A_i \cap C$ from the assumption that the corresponding property holds for the previous terms in the sequence.

When A_i comes from rule 1 or 2, $A_i^C = A_i \cap C$ by Definition 7.3.8. If $A_i = A_j \setminus A_k$ and we assume that $A_j^C = A_j \cap C$ (and the same for A_k), then $A_i^C = (A_j \cap C) \setminus (A_k \cap C) = (A_j \setminus A_k) \cap C = A_i \cap C$. Similarly, when $A_i = A_j \times A_k$, then $A_i^C = (A_j \cap C) \times (A_k \cap C) = (A_j \times A_k) \cap C$ where the last equality follows because C is an apt class. Then, let T be the class of all triples; if $A_j^C = A_j \cap C$ and $A_i = \sigma[A_j \cap T]$ for some transformation σ by rule 6 or 7, then $A_i^C = \sigma[A_j \cap C \cap T]$ because C is an apt class so $T^C = T \cap C$. The transformation σ and its inverse send elements of C to elements of C, again by the apt hypothesis; therefore $A_i^C = \sigma[A_j \cap T] \cap C = A_i \cap C$.

Finally, let A_i be obtained by rule 5 so that $A_i = \mathrm{D}(A_r)$ for some index r, and assume the sequence is C-absolute up to A_i. By hypothesis, A_r is C-reliable, whence $\mathrm{D}(A_r) \cap C = \mathrm{D}(A_r \cap C) = \mathrm{D}(A_r^C)$ by the inductive hypothesis. Therefore $A_i^C = \mathrm{D}(A_r)^C = \mathrm{D}(A_r^C) = \mathrm{D}(A_r) \cap C = A_i \cap C$, and we are done. □

A simple criterion for absoluteness of a class sequence is an immediate consequence of the preceding result.

Corollary 7.3.14. *Let C be an extensional apt class and $A_1, \ldots, A_n$ a U-class sequence over C. If rule 5 is not applied in the class sequence, then it is C-absolute.*

The correspondence between class sequences and formulas presented in Section 3.2 entails that the class sequences of Corollary 7.3.14 give classes identified through formulas without quantifiers. So we may view Corollary 7.3.14 as saying that a class which is defined by a formula without quantifiers is C-absolute whenever C is an extensional apt class. The next criterion refines this one.

Proposition 7.3.15. *Let $C \subseteq U$ be an extensional apt class and let $A_1, \ldots, A_n$ be a U-class sequence over C such that rule 5 has been applied to obtain a term A_j only in the following situation: if $A_j = D(A_i)$ with $i < j$, then $cD(A_i \cap (C \times U)) \subseteq C$. Then the sequence is C-absolute.*

Proof. According to Proposition 7.3.13 it will suffice to see that when A_i satisfies the condition in the statement, then A_i is C-reliable.

The inclusion $\mathrm{D}(A_i \cap C) \subseteq \mathrm{D}(A_i) \cap C$ is general for an apt class C. For the converse inclusion, we assume the hypothesis in the statement and let $x \in \mathrm{D}(A_i) \cap C$; then $\langle x, y \rangle \in A_i$ for some y, hence $\langle x, y \rangle \in A_i \cap (C \times U)$ and $y \in C$ by the hypothesis. Hence $\langle x, y \rangle \in A_i \cap C$, so $x \in \mathrm{D}(A_i \cap C)$ and A_i is C-reliable. □

Let us see an example where this criterion is applied.

Proposition 7.3.16. *Let A be the class of all transitive sets, and let C be a transitive apt class of the ZF$^-$-universe U. Then A is C-absolute.*

Proof. It will suffice to prove that the class A' of sets that are not transitive is C-absolute, because A is obtained as $U \setminus A'$ and U, A' are C-absolute (U can be shown to be C-absolute when seen as $\mathrm{D}(\mathcal{E})$ by Proposition 7.3.13), so A is then C-absolute as seen in the proof of that same proposition.

We may describe A' as $\{x | \exists y, z\ (y \in x \wedge z \in y \wedge z \notin x)\}$. If we call $B = \{\langle x, \langle y, z \rangle\rangle | y \in x \wedge z \in y \wedge z \notin x\}$, then $A' = \mathrm{D}(B)$ and $\mathrm{cD}(B \cap (C \times U)) \subseteq C$: this is clear by the transitivity of C, since $x \in C, y \in x, z \in y$ imply $y, z \in C$. Since a class sequence leading to B does not need any application of rule 5, A' is C-absolute by Proposition 7.3.15. □

7.4 Exercises

Except otherwise stated, in the exercises that follow the universe U is assumed to be a ZF$^-$-universe.

1. A tree is an ordered set T with the property that for every $x \in T$, the set of all $y \in T$ such that yRx is well-ordered by the relation of the tree. We spoke already of this concept in Section 3.2.2. As we then did, we set the root on top, so the tree grows downwards. We draw a line from a to b (with $a \neq b$) when aRb and there is no intermediate element; i.e., no c satisfies $c \neq a, b$ and aRc, cRb.

   For the following sets, find a tree (perhaps with the same element repeated in different nodes of the tree) whose root is the given set and with an object x right below y when $x \in y$. Determine the transitive closure of each of the three sets.

   $a = \{\emptyset, \{\{\emptyset\}, \{\emptyset, \{\emptyset\}\}\}\}.$

   $b = \{\{\emptyset, \{\emptyset, \{\emptyset\}\}\}, \{\{\{\emptyset\}\}\}\}.$

   $c = \{\{\{\emptyset, \{\{\emptyset\}\}\}\}\}.$

2. What is the transitive closure of $\{\omega\}$?

3. Find the transitive closure of the set $\mathbb{Z}$.

4. In a 6-universe where there is an object x which is not a set, give an example of a set with more than four elements, all of whose elements are sets but which is not a pure set.

5. In a ZF$^-$-universe, objects and sets are the same and hence the idea of a set is not really needed. So, U can be seen as an *abstract universe*: a collection of objects with a given undefined relationship $\in$ between those objects. The basic concepts and the axioms may be given without mention of sets. For instance, $x \subseteq y$ is the implication $\forall z\,(z \in x \Rightarrow z \in y)$, and this says that every object that has the relation $\in$ to x has this relation to y too.

   Describe in an abstract universe U, a pair $\{a, b\}$, an ordered pair $\langle a, b\rangle$, and the cartesian product of two objects.

6. Describe, also without using sets and as in the preceding exercise, what is the meaning in an abstract universe of an order relation in an object u. Do the same for a well-ordering of an object.

7. State axioms 1 to 4 for an abstract universe, by referring only to objects as in the preceding exercise.

8. Let C be a U-class seen as an abstract universe. Characterize the objects $c \in U$ that belong to C and, in that universe C, are a pair.

9. Prove the assertions about ordered pairs and cartesian product for an extensional apt class C, made just before Proposition 7.3.6.
10. Let C be an apt transitive class and let $R, s \in C$ such that R is, in the universe U, a relation in the set s which is a strict well-ordering (that is, it is a strict order relation so that the corresponding ordering of Proposition 4.2.7 is a well-ordering). Prove that R is also a strict well-ordering of the object s in the universe C. Discuss the questions whether R is also an ordering and a well-ordering of s when C is an apt extensional class.
11. Suppose C is an extensional apt class. Prove that the axiom of unions holds in C if and only if for every $x \in C$ there is $y \in C$ such that $C \cap y = C \cap (\cup(C \cap x))$. Relate this result to Proposition 7.1.9.
12. Suppose C is an extensional apt class. Prove that the axiom of the power set holds in C if and only if for every $x \in C$ there is $y \in C$ such that $C \cap y = \{u \in C | u \cap C \in \mathbb{P}(x \cap C)\}$. Relate this result to Proposition 7.1.9.
13. A system of axioms is said to be *inconsistent* when a contradiction (i.e., a certain assertion P and its contrary $\neg(P)$) can be deduced from the axioms. A system of axioms which is not inconsistent is called *consistent*. Prove that if the axioms for 6-universes form a consistent system, then so do the axioms for ZF^--universes – this result means that axiom number 7 is consistent with the rest of the axioms.
14. Let C be an extensional apt class and $x \in C$. What can be said on the question whether $\cup x$ (viewed as a class in the natural way) is C-absolute or not? (i.e., is $(\cup x)^C = C \cap (\cup x)$?). What if C is moreover transitive?
15. For an extensional apt class C and $x \in C$, prove that the successor of x, $s(x)$, with the usual definition as a class, is C-absolute. Is the class of all ordered pairs of the form $\langle x, s(x)\rangle$, C-absolute?
16. Let C be a transitive U-class which is a 6-universe. What is wrong with the following argument that "proves" that every finite set belongs to C?

    We use natural induction to see, in U, that each $x \in \omega$ satisfies the property: if t is a set and $f : x \to t$ is a bijective map, then $t \in C$. First, if $x = 0$, t has to be the empty set, hence it belongs to C. For the inductive step, suppose the property for each $y < s(x)$ and that $g : s(x) \to t$ is some bijective map. Let $g(x) = u \in t$. Then t can be partitioned as the union of $\{u\}$ and $t \setminus \{u\}$. Now, $g|_x : x \to t \setminus \{u\}$ is bijective, hence $t \setminus \{u\} \in C$; and $1 \to \{u\}$ is also bijective, hence $\{u\} \in C$. Since C satisfies the axiom of unions, $t \in C$.
17. Transform the "argument" of the preceding exercise so that the following correct result is proved: if C is a transitive U-class which

is a 6-universe and $t \subseteq C$ is a finite set of U, then $t \in C$ (the definition of finite sets is Definition 6.3.11).

18. Let C be a transitive class which is a 6-universe. Let F be the U-class of finite sets. Prove that if $\omega \in C$, then the class F is C-absolute.

In the exercises that follow, C is an apt transitive class. In each of the instances of a class below, it is asked whether it is C-absolute or not (by taking in each case the most natural class sequence, or equivalently, defining formula that leads to that class). Give a proof when it is C-absolute and reason why it may fail to be C-absolute in the other cases (concrete counterexamples are not required). Study the same problem when C is only supposed to be an extensional apt class.

19. For given $a, b \in C$, the class $a \setminus b$.
20. The class of all triples $\langle x, y, z\rangle$ such that $z = x \setminus y$.
21. For a given $a \in C$ with $a \neq \emptyset$, the class $\bigcap a$.
22. The class of all ordered pairs $\{\langle x, y\rangle | x \neq \emptyset \wedge y = \bigcap x\}$.
23. The class of all relations, i.e., all ordered pairs $\langle R, x\rangle$ where R is a relation in x (see Definition 5.2.15).
24. For a fixed object $s \in C$, the class of pairs $\langle R, \mathrm{D}(R)\rangle$ with R any relation in s.

Detection of C-absolute classes is frequently made easier through the use of formulas for the description of classes. Let us define the collection of Δ_0-formulas as those formulas obtained through the following principles.

(a) *Any simple formula is a Δ_0-formula.*

(b) *If F is a formula of level $k > 1$ of the form $\neg P$ or $P \wedge Q$ or $P \vee Q$ or $P \Rightarrow Q$ or $P \Leftrightarrow Q$, then F is a Δ_0-formula if P, Q are Δ_0-formulas.*

(c) *If F is a formula of level $k > 1$ of the form $\exists y$ (P) and P is $y \in x_1 \wedge Q(x_1, \ldots, x_n, y)$ for a Δ_0-formula Q, then F is a Δ_0-formula.*

(d) *If F is a formula of level $k > 1$ of the form $\forall y$ (P) and P is $y \in x_1 \Rightarrow Q(x_1, \ldots, x_n, y)$ for a Δ_0-formula Q, then F is a Δ_0-formula.*

25. Prove that if C is an apt transitive class and A is a class over C that is described by a Δ_0-formula, then A is C-absolute (Hint: Use finite induction on the level of the formula, starting with simple formulas, and apply Proposition 7.3.15.)

8

Ordinals

In this chapter, U is supposed to be a ZF$^-$-universe. Accordingly, all objects of U are sets.

8.1 Well-Ordered Classes and Sets

In Definition 6.3.4 we have introduced the notion of a well-ordered set and noted in Remark 6.3.5 that any well-ordering is a total ordering. It follows also from Lemma 4.2.2 that if a is a subset of a well-ordered set c, then the relation restricted to a is an ordering; it is a well-ordering too, because any subset of a is a subset of c. As usual, the order relation will be written as $\leq$ and the corresponding strict order (Proposition 4.2.7), as $<$.

Definition 8.1.1. Let X be an ordered set and $Y \subseteq X$ a subset. Y is said to be a *segment* of X if the following holds: $(x \in Y, u < x) \Rightarrow u \in Y$.

Definition 8.1.2. Let X be an ordered set and $u \in X$. Then $S_X(u) = \{x \in X | x < u\}$ is called the *initial segment* of X determined by u.

$S_X(u)$ is just another name for $[u]_<$ (see Definition 4.4.6). $S_X(u)$ is a segment of X. The connection between these two definitions is quite strong for well-ordered sets.

Proposition 8.1.3. *Let X be a well-ordered set, Y a subset of X with $Y \neq \emptyset$. Y is a segment of X if and only if either $Y = X$ or there exists $u \in X$ such that $Y = S_X(u)$.*

Proof. One direction is obvious. For the other, suppose Y is a non-empty segment and $Y \neq X$. Then $X \setminus Y \neq \emptyset$ and thus there is a minimum element $u \in X \setminus Y$. Consider $S_X(u)$ which is an initial segment of X. $v \in S_X(u)$ implies $v \in Y$, hence $S_X(u) \subseteq Y$. If $v \notin S_X(u)$, then $u \leq v$ and $v \notin Y$, because Y is a segment. Thus $Y = S_X(u)$. □

We will need the following observation. Suppose X is a well-ordered set and $v < u \in X$. Then $S_X(v) = S_{S_X(u)}(v)$. This is obvious because $S_X(v) \subseteq S_X(u)$.

DOI: 10.1201/9781003449911-8

Proposition 8.1.4. *If X is an ordered set, then the collection of pairs $\langle u, S_X(u)\rangle$ with $u \in X$ is a set.*

Proof. Exercise. □

8.2 Morphisms Between Ordered Sets

Definition 8.2.1. Let X_1, X_2 be ordered sets, and let us use the same symbols $\leq$ or $<$ for the relations in both sets, following the convention for order relations. A map $f : X_1 \to X_2$ is *monotone* (respectively, a *morphism*) when for any $x, y \in X_1$, $x \leq y \Rightarrow f(x) \leq f(y)$ (resp., $x < y \Rightarrow f(x) < f(y)$). An *isomorphism* is a bijective morphism such that its inverse is a morphism too.

Remarks and Examples 8.2.2. By the definitions, any morphism is a monotone map. In the following examples, we assume property NNS. If $f : \omega \to \omega$ is the map given as $f(n)$ = number of digits in the decimal expansion of n, this is a monotone map (for the natural order in ω), but not a morphism ($f(10) = f(12) = 2$). A morphism which is not an isomorphism (with the usual ordering of $\mathbb{R}$) is $g : \mathbb{R} \to \mathbb{R}$ where $g(x) = e^x$ – because $\mathrm{cD}(g) = \mathbb{R}^+ \neq \mathbb{R}$. A bijective morphism which is not an isomorphism is the identity map $\mathbb{P}(\omega) \to \mathbb{P}(\omega)$ if we take the order given by the inclusion in the initial set; and in the final set, the ordering given thus: $X \preceq Y$ when $X = Y$ or the smallest number which belongs to the symmetric difference of X and Y, belongs to Y. One can see easily that it is a morphism, but not when taken in the opposite direction.

An alternative terminology calls *non-decreasing* maps to the monotone maps, and *strictly increasing* maps to morphisms. Since the property of being monotone or a morphism depends exclusively on the function of the map, we also call the corresponding function *monotone* or *morphism* when the map is such.

Compositions of monotone maps (respectively, of morphisms, of isomorphisms) are monotone maps (resp., morphisms, isomorphisms). When $f : X_1 \to X_2$ is an isomorphism, then f^{-1} is also an isomorphism. We write $X_1 \cong X_2$ to indicate that there is an isomorphism $X_1 \to X_2$. An example of an isomorphism is $h : (-\pi/2, \pi/2) \to \mathbb{R}$ with $h(x) = \tan(x)$; and its inverse isomorphism is $\tan^{-1} : \mathbb{R} \to (-\pi/2, \pi/2)$.

When X_1 is totally ordered, any morphism $X_1 \to X_2$ is injective, because $x \neq y$ implies $x < y$ or $y < x$, hence $f(x) < f(y)$ or $f(y) < f(x)$; and its inverse function gives again a morphism. Therefore, a surjective morphism of totally ordered sets (in particular, of well-ordered sets) is an isomorphism.

Proposition 8.2.3. *Let $f : X \to Y$ be an isomorphism between totally ordered sets, and suppose $f(u) = v$. Then the restriction of f to the domain $S_X(u)$ induces a map $S_X(u) \to S_Y(v)$ which is an isomorphism.*

Proof. It is plain that $a \in S_X(u) \Rightarrow f(a) \in S_Y(v)$ and hence f can be restricted to a map $f_1 : S_X(u) \to S_Y(v)$. Being the restriction of a morphism, f_1 is a morphism. Finally, if $b \in S_Y(v)$, then $b = f(a) < v = f(u)$ so that $a < u$ and $a \in S_X(u)$; so, f_1 is surjective. □

Proposition 8.2.4. *Let $f : X \to X$ be a morphism for a well-ordered set X. Then, for each $u \in X$ one has $u \leq f(u)$.*

Proof. Let $X_0 = \{u \in X | f(u) < u\}$. It is obtained by projection from the intersection of f and the inverse relation $<^{-1}$, hence it is a set and a subset of X. If $X_0 \neq \emptyset$, then there is a minimum $v \in X_0$. Thus $f(v) < v$ and $f(f(v)) < f(v)$ whence $f(v) \in X_0$, contradicting the minimality of v in X_0. Therefore $X_0 = \emptyset$ as we had to show. □

Corollary 8.2.5. *The only isomorphism $X \to X$ of a well-ordered set X is the identity. More generally, if there is an isomorphism $f : X \to Y$ of well-ordered sets, then it is the only isomorphism $X \to Y$.*

Proof. Given the isomorphism $f : X \to X$, we have $x \leq f(x)$ for every $x \in X$ by Proposition 8.2.4. Since f^{-1} is also an isomorphism, $f(x) \leq f^{-1}(f(x)) = x$, and therefore $x = f(x)$. For the second part, different isomorphisms f, g would give the isomorhism $g^{-1} \circ f$, then $g^{-1} \circ f = 1_X$ and consequently $f = g$. □

Proposition 8.2.6. *If X is a well-ordered set and $u \in X$, then there is no isomorphism $X \cong S_X(u)$.*

Proof. Suppose $f : X \to S_X(u)$ is an isomorphism and $v = f(u)$. The composition $j \circ f : X \to X$ where $j : S_X(u) \to X$ is the inclusion, is a composition of morphisms, hence a morphism; by Proposition 8.2.4 $u \leq f(u) = v$. But $v \in S_X(u)$ and thus $v < u$, which gives a contradiction. □

The main result of this introductory part is:

Theorem 8.2.7. *Let X, Y be well-ordered sets. One and only one of the following three possibilities holds.*

(a) There exists an isomorphism $X \cong Y$.

(b) There exists $u \in X$ and an isomorphism $Y \cong S_X(u)$.

(c) There exists $v \in Y$ and an isomorphism $X \cong S_Y(v)$.

Proof. The incompatibility of the three possibilities can be disposed of easily. First $X \cong Y$ and $X \cong S_Y(v)$ would entail $Y \cong S_Y(v)$ which contradicts Proposition 8.2.6. The same reason shows that (a) and (b) are incompatible. Finally, if we had isomorphisms $f : X \to S_Y(v)$, $g : Y \to S_X(u)$ and $f(u) = \mathrm{w} < v$, then $S_X(u) \cong S_Y(\mathrm{w})$ by Proposition 8.2.3 and the observation after Proposition 8.1.3, and it follows that $Y \cong S_Y(\mathrm{w})$ again a contradiction.

From Proposition 8.1.4 we see that $\{\langle u, S_X(u), v, S_Y(v)\rangle | u \in X, v \in Y\}$ is a set and it follows easily that there is a set A of all ordered pairs $\langle u, v\rangle$ such that $S_X(u) \cong S_Y(v)$. We shall now focus on this set A. Note first that $A \neq \emptyset$ (unless $X = \emptyset$ or $Y = \emptyset$, in which case the result is trivial) because if u_0, v_0 are the minimum elements of X, Y, then $\langle u_0, v_0\rangle \in A$.

We show first that A is functional. If $S_X(u) \cong S_Y(v) \cong S_Y(\mathrm{w})$, we would get a contradiction with Proposition 8.2.6 if $v \neq \mathrm{w}$. Let $D = \mathrm{D}(A)$ and $C = \mathrm{cD}(A)$. We observe now that $A : D \to C$ is a morphism: if we had $S_X(u) \cong S_Y(v)$ and $u' < u$ then $S_X(u') \cong S_Y(v')$ for $v' < v$ by Proposition 8.2.3. This also shows that $u' < u \Rightarrow u' \in D$, so that D is a non-empty segment of X.

This has proved that $A : D \to C$ is a surjective morphism and since D, C are well-ordered, this is an isomorphism $D \cong C$. Also, D is a segment of X and C is a segment of Y by the same argument.

Since D is a non-empty segment, we see from Proposition 8.1.3 that either $D = X$ or $D = S_X(u)$ for some $u \in X$. Similarly, $C = Y$ or $C = S_Y(v)$ for some $v \in Y$. By combination, we get the three possibilities of the statement and the case $D = S_X(u)$ and $C = S_Y(v)$. But this is not a possibility, because it would imply that $S_X(u) \cong S_Y(v)$, hence $\langle u, v\rangle \in A$ and $u \in D$, contrary to the assumption that $D = S_X(u)$. This completes the proof. □

Example 8.2.8. The examples of well-ordered sets we have considered up to now are ω and the elements of ω. For these, Theorem 8.2.7 says something that we knew from Proposition 6.3.7: given $x, y \in \omega$, either $x = y$ or one of them is an initial segment of the other. Also, each $x \in \omega$ is an initial segment of ω and the only isomorphism $\omega \to \omega$ is the identity, by Corollary 8.2.5.

8.3 Ordinals

One of the main objectives of set theory is the study of infinity, i.e., infinite sets. In studying usual finite sets, a basic tool is afforded by the natural numbers: we can analyze these finite sets by disposing their elements in a list, and that list corresponds to a number. In set theory we have an equivalent of this process based upon the set ω, so that finite sets in the sense of Definition 6.3.11 correspond to elements of ω. So, if we want to extend this view of sets by arranging their elements in a list, the natural idea is to enlarge the series of the elements of ω by stretching it far beyond ω. It is easy to imagine how the series of the elements of ω can be extended further: ω is already an infinite list, and subsequent elements should be $s(\omega), s(s(\omega))$, etc. However this way of proceeding step by step by repeating the process of the enumeration of the natural numbers has obvious limitations. In fact, the essence of the set theory method for dealing with infinite sets is to convert this intuitive, but limited, idea of infinity into abstract concepts to which concrete reasoning may be applied thus providing a much more effective picture: instead of mimicking usual counting, we extract some of the features of the counting process and of natural numbers (as elements of ω) and rest upon the force of this abstraction to reach a new vision of counting infinities.

In this sense, the key idea is to consider an abstract property that applies to all elements of ω and to ω itself, and to take it as the general concept that we need for our study. That property is the fact that ω (and every element of ω) is a transitive set in which membership is a well-ordering (Proposition 6.3.7). The sets so defined (an idea due to von Neumann) are the ordinals and not only do they perfectly the job of serving as a model for sets which are well-ordered, but also the extremely important tools of induction and recursion seen in the case of ω may be extended to the general concept with equal effectivity.

Definition 8.3.1. An *ordinal* is a transitive set X such that the restriction $(X \times X) \cap \mathcal{M}$ of the membership relation $\mathcal{M}$ is a strict well-ordering of X.

Note that if X is an ordinal and $x \in X$, then x is a transitive set, because $z \in y \in x$ imply $y \in X, z \in X$ by the transitivity of X; and then $z \in x$ because membership is transitive in X. We have already seen that ω (Proposition 6.3.7) and the elements of ω (Proposition 6.3.1) are ordinals. But there are more ordinals. In this section, we shall see that most of the properties of the elements of ω obtained in Chapter 6 may be extended to all ordinals.

Proposition 8.3.2. *If x is an ordinal, then $x \notin x$ and the successor $s(x)$ of x, is an ordinal too.*

Proof. Let x be an ordinal. If $x \in x$ then $x = u \in x$ and thus $u \in u$ which contradicts the hypothesis that membership is anti-reflexive in x; this shows that membership is anti-reflexive also in $s(x)$. Next, we show transitivity of membership in $s(x)$: if $u \in z \in y$ for $u, z, y \in s(x)$, then either $y = x$ and then $u \in x = y$ because x is transitive; or else $y \in x$ and $u \in y$ because membership is transitive in the ordinal x. That $s(x)$ is a transitive set is proven by the same reasons. Thus membership is a strict ordering in the transitive set $s(x)$.

Now, let v be a non-empty subset of $s(x)$. If $v = \{x\}$, then v has an obvious minimum element. Otherwise, $v \cap x$ is a non-empty subset of x and it has a minimum element (with respect to $\in$) because x is an ordinal. It follows immediately that this is the minimum element of v. We conclude that membership well orders $s(x)$, hence $s(x)$ is an ordinal. □

Proposition 6.3.3 states that $x \in y \Leftrightarrow x \subsetneq y$ for elements in ω. We see that the same relation holds for arbitrary ordinals.

Lemma 8.3.3. *If α is an ordinal and $x \in \alpha$, then $x = S_\alpha(x)$. Moreover, any element of α is an ordinal.*

Proof. Because α is transitive, $x \subseteq S_\alpha(x)$. The converse inclusion $S_\alpha(x) \subseteq x$ is direct. We mentioned above that every element of an ordinal α is transitive; moreover, for $x \in \alpha$, x is a subset of α by the first part of this proof, and it is therefore well-ordered by membership, so x is an ordinal. □

Proposition 8.3.4. *Let α, β be ordinals. Then $\beta \in \alpha \Leftrightarrow \beta \subsetneq \alpha$.*

Proof. Suppose $\beta \in \alpha$. By Lemma 8.3.3 and Proposition 8.3.2, $\beta \subseteq \alpha$ and $\beta \neq \alpha$. Conversely, let $\beta \subsetneq \alpha$; there is a smallest element $u \in \alpha$ such that $u \notin \beta$. Then $S_\alpha(u) \subseteq \beta$, so $u \subseteq \beta$ by Lemma 8.3.3. Now, let $x \in \beta$. Both $u \in x$ or $u = x$ imply $u \in \beta$ by transitivity, which is impossible. But $u, x \in \alpha$, hence we must have $x \in u$; it follows that $\beta \subseteq u$ and $\beta = u \in \alpha$. □

Since membership is the strict ordering of an ordinal α, we may see that the corresponding order relation is inclusion: by Proposition 4.2.7, it is the relation $x \in y \vee x = y$; by Proposition 8.3.4, this is $x \subsetneq y \vee x = y$; hence $x \subseteq y$.

Proposition 8.3.5. *If α, β are ordinals and there is an isomorphism $\alpha \cong \beta$, then $\alpha = \beta$.*

Proof. Let $f : \alpha \to \beta$ be the isomorphism. We prove $\alpha = \beta$ by showing that f is the identity. To this end, we use RAA and suppose that f is not the identity. The collection of pairs $\langle x, f(x)\rangle$ such that $x \neq f(x)$ is a non-empty set, and its domain is a non-empty subset c of α. Let $u \in \alpha$ be the minimum of c, so that $x = f(x)$ for any $x \in u$ and so $u = f[u]$. Let $f(u) = y \neq u$ so that $u \subsetneq y$ by Definition 8.2.1 and $u \in y$ by Proposition 8.3.4. By transitivity, $u \in \beta$ and $u = f(v)$ for some $v \in \alpha$. Now, $f(v) = v$ would entail $u = v$, which is impossible because then $f(v) = u = f(u)$. Hence $v \in c$, so that $u \in v$, as u is the minimum of c. Then $y = f(u) \in f(v) = u$, from which $y \in y$ (because $u \subsetneq y$) contradicting Proposition 8.3.2. □

Proposition 8.3.6. *If α, β are ordinals, then exactly one of these three possibilities holds: $\alpha = \beta$ or $\alpha \in \beta$ or $\beta \in \alpha$.*

Proof. The case $\alpha = \beta$ excludes any of the other two by Proposition 8.3.2. If $\alpha \in \beta \in \alpha$, then $\alpha \in \alpha$ by transitivity and we get the same contradiction.

If $\alpha \neq \beta$, then by Propositions 8.3.5 and 8.2.7, either $\alpha \cong S_\beta(y)$ or $\beta \cong S_\alpha(x)$ for elements $y \in \beta$ or $x \in \alpha$. By Lemma 8.3.3 and Proposition 8.3.5, this shows that $\alpha = y \in \beta$ or $\beta = x \in \alpha$. □

The concept of a sequence given in Chapter 6 for ω can be extended to ordinals: if α is an ordinal, a *sequence of length* α is a function f with domain α; and one usually writes x_β to represent $f(\beta)$ for $\beta \in \alpha$. But we may also consider the identity function as a special "sequence", which gives a class, the class of ordinals. We construct this class now.

Let T be the class of all transitive sets, W the class $\{\langle u, R\rangle | R \subseteq u \times u \wedge R$ strictly well orders $u\}$ and $L = \{\langle u, R\rangle | R = \mathcal{M} \cap (u \times u)\}$. Then $T \cap \mathrm{D}(L \cap W)$ is the class of all ordinals, by Definition 8.3.1. We shall write this class as **Ord**.

Proposition 8.3.7. *The class* **Ord** *of all ordinals is a proper class.*

Proof. The class **Ord** is transitive by Lemma 8.3.3, and all its elements are transitive by Definition 8.3.1, thus $\in$ is a strict ordering of **Ord**, and $\subseteq$ is an order in **Ord**. Also, it is a total order by Proposition 8.3.6. Now let $C \subseteq$ **Ord** be a non-empty subset and let $\alpha \in C$. If α is not the minimum of C, then $C \cap \alpha \neq \emptyset$, because the order is total. Since $C \cap \alpha$ is included in the ordinal α, $C \cap \alpha$ has a minimum element $m \in \alpha$. If $\beta \in C \setminus \alpha$, then $\alpha \subseteq \beta$ by Proposition 8.3.6 and hence $m \subseteq \beta$. It follows that m is the minimum of C. Thus, if **Ord** were a set, it would be an ordinal and **Ord** $\in$ **Ord**, which contradicts Proposition 8.3.2. So it is not a set, but a proper class. □

The notion of a well-ordered set can be extended to classes. Specifically, a relation $\preceq$ which is an order in the class C is a *well-ordering* (and C is well-ordered by $\preceq$) if every non-empty subset of C has a minimum. The proof above has shown that **Ord** is well-ordered by the relation $\subseteq$, and we shall write this relation as $\alpha \leq \beta$. Naturally, $\alpha < \beta$ will mean $\alpha \subsetneq \beta$ or, equivalently, $\alpha \in \beta$. Note that the proof above may also be used to show that any given non-empty class C of ordinals has a minimum element, and we will eventually use this strong form of the property. Clearly, the minimum of a class C of ordinals is $\cap C$, as it is included in any element of the class, and an intersection of a non-empty class of ordinals is an ordinal. It is worth looking also at unions of ordinals.

Proposition 8.3.8. *Let S be a set of ordinals. Then $\cup S$ is an ordinal, and it is the supremum, in the ordering of* **Ord**, *of the set S.*

Proof. Any element of $\cup S$ belongs to some $\alpha \in S$, hence it is an ordinal. This entails immediately that $\cup S$ is transitive. Moreover, the ordering of **Ord** restricted to $\cup S$ is a well-ordering, so that $\cup S$, being a set, is an ordinal.

Obviously, $\cup S$ is an upper bound of S, since the ordering is given by inclusion. If λ is an upper bound of S, then it is clear that $\cup S \subseteq \lambda$, hence $\cup S$ is the smallest upper bound of S. □

We end this section with the important result that ordinals can account for all well-ordered sets; i.e., the ordinals furnish representatives for all well-ordered sets, under isomorphism.

Theorem 8.3.9. *If X is a set having a well-ordering, then there exists a (unique) ordinal α with an isomorphism $\alpha \cong X$.*

Proof. We consider the class of triples $\langle\langle x, R\rangle, u, \alpha\rangle$ such that: R is a well-ordering of x, $u \in x$, α is an ordinal and there is an isomorphism $\alpha \cong S_x(u)$. By Proposition 8.3.5, this is a functional class. Thus if X is any set with a well-ordering R, $\{\alpha \in \mathbf{Ord} | \exists u \in X\ (S_X(u) \cong \alpha)\}$ is a set by replacement, and hence, by Proposition 8.3.7 there exists some ordinal α which is not isomorphic to any initial segment of X. By Proposition 8.2.7, either $\alpha \cong X$ or $X \cong S_b(\alpha) = b$ for some ordinal b. □

8.4 Induction and Recursion

Like for ω, there are two forms of an induction principle for ordinals, and both will be called *principle of ordinal induction*. We start with the one that corresponds to the strong form of natural induction, Proposition 6.3.8.

Proposition 8.4.1. *Let A be a class of ordinals. Suppose for each ordinal α the following implication holds:*

$$\alpha \subseteq A \Rightarrow \alpha \in A$$

*Then $A =$ **Ord**.*

Proof. Suppose the condition holds but $A \neq \mathbf{Ord}$. Then $A^c = \mathbf{Ord} \backslash A \neq \emptyset$ and A^c has a smallest element β. This entails that $\beta \subseteq A$ and by our assumption, $\beta \in A$, a contradiction. □

The second form of induction needs as a prerequisite the distinction between successor and limit ordinals. We know from Proposition 8.3.2 that if x is an ordinal, then $s(x)$ is an ordinal. Now, if α is an ordinal and $\alpha = s(u)$ for some u, then α is called a *successor ordinal.* In this case, u is an ordinal too. This is clear by Lemma 8.3.3, because $u \in \alpha$. $\emptyset = 0$ is an ordinal which is not a successor, but $s(0) = 1$ is a successor ordinal. Every other element of ω is a successor ordinal, by Proposition 6.1.5.

An ordinal which is not a successor ordinal is called a *limit ordinal.*

ω is a limit ordinal. This is because if we had $\omega = s(x)$ for some ordinal x, then $x \in \omega$ and consequently $s(x) \in \omega$ because ω is inductive. But this means $\omega \in \omega$, a contradiction to Proposition 8.3.2.

Limit ordinals and successor ordinals can be told apart by the following property.

Proposition 8.4.2. *A nonzero ordinal α is a limit ordinal if and only if for every $\beta \in \alpha$, there exists $\gamma \in \alpha$ such that $\beta \in \gamma$. Also, α is a limit ordinal if and only if $\alpha = \cup\alpha$.*

Proof. First, let α be a successor ordinal. Thus $\alpha = s(\beta)$ and hence $\beta \in \alpha$. Now, $x \in \alpha \Rightarrow x \in \beta$ or $x = \beta$, and it follows from Proposition 8.3.2 that $\beta \notin x$. That is, an ordinal γ with $\beta \in \gamma \in \alpha$ cannot exist. Also, if $x \in \cup\alpha = \cup(\{\beta\} \cup \beta)$, then $x \in \beta$ or $x \in \gamma \in \beta$, and hence $\cup\alpha \subseteq \beta$. Since the other inclusion is clear, $\cup\alpha = \beta \neq \alpha$.

Suppose now that α is a nonzero limit ordinal and $\beta \in \alpha$. It follows that $s(\beta) \subseteq \alpha$. But $s(\beta) \neq \alpha$ since α is limit and therefore $s(\beta) \subsetneq \alpha$ and $s(\beta) \in \alpha$ by Proposition 8.3.4. Since $\beta \in s(\beta)$, $\gamma = s(\beta)$ satisfies the property in the first part of the statement. This also shows that, for α limit, $\beta \in \alpha \Rightarrow \beta \in \cup\alpha$ and thus $\alpha \subseteq \cup\alpha$. The converse inclusion holds because α is transitive and thus $\alpha = \cup\alpha$. □

$s(\omega), s(s(\omega))$, etc. are successor ordinals. Later in this chapter, we will show many limit ordinals greater than ω. Let us now state the second form of the principle of ordinal induction.

Proposition 8.4.3. *Let C be a class of ordinals, and assume that the following properties hold:*

(i) $x \in C \Rightarrow s(x) \in C$; *(ii) for each limit ordinal ρ, $\rho \subseteq C \Rightarrow \rho \in C$.*

*Then $C =$ **Ord**.*

Proof. It is simple to check that, under the given hypotheses, the class C satisfies the condition of Proposition 8.4.1 and then that proposition applies. □

We have also two forms of the theorem of (ordinal) recursion. As with natural recursion, the essence of the result is that if A is a given class, it is possible to define a functional H from **Ord** to the class A, such that each value $H(\alpha)$ is functionally determined by the set of the preceding values (that is, of

$\{H(\beta)|\beta \in \alpha\} = H[\alpha]$) and possibly by α itself. The word "functionally" here means that we need some functional which assigns elements of A to elements of $\mathbb{P}(A)$ (or of $\mathbb{P}(A) \times \mathbf{Ord}$) in order to obtain H.

Theorem 8.4.4. *Let A be a class and F a functional such that $D(F) = \mathbb{P}(A) \times \mathbf{Ord}$ and $cD(F) \subseteq A$. Then there is a unique functional H with domain $\mathbf{Ord}$ satisfying the property, for any $\alpha \in \mathbf{Ord}$:*

$$H(\alpha) = F(\langle H[\alpha], \alpha\rangle)$$

(where $H[\alpha]$ has the meaning explained just after Proposition 4.1.4).

Proof. (For simplicity, we shall write throughout the proof $F(a, b)$ instead of the correct $F(\langle a, b\rangle)$). Let us consider the collection C of all ordered pairs $\langle \alpha, h\rangle$ where $\alpha \in \mathbf{Ord}$, h is a function, $\mathrm{D}(h) = \alpha$, $\mathrm{cD}(h) \subseteq A$ and the following property holds: $(*)$ $x \in \alpha \Rightarrow h(x) = F(h[x], x)$. C is clearly a class, as the conditions for the elements of C may be given through a formula (for the particular classes A, F).

We claim that if $\langle \alpha, g\rangle, \langle \beta, h\rangle \in C$ and $\alpha \leq \beta$, then $g \subseteq h$. To prove the claim, we show that $h(x) = g(x)$ for any $x \in \alpha$. For suppose we had α, β, g, h as above and $\{x \in \alpha | h(x) \neq g(x)\} \neq \emptyset$. This set of ordinals has a minimum m and thus $h[m] = g[m]$. But then $h(m) = F(h[m], m) = F(g[m], m) = g(m)$, and this contradiction proves the claim. In particular, when $\alpha = \beta$, the claim shows that the value h is unique; that is, C is functional.

To see that $\mathrm{D}(C) = \mathbf{Ord}$, we use Proposition 8.4.1 and prove that $\alpha \subseteq \mathrm{D}(C)$ implies $\alpha \in \mathrm{D}(C)$. If $\alpha \subseteq \mathrm{D}(C)$, then let h_β (for each $\beta \in \alpha$) be the functions with $\langle \beta, h_\beta\rangle \in C$. By the claim, these functions satisfy the hypothesis of Lemma 6.4.1 and thus their union is a function from $\cup\alpha$ to A satisfying $(*)$. If α is limit, this is the function h_α, according to Proposition 8.4.2; otherwise, $\alpha = s(\delta)$ and the union of the functions h_β is h_δ. By adding the pair $\langle \delta, F(h[\delta], \delta)\rangle$ we obtain the function h_α. So $\alpha \in \mathrm{D}(C)$ in any case.

Finally, another application of Lemma 6.4.1 and the claim above to the set of all functions h_α gives the functional H. Uniqueness follows from the uniqueness at each α. □

We now follow with the second form of recursion.

Theorem 8.4.5. *Let A be a class and let F, G be functional classes such that $D(G) = A \times \mathbf{Ord}$, $D(F) = \mathbb{P}(A) \times \mathbf{Ord}$ and $cD(F), cD(G) \subseteq A$. Then there is a unique functional class H whose domain is $\mathbf{Ord}$ and such that the following holds, for each ordinal α and for each limit ordinal λ:*

$$H(s(\alpha)) = G(\langle H(\alpha), \alpha\rangle), \quad H(\lambda) = F(\langle H[\lambda], \lambda\rangle)$$

Proof. The proof is similar to that of Theorem 8.4.4 so we will only make some comments about the differences of this case. The class C is constructed as in that proof, but instead of condition $(*)$ for the pairs $\langle x, h\rangle$ of C we use two conditions, one for $y = s(z)$ as: $h(y) = G(\langle h(z), z\rangle)$; and one for y a limit, same as $(*)$. Note that limit ordinals form a class and hence we can define C consistently as a class.

The rest of the proof is similar. □

As with natural recursion and for the same reasons, Theorems 8.4.4 and 8.4.5 are also valid when the domain of the functionals F or G consists only of $\mathbb{P}(A)$ or A respectively. These are the *simplified forms of recursion*. Again as with natural recursion, the theorems are also valid if we limit the domain to some ordinal α; and in that case we only need the functional F to be defined for ordinals up to α.

8.5 Ordinal Arithmetic

8.5.1 Ordinal addition

We extend the ordinary arithmetic with natural numbers to all ordinals. Note that with finite sets, addition of two numbers is performed by taking all the elements of the first set, then adjoining all the elements of the second to make a larger collection. This is essentially the same for infinite ordinals.

Lemma 8.5.1. *Let α, β be ordinals, and let $A = \{0\} \times \alpha, B = \{1\} \times \beta, M = A \cup B$. Define a relation $<$ in M by setting: $\langle 0, x\rangle < \langle 1, y\rangle$ for any x, y; $\langle 0, x\rangle < \langle 0, y\rangle$ whenever $x < y$ for $x, y \in \alpha$; similarly, $\langle 1, x\rangle < \langle 1, y\rangle$ whenever $x < y$ for $x, y \in \beta$ (this is usually called a (strict)* lexicographic order*). Then this is a strict ordering which well orders M.*

Proof. Obviously, $A \cap B = \emptyset$. By considering the different cases one may see easily that the relation is transitive and hence a strict ordering. Let $\emptyset \neq X$ be a subset of M; if $X \cap A \neq \emptyset$, then the smallest element of X is $\langle 0, u\rangle$ with u being the smallest element of $\mathrm{cD}(X \cap A)$; if $X \cap A = \emptyset$, then $X = X \cap B$ and its smallest element is $\langle 1, v\rangle$ with v being the smallest element of $\mathrm{cD}(X \cap B)$. □

Definition 8.5.2. For ordinals α, β we define the *sum* $\alpha + \beta$ as the unique ordinal which is isomorphic to the ordered set M of Lemma 8.5.1, according to Proposition 8.3.9.

Thus, the sum is obtained by taking the elements of the second summand in their order, right after the elements of the first summand in their order. We draw below two examples of ordinal sum. The line on top shows simply the ordinals $5, \omega, s(\omega)$ by representing their elements with dots and in the order from left to right. The other two lines give the sums $\omega+5$ and $(\omega+5)+(\omega+3)$: the values above the line are the elements of the ordinal sum, while the names below the line give the values of the ordinals in the second summand in each case.

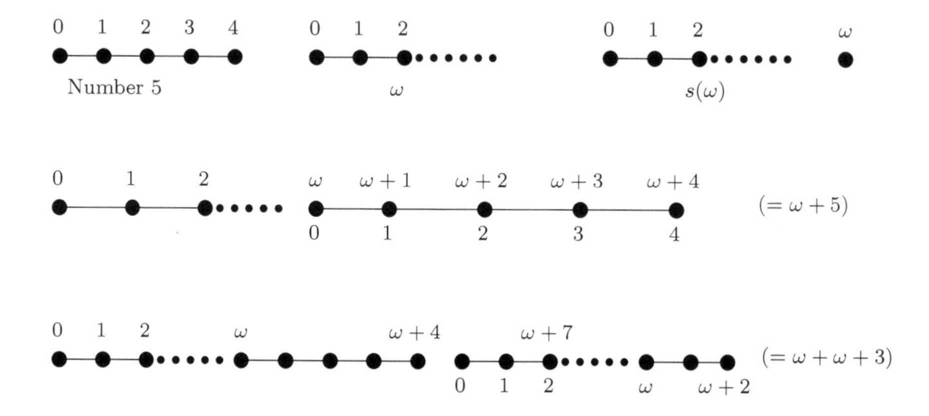

Figure 8.1
Ordinal additions.

Proposition 8.5.3. *The following properties hold for any ordinals α, β and any nonzero limit ordinal λ:*

$$\alpha+0=\alpha, \quad \alpha+s(\beta)=s(\alpha+\beta), \quad \alpha+\lambda=\cup\{\alpha+\beta|\beta\in\lambda\}$$

Proof. As above, we let $A=\{0\}\times\alpha$. The first equation is immediate: $0=\emptyset$ hence $A\cup B=A$ in this case and $\alpha\cong A\cup B$.

Let now $B=\{1\}\times\beta$ and $B'=\{1\}\times s(\beta)$ so that $A\cup B'$ coincides with $A\cup B$ except for its maximum element $\langle 1,\beta\rangle$. The isomorphism $\alpha+\beta\cong A\cup B$ entails then $s(\alpha+\beta)\cong A\cup B'$.

Finally, let λ be a nonzero limit ordinal, and set $C=\{1\}\times\lambda$. By Proposition 8.4.2, $\lambda=\cup\lambda=\cup\{\beta|\beta\in\lambda\}$. For each $\beta\in\lambda$, let us write $B_\beta=\{1\}\times\beta$, so that $C=\cup\{B_\beta|\beta\in\lambda\}$. Therefore $A\cup C=\cup\{A\cup B_\beta|\beta\in\lambda\}$.

Each $A \cup B_\beta$ is an initial segment of $A \cup C$ in the ordering considered in Lemma 8.5.1. By Theorem 8.3.9, $A \cup C$ is isomorphic to a unique ordinal, $\alpha + \lambda$ according to Definition 8.5.2. The restriction of this isomorphism to the initial segment $A \cup B_\beta$ gives an isomorphism which, due to the uniqueness of isomorphisms (Corollary 8.2.5), is $A \cup B_\beta \cong \alpha + \beta$. Since $A \cup C$ is the union of all the initial segments $A \cup B_\beta$, $\alpha + \lambda$ is the union of all the ordinals $\alpha + \beta$, as claimed in the statement of the proposition. □

In particular, $\alpha + 1 = s(\alpha)$. When restricted to natural numbers, which are elements of ω, the addition is the usual one: $n + 0 = n$ and $n + (m + 1) = (n + m) + 1$. Some caution is needed however when other ordinals are taken into account; thus, addition is not commutative in general. For instance, $1 + \omega = \cup(1 + n) = \omega$, but $\omega + 1 = s(\omega)$.

The already known notation $\cup\{\alpha + \beta | \beta \in \lambda\}$ has appeared in Proposition 8.5.3 for the union of a set. An alternative we will use frequently is perhaps more usual and also self-explanatory; namely $\cup_{\beta \in \lambda} \alpha + \beta$. We will be using this in similar cases.

Though the addition of ordinals is not commutative, it possesses other interesting properties. First of all, unlike 1, the number 0 commutes with all ordinals.

Proposition 8.5.4. *If β is any ordinal, then $0 + \beta = \beta$.*

Proof. In this case, $A = \emptyset$ so $A \cup B = B \cong \beta$. □

We have also a monotony property.

Proposition 8.5.5. *Let α, β, γ be ordinals. Then $\beta < \gamma \Leftrightarrow \alpha + \beta < \alpha + \gamma$.*

Proof. For the implication from left to right, let $A = \{0\} \times \alpha, B = \{1\} \times \beta, C = \{1\} \times \gamma$ as in Lemma 8.5.1. Since $\beta < \gamma$, β is an initial segment of γ by Lemma 8.3.3, and hence B is an initial segment of C in the ordering of these sets. Therefore $A \cup B$ is an initial segment of $A \cup C$ and this entails that $\alpha + \beta < \alpha + \gamma$.

The converse is now obvious: if we had $\alpha + \beta < \alpha + \gamma$ and $\gamma \leq \beta$, then $\alpha + \gamma \leq \alpha + \beta$, a contradiction. □

It follows from this result and Proposition 8.4.2 that if λ is limit, then $\alpha + \lambda$ is also limit: any element x of $\alpha + \lambda$ belongs to some $\alpha + \delta$ with $\delta < \lambda$ by Proposition 8.5.3; thus $x < \alpha + \delta < \alpha + \lambda$ and $\alpha + \lambda$ is limit by Proposition 8.4.2. Another simple corollary is the following cancellation property.

Corollary 8.5.6. *Let α, β, γ be ordinals. If $\alpha + \beta = \alpha + \gamma$, then $\beta = \gamma$.*

We have seen above that $1 + \omega = 0 + \omega$ so that we cannot simply change sides in the property of Proposition 8.5.5 or its corollary. Nevertheless, a slightly weakening of the property does hold.

Proposition 8.5.7. *Let α, β, γ be ordinals. Then $\beta \leq \gamma \Rightarrow \beta + \alpha \leq \gamma + \alpha$.*

Proof. We use induction on α (Proposition 8.4.3). The property for $\alpha = 0$ is obvious. Also

$$\beta + s(\alpha) = s(\beta + \alpha) \leq s(\gamma + \alpha) = \gamma + s(\alpha)$$

shows the induction step for successor ordinals. If α is a nonzero limit,

$$\beta + \alpha = \cup_{\mu \in \alpha}(\beta + \mu) \leq \cup_{\mu \in \alpha}(\gamma + \mu) = \gamma + \alpha$$

because each $\beta + \mu \subseteq \gamma + \mu$, by the inductive hypothesis. □

Proposition 8.5.8. *Let α, β be ordinals. Then $\alpha < \beta$ if and only if there exists $\gamma \in Ord$ with $\gamma \neq 0$ and $\beta = \alpha + \gamma$.*

Proof. One direction is a plain consequence of Proposition 8.5.5: if $0 < \gamma$ then $\alpha < \alpha + \gamma = \beta$. For the converse, suppose $\alpha < \beta$ and let $x = \beta \setminus \alpha \neq \emptyset$. x is thus well-ordered by membership, because it is a subset of β. Thus there exists an isomorphism $f : x \cong \gamma$ for some nonzero ordinal γ. If $A = \{0\} \times \alpha$, and $C = \{1\} \times \gamma$, then $A \cup C \cong \alpha \cup x = \beta$ with the ordering of Lemma 8.5.1; so, $\alpha + \gamma = \beta$. □

Proposition 8.5.9. *Let α, β, γ be ordinals. Then $(\alpha+\beta)+\gamma = \alpha+(\beta+\gamma)$.*

Proof. Let $A = \{0\} \times \alpha$, $B = \{1\} \times \beta$, $C = \{2\} \times \gamma$. Consider $A \cup B$ with the ordering of Lemma 8.5.1, and $B \cup C$, $A \cup B \cup C$ with the analogous orderings. It is then straightforward to see that, with similar orderings, $(A \cup B) \cup C \cong A \cup B \cup C \cong A \cup (B \cup C)$. But $(A \cup B) \cup C \cong (\alpha+\beta)+\gamma$ and $A \cup (B \cup C) \cong \alpha+(\beta+\gamma)$, which gives the property by Proposition 8.3.5. □

Proposition 8.5.10. *If α is any ordinal, then there exist a limit ordinal λ and some $x \in \omega$ such that $\alpha = \lambda + x$. This writing is unique.*

Proof. By using Definition 8.5.2, it is immediate that the ordered triples of ordinals $\langle \alpha, \beta, \alpha + \beta \rangle$ form a class A. If L is the class of limit ordinals, the codomain of $(L \times \omega \times \mathbf{Ord}) \cap A$ consists of all ordinals that can be written in the form $\lambda + x$ for some $x \in \omega$ and limit λ. By induction, it is easy to see that these are all the ordinals: this is clear for 0 and for limit ordinals, and also $s(\alpha) = \lambda + s(x)$ if $\alpha = \lambda + x$.

We may use again induction to prove uniqueness. This is obvious for limit ordinals, so assume that α has the unique writing $\alpha = \lambda + x$, but pretend that $s(\alpha) = \rho + y$ (for some limit ρ and $y \in \omega$). Then $y \neq 0$ because $s(\alpha)$ is not limit; hence $y = s(z)$ from which it follows that $\alpha = \rho + z$, so that $\rho = \lambda$ and $y = s(x)$. □

8.5.2 Ordinal multiplication

Again, the concept of multiplication of ordinals generalizes the idea of the multiplication of natural numbers. When explaining what 7×3 is, we might say that it is seven objects three times; that is, seven plus seven plus seven. The definition for infinite ordinals repeats this idea: the product $\alpha \cdot \beta$ is basically a well-ordered set constructed by starting with a copy of the ordinal α, then adjoining a second copy of α, then a third, and repeating the process so that the sequence of the steps corresponds to the sequence forming the ordinal β. But this informal version, where an ordinal is identified with a sequence of actions, has not the precise sense it has for finite multipliers. The actual definition will follow the next lemma.

Lemma 8.5.11. *Let α, β be two ordinals. Let $C = \beta \times \alpha$ and consider the (strict) lexicographic order in C: $\langle x, y\rangle < \langle x', y'\rangle$ if and only if $x < x'$; or $x = x'$ and $y < y'$. Then this gives a (strict) well-ordering of C.*

Proof. It is clear that the relation is a strict total ordering of C. Moreover, if $\emptyset \neq X \subseteq C$, then let $b \in \beta$ be the smallest element of $\mathrm{D}(X)$. If a is the smallest element of α such that $\langle b, a\rangle \in X$, it follows immediately that $\langle b, a\rangle$ is the smallest element of X. □

Definition 8.5.12. For ordinals α, β, we define the *product* $\alpha \cdot \beta$ as the unique ordinal which is isomorphic to the well-ordered set $\beta \times \alpha$ of Lemma 8.5.11.

Proposition 8.5.13. *The following properties hold for ordinals α, β and a nonzero limit λ:*

$$\alpha \cdot 0 = 0, \quad \alpha \cdot s(\beta) = (\alpha \cdot \beta) + \alpha, \quad \alpha \cdot \lambda = \bigcup_{\beta \in \lambda} \alpha \cdot \beta$$

Proof. The first equation is immediate: $0 = \emptyset$ hence $0 \times \alpha = \emptyset$ and $\alpha \cdot 0 = 0$.

Next, $s(\beta) \times \alpha = (\beta \times \alpha) \cup (\{\beta\} \times \alpha)$, and in the lexicographic order, all the elements of the second set are greater than those in the first set. Since $\beta \times \alpha \cong \alpha \cdot \beta$ and $\{\beta\} \times \alpha \cong \alpha$, we get the isomorphism $s(\beta) \times \alpha \cong (\alpha \cdot \beta) + \alpha$, which proves the second equation.

Finally, let λ be a nonzero limit ordinal. For each $\beta \in \lambda$, let us write $\delta_\beta = \alpha \cdot \beta$, so that $\delta_\beta \cong \beta \times \alpha$ with the lexicographic ordering. When $\beta < \beta' \in \lambda$, $\beta \times \alpha$ is an initial segment of $\beta' \times \alpha$ and, due to the uniqueness of isomorphisms (Corollary 8.2.5), the isomorphism $\delta_\beta \cong \beta \times \alpha$ is a restriction of $\delta_{\beta'} \cong \beta' \times \alpha$. So, the isomorphisms extend to an isomorphism $\bigcup_{\beta \in \lambda} \delta_\beta \cong \bigcup_{\beta \in \lambda} (\beta \times \alpha) = (\bigcup_{\beta \in \lambda} \beta) \times \alpha = \lambda \times \alpha$ and this gives the result. □

Multiplication is not commutative: $2 \cdot \omega = \omega$ but $\omega \cdot 2 = \omega + \omega \neq \omega$. But $0 \cdot \alpha = 0 = \alpha \cdot 0$ and $1 \cdot \alpha = \alpha = \alpha \cdot 1$. More generally, the product of elements of ω (as ordinals) is the product of these elements seen in Chapter 6. We now revise some other properties of multiplication.

Proposition 8.5.14. *Let α, β, γ be ordinals and $\alpha \neq 0$. Then $\beta < \gamma \Leftrightarrow \alpha \cdot \beta < \alpha \cdot \gamma$.*

Proof. In the lexicographic ordering of $\gamma \times \alpha$, $\beta \times \alpha$ consists of all elements that are $< \langle \beta, 0 \rangle$. Thus $\alpha \cdot \beta$ is an initial segment of the ordinal $\alpha \cdot \gamma$, from which the inequality follows. □

Corollary 8.5.15. *Let α, β, γ be ordinals with $\alpha \neq 0$. If $\alpha \cdot \beta = \alpha \cdot \gamma$, then $\beta = \gamma$.*

Proposition 8.5.16. *Let α, β, γ be ordinals. Then $\alpha \cdot (\beta \cdot \gamma) = (\alpha \cdot \beta) \cdot \gamma$.*

Proof. These ordinals are both isomorphic to $\gamma \times \beta \times \alpha$ with lexicographic ordering, so they coincide. □

Distributive properties are not generally valid: we have already seen that $(1 + 1) \cdot \omega = 2 \cdot \omega = \omega$, but $1 \cdot \omega + 1 \cdot \omega = \omega + \omega \neq \omega$. However, distributivity holds on one of the sides.

Proposition 8.5.17. *Let α, β, γ be ordinals. Then $\alpha \cdot (\beta + \gamma) = (\alpha \cdot \beta) + (\alpha \cdot \gamma)$.*

Proof. Taking $B = \{\langle 0, b \rangle | b \in \beta\}$ and $C = \{\langle 1, c \rangle | c \in \gamma\}$, we know the ordinals of the statement are isomorphic, respectively, to $(B \cup C) \times \alpha$ and $(B \times \alpha) \cup (C \times \alpha)$ with the orderings given in Lemmas 8.5.1 and 8.5.11. There is then an isomorphism between these two ordered sets with $\langle \langle 0, b \rangle, x \rangle \mapsto \langle 0, \langle b, x \rangle \rangle$ and $\langle \langle 1, c \rangle, x \rangle \mapsto \langle \langle 1, \langle c, x \rangle \rangle$ for $b \in B, c \in C$ and $x \in A$. □

The ordinals extend the usual finite counting to infinity and beyond. But where do we arrive with this process? We know we may count up to ω, which comes after all finite numbers (or their equivalent in set theory, the elements of ω). The next element is $s(\omega) = \omega + 1$, and then come $\omega + 2, \omega + 3, \ldots, \omega + \omega = \omega \cdot 2, \ldots, \omega \cdot 3, \ldots$. Although this process largely surpasses the natural infinite process of counting, it is still not entirely inimaginable, as we can somehow describe these infinitely many infinite sequences. In fact since our means to produce this long sequence are sums and products of ordinals, the sets we reach are obtained through cartesian products or unions of two sets, repeated perhaps countably many times. It turns out that these give always countable sets. Even if we reach $\omega \cdot \omega$, it is still countable (Proposition 6.6.8). Another operation that will be presented in the exercises, exponentiation, allows a quicker way of writing greater ordinals. With this operation $\omega \cdot \omega = \omega^2$, and we can arrive at ω^3, ω^ω or ω^{ω^ω}, for example. But even this operation will not take us off the countable realm. It is through a more abstract view (and a powerful new axiom) that we will find an extremely big, almost unthinkable world of ordinals.

8.6 Exercises

The universe U is assumed to be a ZF^--universe.

1. Prove Proposition 8.1.4 and justify that only one instance of the principle of separation is needed to establish this property for all ordered sets.
2. Explain with full detail why A in the proof of Theorem 8.2.7 is a set.
3. Prove that the set of ordinals of an inductive set is an inductive set.
4. Prove that any transitive set S of ordinals is an ordinal; and (1) if S has a maximum element α, then $S = s(\alpha)$; (2) if S has no maximum and $\beta = \sup(S)$, then $S = \beta$. Deduce that any set of ordinals is bounded by some ordinal.
5. Prove that there is not an infinite strictly descending sequence of ordinals $\alpha_1 > \alpha_2 > \cdots > \alpha_n > \ldots$. More generally, prove that the corresponding property holds for any well-ordered set.
6. Prove the second version of the recursion principle, Theorem 8.4.5.
7. Prove Theorem 8.4.4 in the simplified form which is referred to by the end of that section: i.e., when the domain of the functional F is $\mathbb{P}(A)$.
8. Prove that Theorem 8.4.4 holds also for a well-ordered class W (possibly proper) instead of the class **Ord**, if every initial segment in W is a set. It will be necessary to change accordingly the definition of H, with the adequate term substituted for $H[\alpha]$.
9. State and prove a *parametric version* of Theorem 8.4.4: the domain of the functional F in the statement of the theorem is now $C \times \mathbb{P}(A)\times$ **Ord** (for some class C), and the functional H to be defined has domain $C\times$ **Ord**.
10. Let W be a well-ordered set (with a strict relation $<$) and let $u \in U \setminus W$. For any segment $S \subseteq W$, let us consider the relation in the set $W \cup \{u\}$ which is the union of $<$ and the set $\{\langle x, u\rangle | x \in S\} \cup \{\langle u, y\rangle | y \in W \setminus S\}$. Prove that $W \cup \{u\}$ is also well-ordered with this relation.

11. Construct a functional class F with $\mathrm{D}(F) = \mathbf{Ord}$, $\mathrm{cD}(F) = \{\alpha|\alpha \text{ is a limit ordinal}\}$ and which is strictly increasing (i.e., $\alpha < \beta \Rightarrow F(\alpha) < F(\beta)$). Give explicitly the value $F(s(\alpha))$ in terms of $F(\alpha)$. When α is a successor ordinal, $F(\alpha)$ is called a *successor-limit* ordinal.

12. A functional class F with $\mathrm{D}(F) = \mathbf{Ord}$ and $\mathrm{cD}(F) \subseteq \mathbf{Ord}$ is said to be *normal* if it is strictly increasing and for every limit ordinal λ, $F(\lambda) = \sup(F[\lambda])$. Prove that if F is a normal functional class of that kind, $\alpha \subseteq F(\alpha)$ for every ordinal α; and $F(\lambda)$ is a limit ordinal if λ is a limit ordinal.

13. Let F be a normal functional class (see the preceding exercise), and let $\alpha \in \mathbf{Ord}$. Prove that there is some ordinal β such that $\alpha \leq \beta$ and $F(\beta) = \beta$. (Hint: assume $\alpha \neq F(\alpha)$ and construct explicitly a set with the elements $\alpha, F(\alpha), F(F(\alpha)), \ldots$. Show then that the supremum of this set has the required property).

14. The conditions in Proposition 8.5.3 can be seen as giving a recursive definition (for each ordinal α), based on Theorem 8.4.5, of the sum of ordinals $\alpha + \beta$. Explain how this can be done by using a version of the recursion theorem analogous to that given in Exercise 9. Conclude that there is a unique functional H with domain $\mathbf{Ord} \times \mathbf{Ord}$ such that $H(\alpha, \beta)$ satisfies the conditions in Proposition 8.5.3 for $\alpha + \beta$.

15. Repeat the preceding exercise for Proposition 8.5.13 and the product of ordinals.

16. Find sets of real numbers such that, with the usual inequality ordering, are isomorphic to each of the following ordinals: (i) $\omega + 1$; (ii) $\omega \cdot 3$: (iii) ω^2.

17. Find the smallest ordinal $\alpha > \omega$ with the property that $\beta, \gamma < \alpha \Rightarrow \beta + \gamma < \alpha$. Do the same for the product, instead of the sum.

18. Let λ be a limit ordinal. Prove that the following two conditions are equivalent: (i) for every $\alpha < \lambda$, $\alpha + \lambda = \lambda$; (ii) for any $\alpha, \beta < \lambda$, $\alpha + \beta \neq \lambda$.

19. Prove Proposition 8.5.4 by using the equations of Proposition 8.5.3, and applying induction. Do the same for Proposition 8.5.5.

20. Justify that if $\alpha \leq \beta$, then there exists a unique ordinal γ such that $\alpha + \gamma = \beta$. Show that, on the contrary, it is not possible to obtain a difference operation on the other side. Specifically, for $\alpha < \beta$ it is in general not possible to find γ such that $\gamma + \alpha = \beta$.

21. Let α, γ be ordinals. Prove that $\alpha + \gamma = \alpha \cup \{\alpha + \beta|\beta \in \gamma\}$.

22. Let α, β, γ be ordinals. Prove that if $\beta \leq \gamma$, then $\beta \cdot \alpha \leq \gamma \cdot \alpha$.

23. Give an example to show that multiplication of ordinals is not cancellative on the right: i.e., there are $\alpha \neq 0$, β, γ with $\beta \neq \gamma$ and $\beta \cdot \alpha = \gamma \cdot \alpha$.

24. Define for each ordinal α a functional H (we write $H(\rho) = \alpha^\rho$) with the properties: $\alpha^0 = 1$, $\alpha^{s(\beta)} = \alpha^\beta \cdot \alpha$, $\alpha^\lambda = \cup_{\beta \in \lambda} \alpha^\beta$ for λ nonzero limit. Show that this operation extends the known properties: (i) $\alpha^{\beta+\gamma} = \alpha^\beta \cdot \alpha^\gamma$; and (ii) $(\alpha^\beta)^\gamma = \alpha^{\beta \cdot \gamma}$.

25. Let α, β be ordinals, $\alpha \neq 0$. Prove that each element of the product $\alpha \cdot \beta$ can be written as $\alpha \cdot \gamma + \delta$ with $\gamma \in \beta$ and $\delta \in \alpha$, in a unique way.

26. Let $\alpha > 1$ and $\beta > 0$ ordinals. Show that there exist uniquely determined ordinals $\gamma_0 > \gamma_1 > \cdots > \gamma_k$ and $\delta_0, \delta_1, \ldots, \delta_k$ with $\delta_i < \alpha$ so that

$$\beta = \alpha^{\gamma_0} \cdot \delta_0 + \alpha^{\gamma_1} \cdot \delta_1 + \cdots + \alpha^{\gamma_k} \cdot \delta_k$$

    This is called the expansion of β to the base α.

    (Hint: Look for the smallest ordinal γ such that $\alpha^\gamma \geq \beta$ and show that if γ is a limit ordinal then $\alpha^\gamma = \beta$; use also the preceding exercise).

27. List the following ordinals from smaller to greater:

$$\omega, \omega + \omega + \omega, 3 \cdot \omega, \omega \cdot (\omega + 2), \omega^2, 3^\omega, \omega \cdot 3, \omega^\omega, 2^{\omega^\omega}, (\omega + 1) \cdot \omega$$

28. We use the definition of a successor-limit ordinal given in exercise 11. Prove that a limit ordinal λ is not a successor-limit if and only if the expansion to the base ω of λ has no terms with ω^1 or ω^0.

29. Prove that ω^ω and ω^{ω^ω} are denumerable (Hint: Find a systematic way for obtaining bijections $\omega \to \omega^k$ for each natural k; and use Proposition 6.6.9 to prove first that ω^ω is denumerable).

30. Use natural recursion to show that there is a function with domain ω whose codomain contains all ordinals of the form $\omega^{\omega^{\omega^{\cdots^{\omega}}}}$ with finitely many (in the sense of Definition 6.3.11) exponents. Prove that the union of the codomain of that function is an ordinal exceeding all the powers of the form $\omega^{\omega^{\omega^{\cdots^{\omega}}}}$.

9

ZF-Universes

The central development of this chapter parallels that of Chapter 7, with the concepts of well-founded set and regular set corresponding respectively to those of pure set and set of that chapter: it was seen in Chapter 7 that the class of pure sets in a 6-universe $\mathcal{U}$ is again a 6-universe; and axiom 7 was introduced to the effect that every object is a (pure) set. Now, we see that the class of well-founded sets in a ZF$^-$-universe U is a ZF$^-$-universe and then axiom 8 establishes that every set is regular (and also well-founded). It turns out that the consecutive restrictions of the universe provided by these two axioms, first to the class of pure sets, then to the class of well-founded sets, do the theory no harm in the sense that natural numbers (or the like of them, i.e., ω), integer, rational, real or complex numbers are the same in the restricted universes and in the initial 6-universe, due to the fact that the class of well-founded sets, as the class of pure sets, is closed under subsets. However, we cannot be sure that those sets of numbers remain the same when changing the initial universe. Not only because ω may be different in different universes; but, more importantly, because the subsets of $\mathbb{Q}$ may be different in one or the other universe. And these subsets determine which are the real numbers.

In the first part of this chapter, U will be assumed to be a ZF$^-$-universe.

9.1 Well-Founded Sets

Definition 9.1.1. Let A be a set. We say that A is *regular* if either $A = \emptyset$ or there exists an element $x \in A$ such that $x \cap A = \emptyset$.

The idea that every set should be regular has undoubtedly some merit. At least, the intuition of a set as represented in the well-known [1] Cantor's definition[1] suggests this: a set A has to be constructed from objects, hence it

[1] For Cantor, a set is "the reunion in a totality, of objects determined and well distinguished, of our perception or of our thinking".

DOI: 10.1201/9781003449911-9

seems plausible that those objects must be known previously to the formation of A. But if A is not regular, each of its elements contains objects of A; and this implies that describing or thinking of any element of A necessarily involves consideration of other elements of that same set, so there is no element of that set with a description independent of A. This comes near of creating a vicious circle: to define such a set we need to identify its elements, but whichever element we try to define, we need to know other elements of A to construct it. So, it is no surprise that mathematicians may consider regularity as a natural property for sets. Formally, the next definition reminds that of a pure set (see Definition 7.1.5).

Definition 9.1.2. Let u be a set. We say that u is a *well-founded set* if all the subsets of $\mathrm{tc}(u)$, the transitive closure of u, are regular sets.

As a first example besides the trivial one of $\emptyset$, ω is regular because $\emptyset \in \omega$ and $\emptyset \cap \omega = \emptyset$. Moreover, any non-empty subset of ω has a minimum element for the membership relation and hence it is regular. Since ω is transitive we see that ω is well-founded. More generally, every ordinal is a well-founded set by the same reason. Also, every subset x of a well-founded set u is well-founded because $\mathrm{tc}(x) \subseteq \mathrm{tc}(u)$.

9.2 The von Neumann Universe of a Universe

Our objective now is to show that the collection W of all well-founded sets is by itself a ZF$^-$-universe. But we will do this in an indirect way, by describing a hierarchy of objects of U giving a class that will turn out to coincide with W. We thus obtain a classification of the objects of W, and it is through this description that we will infer that W forms a ZF$^-$-universe.

9.2.1 The construction of V

Proposition 9.2.1. *There is a functional H with domain* **Ord** *such that $H(\emptyset) = \emptyset$, and for each ordinal α and each limit ordinal λ:*

$$H(\alpha + 1) = \mathbb{P}(H(\alpha)), \quad H(\lambda) = \cup H[\lambda]$$

Proof. With reference to Theorem 8.4.5, let the class A be the universe U and F, G be functionals defined on U: $F(x) = \mathbb{P}(x)$ and $G(x) = \cup x$. We know these are functional classes, hence there exists H with the stated properties by that theorem. □

The usual and useful writing of these sets is: V_α for $H(\alpha)$. Like this, we have:

$$V_0 = \emptyset, \quad V_{\alpha+1} = \mathbb{P}(V_\alpha), \quad V_\lambda = \cup_{\beta\in\lambda} V_\beta \text{ for } \lambda \text{ limit}$$

We consider some easy properties of these sets.

Proposition 9.2.2. *Each set V_α is transitive, and $V_\alpha \subseteq V_{\alpha+1}$. Consequently, $\alpha \leq \beta \Rightarrow V_\alpha \subseteq V_\beta$.*

Proof. We prove both properties by ordinal induction (Proposition 8.4.3). Let C be the class of ordinals α such that V_α is transitive and $V_\alpha \subseteq V_{\alpha+1}$. $\emptyset \in C$ is obvious. Let $\alpha \in C$. If $x \in V_{\alpha+1}$, $x \subseteq V_\alpha$ and $x \subseteq V_{\alpha+1}$ by the induction hypothesis; hence $V_{\alpha+1}$ is transitive and $x \in V_{\alpha+2}$. We have thus shown that $\alpha + 1 \in C$.

Suppose now that λ is a nonzero limit ordinal and $\beta \in C$ for all $\beta \in \lambda$. A union of transitive sets is transitive hence V_λ is transitive. Therefore $x \in V_\lambda \Rightarrow x \subseteq V_\lambda$ and $x \in V_{\lambda+1}$.

The last assertion follows easily, again by induction. □

Proposition 9.2.3. *For each $\alpha \in$ **Ord**, $\alpha \subseteq V_\alpha$.*

Proof. We use again Proposition 8.4.3. The induction hypothesis for α gives $\alpha \in V_{\alpha+1}$ hence $\{\alpha\} \subseteq V_{\alpha+1}$. Since also $\alpha \subseteq V_\alpha \subseteq V_{\alpha+1}$ by Proposition 9.2.2, we get $\alpha + 1 = \alpha \cup \{\alpha\} \subseteq V_{\alpha+1}$.

Let λ be a limit ordinal, so that $\lambda = \cup\lambda$ (Proposition 8.4.2). By the induction hypothesis again, each element of λ belongs to V_λ hence $\lambda \subseteq V_\lambda$. □

We shall denote by V the class $\cup_{\alpha\in\mathbf{Ord}} V_\alpha$, which is indeed a class because it is the union of the codomain of the functional H of Proposition 9.2.1; and a transitive class by Proposition 9.2.2. V will be called the *von Neumann universe* of the given universe U.

Let $x \in V$. Then the ordinals α such that $x \in V_\alpha$ form a non-empty class. Consequently there is a smallest $\alpha \in$ **Ord** such that $x \in V_\alpha$; and such α cannot be a limit ordinal, because $V_\alpha = \cup_{\beta\in\alpha} V_\beta$ for α limit. We call *rank of* x, $\mathrm{rk}(x)$ to the smallest ordinal α such that $x \in V_{\alpha+1}$. The ordered pairs $\langle x, \mathrm{rk}(x)\rangle$ form a class, which is a functional called the *rank functional.*

Corollary 9.2.4. *If $x \in U$ and $x \subseteq V$, then $x \in V$. Moreover, if $x, y \in V$ then $x \in y \Rightarrow rk(x) < rk(y)$.*

Proof. Let R denote the rank functional. Since $x \subseteq V$, $R[x]$ is a set by the simplified replacement axiom (Proposition 5.1.4), and its union is an ordinal

μ which is an upper bound of $R[x]$ by Proposition 8.3.8. This entails that $x \subseteq V_\mu$ and hence $x \in V_{\mu+1}$, so $x \in V$.

For the final assertion, let $\alpha = \mathrm{rk}(x)$ so $x \in V_\beta$ and $\beta = s(\alpha)$. If $x \in y$ and $\mathrm{rk}(y) \leq \alpha$, then $y \in V_\beta = \mathbb{P}(V_\alpha)$, and $y \subseteq V_\alpha$. But then $x \in V_\alpha$ which contradicts $\mathrm{rk}(x) = \alpha$. □

9.2.2 V and the axioms

Our purpose is to show that the class V fulfills all the conditions of a ZF$^-$-universe.

Proposition 9.2.5. *The von Neumann universe V is a transitive class of U, closed under subsets; and a 4-universe which satisfies axiom* 7.

Proof. V is transitive because it is a union of transitive sets, by Proposition 9.2.2. Transitivity also entails that axiom 7 holds in V, by Proposition 7.3.3.

It is obvious that $\emptyset \in V$ and if $x, y \in V$, then $x, y \in V_\alpha$ for some α, and thus $\{x, y\} \in V$ by Corollary 9.2.4. So, V satisfies axioms 1 and 2.

If $x \in V_\alpha$, then $x \subseteq V_\alpha$ by Proposition 9.2.2 and $\cup x \subseteq V_\alpha$, again by transitivity. Then $\cup x \in V_{\alpha+1} \subseteq V$ and axiom 3 holds, by Proposition 7.1.9.

Finally, let $x \in V$. For any $y \subseteq x$, transitivity of V implies $y \subseteq V$ and $y \in V$ by Corollary 9.2.4. Thus V is closed under subsets and hence $\mathbb{P}(x) = \mathbb{P}(x) \cap V$. Let $\mathrm{rk}(x) = \alpha$ so $x \subseteq V_\alpha$. Thus $\mathbb{P}(x) \subseteq V_{\alpha+1}$ and $\mathbb{P}(x) \in V_{\alpha+2}$. Therefore $\mathbb{P}(x) \in V$ and axiom 4 holds in V by Proposition 7.1.9. □

Proposition 9.2.6. *The von Neumann universe V satisfies axiom* 6.

Proof. By Proposition 9.2.3 and Corollary 9.2.4, every ordinal belongs to V. So $\omega \in V$. Then ω is an inductive object of V by Proposition 7.1.11, and thus axiom 6 holds in V. □

To prove that V is a ZF$^-$-universe, we are only left with the replacement axioms.

Proposition 9.2.7. *Each instance of Axiom* 5 *is satisfied in V, and V is a ZF$^-$-universe.*

Proof. Let C be a V-class of triples, take $u, s \in V$ and suppose that the restriction of C to $\{u\} \times s$ is functional (in V). We know from Proposition 7.1.12 that C is a U-class and it is immediate that $C|_{\{u\}\times s}$ is also functional in U. Then the codomain of that restriction is a set $y \in U$ by the corresponding instance of the axiom of replacement in U. But $y \subseteq V$ because $C \subseteq V$ and V is transitive. Therefore $y \in V$ by Corollary 9.2.4, and it is also the codomain of that restriction of C in the universe V. So the C-instance of replacement holds in V. □

This shows that V is a ZF$^-$-universe (note, however, that we have no (finite) proof of this fact, but we have a proof of the axiom of replacement for each particular class). We remark that $\omega \in V$, and so do all the elements of ω. Moreover, V is closed under the usual operations made in U with objects of V: unions, ordered pairs, cartesian products, subsets; consequently, all the constructions giving the imitations of the different classes of numbers, their functions and the like, can all be carried on V with exactly the same values; so this universe is, as U, enough to serve as a foundation for the basic constructions of mathematics. From this point of view then, $V \subseteq U$ is as good a universe as U itself.

9.2.3 The axiom of regularity

Proposition 9.2.8. *If $x \in V$, then $tc(x) \in V$.*

Proof. If $x \in V$, then $x \subseteq V_\alpha$ for some α; since V_α is transitive by Proposition 9.2.2, $\mathrm{tc}(x) \subseteq V_\alpha \subseteq V$ and $\mathrm{tc}(x) \in V$ by Corollary 9.2.4. □

Proposition 9.2.9. *Every element of V is a regular set.*

Proof. Let $x \neq \emptyset$ and $x \in V$. The rank functional contains all pairs $\langle u, \mathrm{rk}(u)\rangle$; so, when u moves on x, its codomain is a set (by replacement) containing all ranks of elements of x. Being a non-empty set of ordinals, it has a minimum element μ. Let $y \in x$ with $\mathrm{rk}(y) = \mu$, and suppose $z \in y \cap x$. Then $\mathrm{rk}(z) < \mu$ by Corollary 9.2.4, and this contradicts the choice of μ. Therefore $y \cap x = \emptyset$ and x is regular. □

Proposition 9.2.10. *If $A \in U$ is a well-founded transitive set, then $A \in V$.*

Proof. Let A be well-founded and transitive, but $A \notin V$. By Corollary 9.2.4, $A \nsubseteq V$ hence $A \setminus V = A'$ is not empty. By hypothesis, A' is regular and there exists $x \in A'$ such that $x \cap A' = \emptyset$ which means that $x \subseteq V$, by the transitivity of A. But then $x \in V$ again by Corollary 9.2.4, contrary to $x \in A'$. □

Corollary 9.2.11. *For any set $x \in U$, x is well-founded if and only if $x \in V$.*

Proof. If $x \in V$, $\mathrm{tc}(x) \in V$ by Proposition 9.2.8, hence every subset of $\mathrm{tc}(x)$ is regular by Proposition 9.2.9 because V is closed under subsets. By Definition 9.1.2, x is well-founded.

Conversely, let x be well-founded so $\mathrm{tc}(x)$ is transitive and well-founded. By Proposition 9.2.10, $\mathrm{tc}(x) \in V$ and $x \in V$ because $x \subseteq \mathrm{tc}(x)$. □

We have thus found that the class W of all well-founded sets of U is a ZF$^-$-universe; at the same time, we have a description which classifies the elements of W (or V) according to their rank. Looking at the definition of the sets V_α we see that there is just one element $\emptyset$ with rank 0, only one element again, $\{\emptyset\}$ of rank 1, two elements of rank 2: $\{\{\emptyset\}\}$ and $\{\emptyset, \{\emptyset\}\}$; and finitely many elements of rank n. But these are only the very first steps of the story. We find already infinite sets with $V_{\omega+1}$ (we recall that $\omega \in V_{\omega+1}$ since $\omega \subseteq V_\omega$ by Proposition 9.2.3) and in fact $\mathbb{P}(\omega) \in V_{\omega+2}$ and we know that $\mathbb{P}(\omega)$ is uncountable.

As we did with pure sets, it makes sense to start now with the ZF^--universe V and construct the class of well-founded sets of the universe V. Once more, the result is quite simple.

Proposition 9.2.12. *If $W = V$ is the class of well-founded sets of U and $x \in V$, then x is a well-founded set as a member of the universe V. Thus the relativized class of well-founded sets of the universe V, is $W^V = V$.*

Proof. Note first that the class of regular sets is V-absolute, because all elements of $x \in V$ are V-visible by the transitivity of V. Also the transitive closure $\mathrm{tc}(x)$ is the same in V or in U, since V is closed for unions as seen in Proposition 9.2.5. Moreover, the subsets of $\mathrm{tc}(x)$ are the same in V or U, again by Proposition 9.2.5. Consequenty, the class of well-founded sets W^V of the universe V coincides with V; that is, every object of V is well-founded in V. □

Corollary 9.2.13. *Let U be a ZF^--universe. The following conditions are equivalent.*

(i) $U = V$, the von Neumann universe of U.

(ii) Every element of U is a regular set.

(iii) Every element of U is a well-founded set.

Proof. (i) $\Rightarrow$ (ii) follows by Proposition 9.2.9. (ii) $\Rightarrow$ (iii) is obvious by the definition of well-founded sets. (iii) $\Rightarrow$ (i) is a consequence of Corollary 9.2.11. □

As observed above, the class V of well-founded sets is a good candidate for being taken as the universe, since all basic constructions of mathematics have an equivalent inside that universe. This suggests that adding the axiom $V = U$ is reasonable (and there is the additional reason mentioned in the comment after Definition 9.1.1). This will be indeed the next axiom of the theory.

Axiom 8 (of regularity). Every set is regular.

Equivalently, $U = V = W$. We shall say that a universe that satisfies all axioms from 1 to 8 is a ZF-universe, that is, a *Zermelo-Fraenkel universe.* Zermelo-Fraenkel set theory is the theory of the universe of all sets when one accepts all axioms from 1 to 8. From now on we will write V to denote a ZF-universe. Some consequences of axiom 8 now follow.

Proposition 9.2.14. *If U is a 2-universe which satisfies axiom 8, then $x \notin x$ for every set $x \in U$.*

Proof. Suppose $x \in U$ and $x \in x$. By the axiom of pairs there is a set $y = \{x\}$. Since it is not empty, the axiom of regularity implies that $x \cap y = \emptyset$. But $x \in x \cap y$, which is a contradiction. □

Besides this, axiom 8 forbids also the existence of sets x, y such that $x \in y$ and $y \in x$; or sets x, y, z such that $x \in y, y \in z, z \in x$. More generally, we have the following.

Proposition 9.2.15. *If V is a ZF-universe, then there is no infinite sequence of objects x_i with*

$$x_1 \ni x_2 \ni \dots$$

Proof. Let $x_1 \ni x_2 \ni \dots$ be such an infinite sequence. By definition, it is a function with domain ω, whose codomain is a set s such that $s \cap x_i \neq \emptyset$ for every $i \in \omega$; so it is a non-regular set, contrary to axiom 8. □

9.2.4 Absoluteness in ZF-universes

We studied absolute classes in Chapter 7. It happens that the axiom of regularity has some interesting consequences for these classes. The following is a key result.

Proposition 9.2.16. *Let V be a ZF-universe. An element $x \in V$ is an ordinal if and only if x is transitive and the membership relation is a strict total ordering of x.*

Proof. The condition is necessary by the definition of ordinal. Conversely, assume that x satisfies the conditions of the statement and let us show that membership is a well-ordering so that x will be an ordinal. Let $A \neq \emptyset$ and $A \subseteq x$. By regularity, there exists $y \in A$ with the property $y \cap A = \emptyset$. Now, if $u \in A$ then $u \notin y$, hence $y \in u$ or $y = u$ because the ordering is total. But then y is the minimum of A. □

Proposition 9.2.17. *Let V be a ZF-universe and C a transitive class which is a ZF$^-$-universe. Then the following classes are C-absolute: (i) the class **Ord** of ordinals; (ii) the class of limit ordinals. Moreover, $\omega \in C$.*

Proof. (i) Recall that the class of transitive sets is C-absolute by Proposition 7.3.16. We use the concept of ordinal of Definition 8.3.1. If $x \in C$ is an ordinal viewed in V, then $\in$ is a well-ordering of x, and clearly it is also so if x is viewed in C. Conversely, suppose x is an ordinal in C; if $a, b \in x$, we know that $\{a, b\}$ is a subset of x in C, from which it follows that x is totally ordered by $\in$ in V, hence x is an ordinal of V by Proposition 9.2.16.

(ii) This is a consequence of (i) combined with Proposition 8.4.2, because $\cup x$ is the same in V or in C for ordinals $x \in C$ by Proposition 7.1.9.

For the final assertion, C being a ZF$^-$-universe contains a limit ordinal all of whose elements are non-limit ordinals. Viewed in V, this is still a limit ordinal all of whose elements are non-limits, by item (ii). This is $\omega \in V$, and hence $\omega \in C$. □

The following corollary will be of great use.

Corollary 9.2.18. *Let V be a ZF-universe and C a transitive class which is a ZF$^-$-universe. Either all ordinals of V are ordinals of C or else there is some limit ordinal $\lambda \in V$ such that λ is the class of all ordinals of C.*

Proof. Let us assume that the class of ordinals of C is $\neq$ **Ord**. Therefore there is a minimum ordinal $\lambda \in V$ such that $\lambda \notin C$ because **Ord** is well-ordered by the membership relation (see the observations after the proof of Proposition 8.3.7). By Proposition 9.2.17 and the transitivity of C, all elements of λ are ordinals of C and C does not contain any ordinal $> \lambda$. If we had $\lambda = s(\alpha)$, then $\alpha \in C$ whence $\lambda = \alpha \cup \{\alpha\} \in C$. □

Proposition 9.2.19. *Let V be a ZF-universe and C a transitive class which is a ZF$^-$-universe. Then: (i) C is a ZF-universe; (ii) for each ordinal $\alpha \in C$, $V_\alpha^C = V_\alpha \cap C$; (iii) the class of pairs $\langle x, rk(x)\rangle$ is C-absolute.*

Proof. (i) Since every set of C is regular in V, it is also regular in C because C is transitive. Thus C satisfies axiom 8.

(ii) We know from Proposition 7.1.9 and its proof, that $\mathbb{P}(x)^C = \mathbb{P}(x) \cap C$. Then (ii) is proved by ordinal induction (Theorem 8.4.3): if the property holds for α, then $V_{\alpha+1}^C = (\mathbb{P}(V_\alpha^C))^C = \mathbb{P}(V_\alpha^C) \cap C = ((\mathbb{P}(V_\alpha) \cap C) \cap C = \mathbb{P}(V_\alpha) \cap C = V_{\alpha+1} \cap C$.

If $\mu \in C$ is a limit ordinal, then $V_\mu^C = \cup_{\delta \in \mu} V_\delta^C = \cup_{\delta \in \mu}(V_\delta \cap C) = V_\mu \cap C$.

(iii) By Proposition 9.2.17, the ordinals of C form a segment of **Ord** and thus the von Neumann hierarchy of the universe C is indexed by a segment of ordinals of V. If $x \in C$, item (ii) entails that $\mathrm{rk}(x) = \mathrm{rk}^C(x)$ (the C-rank of x), hence the class of pairs $\langle x, rk(x)\rangle$ is C-absolute. □

9.3 Well-Founded Relations

9.3.1 Induction and recursion for well-founded relations

In this paragraph, it is enough to assume that the universe is a ZF^--universe.

Let R be a relation with field A. We may define the following class $p_R(x)$ for each $x \in A$:

$$p_R(x) = \{z \in A | zRx\}$$

When the relation R may be understood, we write simply $p(x)$; $p(x)$ is called the class of *predecessors* of x. $p_R(x)$ is an old acquaintance: it is nothing else than $[x]_R$ with the notation of Definition 4.4.6, or the initial segment $S_A(x)$ of Definition 8.1.2 for order relations; however, usage ties these other notations to equivalences or to orderings, hence we follow this new notation for our current purpose.

When $p(x)$ is a set for every $x \in A$, we may use natural recursion (Theorem 6.4.2) to define another set for each x: take $g(0) = p(x)$ and $g(t+1) = \cup_{z \in g(t)} p(z)$. Then $\mathrm{cD}(g)$ is a subset of $\mathbb{P}(A)$ by replacement and we set $a_R(x) = \cup\, \mathrm{cD}(g)$, the set of *ancestors* of x. Again, we usually write simply $a(x)$.

Examples 9.3.1. If R is the membership relation in the universe U, then $p(x) = x$ and $a(x)$ is the transitive closure of x, $\mathrm{tc}(x)$.

When the relation R is transitive, if $y \in p(x)$ then $p(y) \subseteq p(x)$, from which it follows that $a(x) = p(x)$.

For the next example we admit property NNS about ω. Suppose that the population of the Earth could be seen as a universe and consider the parenthood relationship; that is, xRy means that x is the father or mother of y; then $p(y)$ is a set of two persons, $\cup\{p(z)|z \in p(y)\}$ is the set of the grandparents of y, and so on. In this case $a(x)$ is really the set of ancestors of x. But of course, the population of the Earth cannot be made into a ZF^--universe – if not for other reason, because it is not infinite.

Lemma 9.3.2. *Let R be a relation with $Fd(R) = A$ and suppose that $p(x)$ is a set for every $x \in A$. Then $p(x) \subseteq a(x)$ for any $x \in A$; and if $u \in a(x)$ and yRu, then $y \in a(x)$.*

Proof. The first assertion is obvious. If $u \in a(x)$, then $u \in g(t)$ (with g as above) for some $t \in \omega$; since yRu, we have $y \in p(u)$ hence $y \in g(t+1)$ and $y \in a(x)$. □

Definition 9.3.3. Let R be a relation with field A. We say that R is *well-founded* (in A) if the following two conditions hold.

(i) For each $x \in A$, $p_R(x)$ is a set.

(ii) Every non-empty subset $y \subseteq A$ contains a *strict* minimal element; that is, there is $m \in y$ such that no $u \in y$ satisfies uRm.

Remark 9.3.4. The second condition of this definition implies that R has to be anti-reflexive because each $\{x\}$ must contain a strict minimal element. This definition generalizes strict well-orderings in sets, since the minimum element of a non-empty subset is strict minimal because of the anti-reflexivity. In fact, a well-founded relation in A is a strict well-ordering of A when it is transitive and a total order. Also, there is a connection between well-founded relations and well-founded sets: a set A is well-founded precisely if the membership relation of $\mathrm{tc}(A)$ is well-founded.

Proposition 9.3.5. *Let R be a well-founded relation with $Fd(R) = A$. Any non-empty given subclass $L \subseteq A$ contains a strict minimal element.*

Proof. Take $x \in L$ and suppose x is not a strict minimal element of L so that $p(x) \cap L \neq \emptyset$ and therefore $a(x) \cap L$ is a non-empty subset of A. By Definition 9.3.3, it has a strict minimal element u. If we had vRu for some $v \in L$, then $v \in a(x) \cap L$ by Lemma 9.3.2, which contradicts the minimality of u. Thus u is a strict minimal element of L. □

The analogy with well-ordered sets goes as far as producing an induction and a recursion principle for well-founded relations.

Proposition 9.3.6. *(Induction principle for well-founded relations) Let R be a well-founded relation with $Fd(R) = A$ and let $L \subseteq A$ be a non-empty subclass satisfying: $p(x) \subseteq L \Rightarrow x \in L$.*

Then $L = A$.

Proof. Use RAA and suppose that L satisfies the condition but $L \neq A$, so that $A \setminus L$ is a non-empty subclass of A. By Proposition 9.3.5, $A \setminus L$ contains a strict minimal element, say y. Then $p(y) \subseteq L$ because of the minimality, and thus $y \in L$ by hypothesis. This is a contradiction. □

Proposition 9.3.7. *Let R be a well-founded relation with $Fd(R) = A$. Let X be a class and F a functional such that $D(F) = \mathbb{P}(X)$ and $cD(F) \subseteq X$. There exists a unique functional H with domain A such that the following condition holds, for every $x \in A$: $H(x) = F(H[p(x)])$.*

Proof. Obviously, we may assume that $A \neq \emptyset$. For $x \in A$, let us write $\sigma(x)$ to denote the set $a(x) \cup \{x\}$, a hybrid between the successor of x and the ancestor set of x. Let C be the class of all pairs $\langle u, h\rangle$ where: $u \in A$, h is a function, $\mathrm{D}(h) = \sigma(u)$ and $\mathrm{cD}(h) \subseteq X$, and the following condition $(*)$ holds: $\forall x \in \mathrm{D}(h)\ \ (h(x) = F(h[p(x)]))$.

Let $\langle u, h\rangle$ and $\langle u', h'\rangle$ belong to C. We claim that h, h' coincide on the elements of their common domain $\sigma(u) \cap \sigma(u')$. Suppose, to the contrary, that the set $\{x \in \sigma(u) \cap \sigma(u') | h(x) \neq h'(x)\}$ is not empty. Because R is well-founded, there is a strict minimal element m in that set, and obviously $p(m) \subseteq \sigma(u) \cap \sigma(u')$. By the choice of m, the functions h, h' coincide on $p(m)$ hence $h(m) = F(h[p(m)]) = F(h'[p(m)]) = h'(m)$, a contradiction. In particular, C is functional.

Let now H be the class $H = \{\langle x, y\rangle | x \in A \wedge y \in X \wedge (\exists \langle u, h\rangle \in C\ (h(x) = y))\}$. It follows from the above claim that H is a functional.

Let $A_0 = \{x \in \mathrm{D}(H) | H(x) = F(H[p(x)])\}$. Note that if $x \in A_0$ and $y \in a(x)$, then $y \in A_0$. Because $x \in \sigma(u)$ for some $\langle u, h\rangle \in C$ and $y \in a(x) \Rightarrow y \in a(u)$ so that $y \in \mathrm{D}(h)$ and $h(y) = F(h[p(y)]) = F(H[p(y)])$. We show that $A = A_0$. Suppose instead that $A \setminus A_0 \neq \emptyset$, so that there is a strict minimal element $z \in A \setminus A_0$. This entails that $p(z) \subseteq A_0$, and by the above observation, $a(z) \subseteq A_0$. Therefore, $g = H|_{a(z)}$ is a function satisfying $g(t) = F(g[p(t)])$ everywhere in $a(z)$. If we add the pair $\langle z, F(g[p(z)])\rangle$ to g we obtain h with $\langle z, h\rangle \in C$, whence $z \in \mathrm{D}(H)$ and $z \in A_0$, a contradiction that proves $A = A_0$.

Uniqueness follows by the same reason that proves the claim in the second paragraph of this proof. □

9.3.2 Mostowski's theorem

Recall the concept of extensional class given in Definition 7.3.1. The version of this definition for an arbitrary binary relation in the place of the membership relation gives the following definition.

Definition 9.3.8. In a ZF-universe V, let R be a relation with $\mathrm{Fd}(R) = A$. We say that R is *extensional* if the following holds for every pair of elements $x, y \in A$:

$$p(x) = p(y) \Rightarrow x = y$$

When C is a V-class such that $\emptyset \in C$, then C is extensional (in the sense of Definition 7.3.1) if and only if every non-empty $x \in C$ contains some element of C, and the membership relation $(C \times C) \cap \mathcal{M}$ is extensional in the sense of Definition 9.3.8.

Theorem 9.3.9. *(Mostowski's theorem) In a ZF-universe V, let R be a relation with $Fd(R) = A$ that is extensional and well-founded. Then there exist a transitive class C and an injective functional H with $D(H) = A$ and $cD(H) = C$, such that H is an isomorphism with respect to the relations R in A and membership in C; that is, $xRy \Leftrightarrow H(x) \in H(y)$ for any $x, y \in A$.*

Proof. We apply Proposition 9.3.7. With reference to its statement, the class X will be the universe V and the functional F is the identity. The conclusion is that there exists a functional H with domain A satisfying $H(x) = H[p(x)]$. We shall now define $C = \mathrm{cD}(H)$ and hence we must show that C is transitive, H is injective and an isomorphism in the sense above.

To see that C is transitive, let $x \in C$ and $y \in x$. By hypothesis, $x = H(u)$ for $u \in A$, hence $x = H[p(u)]$ and $y = H(v)$ for some $v \in p(u)$. Thus $y \in C$.

As for injectivity, suppose that $L = \{u \in A | \exists v \in A\ (v \neq u \wedge H(v) = H(u))\}$ is not empty. Then L has a strict R-minimal element m by Proposition 9.3.5, and there is $u \neq m$ with $H(u) = H(m)$, so $H[p(u)] = H[p(m)]$. By the choice of m, H is injective on the elements of $p(m)$, hence $p(m) = p(u)$. But extensionality of R implies then that $u = m$, a contradiction. Therefore $L = \emptyset$ and H is injective.

Let $x, y \in A$. If xRy then $x \in p(y)$ so that $H(x) \in H[p(y)] = H(y)$. Conversely, if $H(x) \in H(y) = H[p(y)]$, then $H(x) = H(u)$ for some $u \in p(y)$. But injectivity entails that $x = u$ and xRy. □

It is usual to call π to the isomorphism of the theorem; thus $\pi(x) = \pi[p(x)]$ for each $x \in A$, and when the relation R is the membership relation, $\pi(x) = \pi[x \cap A]$. $\pi[A]$ is called the *transitive collapse* of $\langle R, A \rangle$.

Example 9.3.10. In any class A of ordinals, membership is extensional and well-founded, because if $\alpha < \beta$ belong to A, $\alpha \in \beta$ but $\alpha \notin \alpha$, so that they have different predecessors (in A). Suppose that, for instance, A contains all the elements of ω, but the next ordinal in A is ω^2; then all $\omega^2 + n$ belong to A, but again $\omega^2 + \omega \notin A$, and $\omega^2 + \omega^2 = \omega^2 \cdot 2$ is the next ordinal in A. We may define the Mostowski's collapse for this class and describe its first elements. $\pi(n) = \pi[n \cap A] = \pi[n] = n$, so π is the identity on ω. Then $\pi(\omega^2) = \pi[\omega^2 \cap A] = \pi[\omega] = \omega$. Then $\pi(\omega^2 + n) = \omega + n$ and $\pi(\omega^2 \cdot 2) = \omega + \omega = \omega \cdot 2$. Basically, the collapse of a class A in this type of example is obtained by numbering in order the elements of A (by the ordinals) and jumping over the gaps in that series.

9.4 Exercises

1. Let U be a ZF$^-$-universe and V the von Neumann universe of U. Prove that U and V have exactly the same ordinals and exactly the same limit ordinals.

2. Prove that, in a ZF$^-$-universe, the sets $\mathbb{P}(\omega)$ and V_ω are well-founded sets, and find their ranks.

Except when stated otherwise, the universe V in the exercises that follow is a ZF-universe.

3. Let A be a class of the ZF-universe V and let $\widetilde{A} = \{x \in A | \forall y \in A \ (\mathrm{rk}(x) \leq \mathrm{rk}(y)\}$. Prove that $\widetilde{A}$ is a set.
4. Let R be a relation which is an equivalence in its field $A = \mathrm{Fd}(R)$. For $A' \subseteq A$ a subclass, let $R' = R_{A'}$, the restriction of R. Check that R' is an equivalence in A'.

   Given the equivalence relation R with $\mathrm{Fd}(R) = A$, construct a subclass A' such that, R' being as above: (1) each equivalence class of R' is a set, so that there is a quotient class A'/R'; and (2) to each $x \in A$ there is $x' \in A'$ with $[x]_R = [x']_R$ (and hence $[x']_{R'} \subseteq [x]_R$).

   (This shows that, even if an equivalence relation has not properly an associated partition, we may obtain from that relation a true partition that faithfully represents the equivalence classes of the starting relation).
5. Let R be a relation. Prove that there exists a restriction R' of R with $\mathrm{D}(R') = \mathrm{D}(R)$ and such that for every set x we have that the codomain of $R'|_x$ is a set.

   This property for the relation R is called the R-instance of the *collection principle.*
6. Let U be a universe satisfying all axioms from 1 to 8 except possibly axiom 5. Find a class X of ordered pairs with the following property. If the X-instance of the principle of separation holds and C is a U-class of triples, then the C-instance of the collection principle (viewing C as a class of ordered pairs) implies the C-instance of the axiom of replacement (Hint: Show that Proposition 5.1.17 follows from just one instance of the principle of separation).
7. Let C be a transitive class of V which is a ZF^--universe. For ordinals $\alpha, \beta \in C$, prove that $\alpha + \beta$ and $\alpha \cdot \beta$ are the same ordinals both in C and in V.
8. Consider the class R of all triples $\langle \alpha, \beta, f \rangle$ with f being a bijection $\alpha \to \beta$, and α, β ordinals. Is this class C-absolute for C a transitive class which is a ZF^--universe? What can be said about the same question for the class R_1 of ordered pairs of ordinals $\langle \alpha, \beta \rangle$ such that there is a bijective map $\alpha \to \beta$?
9. For any $x \in V$, prove that $\mathrm{rk}(x)$ is the supremum of all ordinals of the form $\mathrm{rk}(y) + 1$ when y ranges over the elements of x.
10. Prove that for every ordinal α, $\mathrm{rk}(\alpha) = \alpha$.
11. Find two objects x, y such that $\mathrm{rk}(x) = \mathrm{rk}(\cup x)$ and $\mathrm{rk}(y) > \mathrm{rk}(\cup y)$.
12. Show that if $\mathrm{rk}(x)$, $\mathrm{rk}(y) \leq \alpha$, then the sets $\{x, y\}, \langle x, y \rangle, \cup x, \mathbb{P}(x)$ have rank $< \alpha + \omega$.
13. Prove that the sets $\mathbb{Z}, \mathbb{Q}, \mathbb{R}$ belong to $V_{\omega+\omega}$.

14. Let A be a set in a ZF$^-$-universe. Prove that every subset of A is regular if and only if the membership relation is well-founded in A. Deduce that a set A is well-founded if and only if the membership relation is well-founded in $\text{tc}(A)$.

15. In ZF$^-$-universes, sets and objects are the same thing and we may dispense with using the term (and the idea of) set. Recall that exercises 5 and 7 of Chapter 7 asked for describing properties of objects in that form. Define regular objects and well-founded objects, and give axiom 8 by following those same rules.

16. Let R be a relation in a V-class C which is apt. Prove that R is extensional as a class, but not necessarily an extensional relation.

17. In a ZF-universe V, let R be a relation with $\text{Fd}(R) = A$. Prove the converse of Mostowski's theorem; i.e., if there is an isomorphism F between $\langle A, R\rangle$ and $\langle C, \in\rangle$ with C a transitive class, then R is extensional and well-founded.

18. Let C be the set defined by natural recursion as $C = \{a_0, a_1, \ldots\}$ with $a_0 = \emptyset$ and $a_{t+1} = a_t$. Prove that $C \in V_{\omega+1}$ and deduce that membership is a well-founded relation in C but is not a strict ordering of C.

19. For this and the next exercise we accept property NNS for ω. Let $P \subseteq \omega$ be the set of even natural numbers with the relation $<$. Show that this is a strict well-ordering of P (and hence well-founded), but P is not transitive. Find the Mostowski's collapse $\pi[P]$ and each of the values $\pi(2n)$.

20. For the same set P of the preceding exercise, consider the relation formed with all ordered pairs $\langle n, n+2\rangle$; show that it is not an order relation, but it is extensional and well-founded. Find also the Mostowski's collapse of P for this relation.

21. Let R be an extensional and well-founded relation with field A, and let B be a transitive class such that there is an isomorphism between A and B with respect to the relations R and $\in$. Prove that B is the transitive collapse of $\langle R, A\rangle$.

22. Show that if a contradiction can be deduced from the axioms of ZF-universes, then a contradiction can be deduced from the axioms for 6-universes.

Let U be a ZF$^-$-universe. Let us say that a pointed directed graph *of U is a triple $\langle A, R, u\rangle$ where $u \in A$, R is a binary relation in A. The elements of A are the* nodes *of the graph, the pairs $\langle a, b\rangle \in R$ (written with an arrow $a \to b$) are the* edges *of the graph, u is the* root *of the graph. The pointed directed graph $\langle A, R, u\rangle$* represents *the set X in case there is a surjective map $r : A \to tc(X) \cup \{X\}$ (the representation) with the following properties for any $x, y \in A$:*

(1) $xRy \Leftrightarrow r(y) \in r(x)$*; and (2)* $r(u) = X$*. In the coming exercises,* graph *means pointed directed graph.*

23. In the class of all graphs, we consider the relation $\langle A, R, u\rangle \sim \langle A', R', u'\rangle \Leftrightarrow$ there is a bijective map $F : A \to A'$ such that $F(u) = u'$ and for every $x, y \in A$, $xRy \Leftrightarrow F(x)R'F(y)$. Show that this is an equivalence relation such that if $\langle A, R, u\rangle$ represents the set X, then every graph equivalent to $\langle A, R, u\rangle$ represents the same set X.

24. Justify that for any set X there is a graph which represents the set X. Prove that there exist non-equivalent graphs that represent the same set (one may find such an example with a finite and small set).

25. In a graph $\langle A, R, u\rangle$, a *path* is a countable (denumerable or finite) sequence of elements a_k of A satisfying $a_k R a_{k+1}$ for any a_{k+1}. We shall call a graph $\langle A, R, u\rangle$ *well-founded* when R^{-1} is a well-founded relation. Show that a well-founded graph has no infinite paths (a circular path as $a_1 R a_2 R a_3 R a_1$ counts as infinite). By using recursion for well-founded relations (Proposition 9.3.7, recall that this holds in ZF$^-$-universes), show that a well-founded graph $\langle A, R, u\rangle$ cannot be a representation of more than one set (Hint: Consider the uniqueness part of Proposition 9.3.7; and take into account that $r(x) = \{r(y) | xRy\}$ for each node of the graph and a representation r).

26. Let X be a set represented by the graph $\langle A, R, u\rangle$. Prove that X is a well-founded set if and only if $\langle A, R, u\rangle$ is a well-founded graph. Conclude that, in a ZF$^-$-universe, axiom 8 is equivalent to the assertion: any graph which represents a set is well-founded.

27. Create three non equivalent and not well-founded graphs in the universe U (for instance, a loop). By assuming that these graphs represent "sets", write which would these sets possibly be, adequately labeling their elements (since these sets have to be non-well-founded by the preceding exercise, one would have to accept relations like $x \in x$ or $x \in y, y \in x$).

By deleting Axiom 8 and considering ZF$^-$-universes where graphs that are not well-founded represent sets, it is possible to create a different theory, ZFA$^-$-theory, which has been studied by P. Aczel.

10

Cardinals and the Axiom of Choice

In the first part of this chapter, we work in a ZF-universe V. Since we are not assuming property NNS, we are forced to distinguish, when speaking of finite sets, between the intuitive ordinary sense of *finite* and the meaning of *finite* in the sense of Definition 6.3.11, where a set s is finite when there is a bijective map from some element of ω to s.

10.1 Equipotent Sets and Cardinals

Set theory affords a double generalization of the arithmetic of natural numbers. On the one hand, ordinals, as we have already seen, provide a far-reaching extension of the usual ordered series of the natural numbers. But natural numbers are also used to count the elements of a (finite) collection and to compare the sizes of different collections, and ordinals are not the right tool to address this issue in the infinite case. Sizes of finite sets are measured through natural numbers following two rules: if the elements of A and of B can be paired (in set theory terms, if there is a bijective map $A \to B$) then A and B have equal size, i.e., equal number of elements; and if there is an injective map $A \to B$ which is not bijective, then A is smaller than B, i.e., A has less number of elements than B. But for infinite sets, these two rules cannot stand together: we have seen for instance that the set of even natural numbers is properly included in the set of all numbers, but there is a bijection between both sets. Mathematicians have found it useful to focus on the condition of the existence of bijections as a measure of "equal size". The concepts corresponding to "bigger" and "smaller" for infinite sets will be introduced later.

Definition 10.1.1. Let A, B be sets. We say that A and B are *equipotent* when there is a bijective map $f : A \to B$. If this happens, we write $A \simeq B$ and this is a symmetric, reflexive and transitive relation in V, as can be easily seen from the properties of bijective maps.

DOI: 10.1201/9781003449911-10

We have already met, without introducing the term, equipotency of sets in Chapter 6. We saw there that the set ω is equipotent to the set of integers or to the set of rational numbers. On the contrary, ω is not equipotent to the set of real numbers which, in turn, is equipotent to the power set $\mathbb{P}(\omega)$. We remark that equipotency is not an absolute notion: for a transitive class M, it might happen that two sets $x, y \in M$ are equipotent in V but not in M, since there could exist a bijection $f \in V$ with $f \notin M$, and such that no other bijection for x, y exists in M.

Lemma 10.1.2. *Let $f : A \to B$, $g : B \to C$ be injective maps between sets and suppose that $A \simeq C$. Then $B \simeq C$.*

Proof. Let $h : C \to A$ be a bijection. Then $g : B \to C$ and $f \circ h : C \to B$ are injective maps. By Corollary 6.6.6, $B \simeq C$. □

As a first step in studying equipotency of sets, we start by studying equipotency of ordinals. The key concept in this study is the following.

Definition 10.1.3. A *cardinal* is an ordinal μ such that there is no element $\alpha \in \mu$ with $\alpha \simeq \mu$.

Remarks and Examples 10.1.4. By the definition, $\emptyset$ is a cardinal. Every natural number and, more generally, every $x \in \omega$ is a cardinal by Proposition 6.6.2. Then ω is a cardinal because for no $x \in \omega$ is there a bijection $x \simeq \omega$. But we know that $\omega + 1 \simeq \omega$ and also $\omega + \omega \simeq \omega$ or even $\omega \cdot \omega \simeq \omega$, according to Proposition 6.6.8. So, none of these is a cardinal, as they are equipotent to a smaller ordinal. On the other hand, we know from Theorem 6.6.11 that $\mathbb{P}(\omega)$ is infinite and not equipotent to ω. Hence if we could find a well-ordering of this set (or, equivalently, a well-ordering of $\mathbb{R}$, because $\mathbb{R}$ is equipotent to $\mathbb{P}(\omega)$), then there would exist an ordinal isomorphic (and therefore equipotent) to $\mathbb{P}(\omega)$, by Theorem 8.3.9; and it could be deduced from this that there exists some cardinal which is greater than ω. But a new axiom is needed to complete this idea.

Recall that the class of ordinals **Ord** is well-ordered by inclusion with the corresponding strict order being membership. This ordering gives by restriction a well-ordering of the class of cardinals, written also as $\leq$, respectively $<$. Thus, for cardinals μ and ρ, $\mu < \rho$ means $\mu \in \rho$, $\mu \leq \rho$ means $\mu \subseteq \rho$; and $\rho \leq \mu, \mu \leq \rho$ imply $\mu = \rho$.

Comment 10.1.5. The problem just mentioned of finding a well-ordering of the set $\mathbb{R}$ of real numbers was proposed by Hilbert in 1900. In the list of 23 problems he presented in the *Deuxième Congrès International des Mathématiques*, it was included as a part of the first problem about the power of the continuum. It had been conjectured by Cantor, and solved by Zermelo in 1904 with the aid of the axiom of choice, an aid which was highly controversial at the time.

Proposition 10.1.6. *(i) If μ, ρ are cardinals, then $\mu \simeq \rho \Leftrightarrow \mu = \rho$.*

(ii) If A is a set, μ a cardinal and $A \simeq \mu$, μ is the unique cardinal equipotent to A. We say then that μ is the cardinal of A *and write $\mu = |A|$.*

(iii) Any ordinal α is equipotent to a unique cardinal $|\alpha|$, which is the greatest cardinal that is $\leq \alpha$.

(iv) If $\alpha \leq \beta$ are ordinals, then $|\alpha| \leq |\beta|$.

(v) Every set with a well-ordering has a cardinal.

(vi) Let A, B be sets which have a cardinal. Then $|A| \leq |B|$ if and only if there is an injective map $f : A \to B$.

Proof. (i) If $\mu \neq \rho$ and $\mu \simeq \rho$, then $\mu \in \rho$ or $\rho \in \mu$ by Proposition 8.3.6. But each of these possibilities leads to a contradiction by the definition of cardinal.

(ii) If $A \simeq \mu$ and $A \simeq \rho$, then $\mu \simeq \rho$ by transitivity and $\mu = \rho$ by (i).

(iii) Since $\alpha \simeq \alpha$, the class of ordinals equipotent to α is not empty and has a smallest element μ, which is necessarily $\leq \alpha$. μ is a cardinal, because if $\rho < \mu$ and $\rho \simeq \mu$, then $\rho \simeq \alpha$ contrary to the choice of μ. Finally, if a cardinal κ satisfies $\mu \leq \kappa$ and $\kappa \leq \alpha$, then $\mu \simeq \kappa$ by Lemma 10.1.2, and $\mu = \kappa$ by item (i).

(iv) If $\alpha \leq \beta$, then $|\alpha| \leq \beta$ and $|\alpha| \leq |\beta|$ by (iii).

(v) If A is a well-ordered set, then $A \cong \alpha$ for some ordinal α by Proposition 8.3.9. If $\mu = |\alpha|$, then $A \simeq \mu$ and $\mu = |A|$.

(vi) Let $|A| = \mu$ and $|B| = \rho$. If $\mu \leq \rho$, then the composition $A \to \mu \hookrightarrow \rho \to B$ obtained from the assumed bijections is an injective map $A \to B$. Conversely, suppose there is an injective map $f : A \to B$. By composition with the bijections, we obtain an injective map $g : \mu \to \rho$. If we had $\rho < \mu$, there is an injective map $\rho \to \mu$, and by the CSB-theorem (Corollary 6.6.6), $\rho \simeq \mu$ and $\rho = \mu$ by (i), which contradicts the assumption $\rho < \mu$. Thus $\mu \leq \rho$ since the ordering of the ordinals is a total ordering. □

We give now the first cardinals. An ordinal α is *finite* when $\alpha \in \omega$.

Proposition 10.1.7. *Every finite ordinal is a cardinal. ω is also a cardinal.*

Proof. The first part has already been shown in Remark 10.1.4. As for ω, suppose there is a bijection $x \to \omega$ with $x \in \omega$. Since $x \subseteq s(x) \subseteq \omega$, Lemma 10.1.2 applies and $x \simeq s(x)$, a contradiction to the fact that $s(x) \in \omega$ and is a cardinal. □

The elements of ω are finite sets in the sense of Definition 6.3.11; they are called *finite cardinals*. The other cardinals are infinite sets, as they cannot have a bijection with any smaller cardinal. They are then called *infinite cardinals*. Unlike ω, the ordinals $\omega \cdot 2$ or ω^2 are not cardinals.

Proposition 10.1.8. *Any infinite cardinal is a limit ordinal.*

Proof. Let $\alpha + 1$ be an infinite successor ordinal, so that $\omega \subseteq \alpha$. There is a bijective map $f : \alpha + 1 \to \alpha$ obtained with the ordered pairs $\langle x, x + 1\rangle$ for $x \in \omega$, $\langle \beta, \beta\rangle$ for all infinite $\beta \in \alpha$; and $\langle \alpha, 0\rangle$. Thus $\alpha+1$ cannot be a cardinal by Definition 10.1.3. □

10.2 Operations with Cardinals

10.2.1 Addition

We define operations with cardinals through the corresponding operations with ordinals. Since cardinals are ordinals, we must distinguish between the operations as cardinals and as ordinals. For this reason, we shall now write $\alpha +_o \beta$ or $\alpha \cdot_o \beta$ to refer to the respective operations as ordinals, be they cardinals or not. The following observation will be useful to our purpose.

Lemma 10.2.1. *Let α, β, γ be ordinals. Then $\alpha \simeq \beta \Rightarrow \alpha +_o \gamma \simeq \beta +_o \gamma$ and $\gamma +_o \alpha \simeq \gamma +_o \beta$.*

Proof. This is an easy exercise. □

Definition 10.2.2. Let ρ, μ be cardinals. We define the sum as:

$$\rho + \mu = |\rho +_o \mu|$$

Proposition 10.2.3. *Let A, B be disjoint sets (i.e., $A \cap B = \emptyset$). If A, B have cardinals, then $A \cup B$ has a cardinal and $|A \cup B| = |A| + |B|$.*

Proof. Let $A \simeq \mu$, $B \simeq \rho$. By combining both bijections we get a bijective map $A \cup B \to (\{0\} \times \mu) \cup (\{1\} \times \rho)$. In turn, this last set is equipotent to $\mu +_o \rho$ by Definition 8.5.2. Thus $\mu + \rho = |A \cup B|$ by Definition 10.2.2. □

Proposition 10.2.4. *The sum of cardinals is associative and commutative.*

Proof. Since $\kappa + \mu \simeq \kappa +_o \mu$, we see from Lemma 10.2.1 that $(\kappa +_o \mu) +_o \rho \simeq (\kappa + \mu) +_o \rho \simeq (\kappa + \mu) + \rho$. Similarly, $\mu + \rho \simeq \mu +_o \rho$ and hence $\kappa +_o (\mu +_o \rho) \simeq \kappa +_o (\mu + \rho) \simeq \kappa + (\mu + \rho)$. Then associativity follows from Proposition 8.5.9.

For the commutativity, note that $\mu +_o \rho \simeq (\{0\} \times \mu) \cup (\{1\} \times \rho)$ by Definition 8.5.2, and analogously for $\rho +_o \mu$. Thus $\mu +_o \rho \simeq \rho +_o \mu$ through the bijection where $\langle 0, x\rangle \mapsto \langle 1, x\rangle$ and $\langle 1, y\rangle \mapsto \langle 0, y\rangle$. The conclusion follows. □

Sum of finite cardinals coincides by definition with sum of finite ordinals. As for infinite cardinals, the definitive result is:

Proposition 10.2.5. *Let $\mu \leq \kappa$ be cardinals such that κ is infinite. Then $\mu + \kappa = \kappa$.*

Proof. First, $\kappa \leq \mu +_o \kappa$ by Proposition 8.5.5, and then $\kappa \leq \mu + \kappa$ by Lemma 10.1.6 (iv). Also $\mu +_o \kappa \leq \kappa +_o \kappa$ by Proposition 8.5.7 ; and thus $\mu + \kappa \leq \kappa + \kappa$. So, it will be enough to show that $\kappa +_o \kappa \simeq \kappa$ when κ is infinite.

From the construction of $\kappa +_o \kappa$ in Lemma 8.5.1, we know that $\kappa +_o \kappa \simeq \{0,1\} \times \kappa$. By Propositions 8.5.10 and 10.1.8, the elements of κ are precisely all sums $\lambda +_o t$ with $\lambda < \kappa$ a limit ordinal; and $t \in \omega$ arbitrary. Then there is a bijective map $f : \{0,1\} \times \kappa \to \kappa$ with $f(0, \lambda + t) = \lambda + 2t$ and $f(1, \lambda + t) = \lambda + (2t + 1)$, and this shows the result. □

10.2.2 Multiplication

We introduce now the multiplication of cardinals:

Definition 10.2.6. If μ, ρ are cardinals, then $\mu \cdot \rho = |\mu \cdot_o \rho|$.

Proposition 10.2.7. *Let A, B, C be sets with $A \simeq B$. Then $A \times C \simeq B \times C$. Consequently, if A, B are sets which have a cardinal, then $A \times B$ has a cardinal and $|A \times B| = |A| \cdot |B|$.*

Proof. If $f : A \to B$ is a bijective map, then we get a bijection $g : A \times C \to B \times C$ by setting $g(\langle a, c \rangle) = \langle f(a), c \rangle$. Next, if $|A| = \rho$ and $|B| = \mu$ for certain cardinals, it follows that $A \times B \simeq B \times A \simeq \mu \times \rho$. By the description of the ordinal multiplication in Lemma 8.5.11 and Definition 8.5.12, $|\mu \times \rho| = |\rho \cdot_o \mu| = \rho \cdot \mu = |A| \cdot |B|$, which proves the assertion. □

From this property and the bijection $\mu \times \rho \simeq \rho \times \mu$, commutativity of the product of cardinals follows. Associativity depends on the existence of the natural bijection $A \times (B \times C) \simeq (A \times B) \times C$. Again the product of finite cardinals, being the product of finite ordinals, is the product of elements of ω; in particular, cardinal product generalizes the usual product of natural numbers. Also distributivity holds for cardinals.

Proposition 10.2.8. *If κ, μ, ρ are cardinals, then $\kappa \cdot (\mu + \rho) = \kappa \cdot \mu + \kappa \cdot \rho$; and $(\kappa + \mu) \cdot \rho = \kappa \cdot \rho + \mu \cdot \rho$.*

Proof. Because of the commutativity, it is enough to prove the first equation. Now, $\kappa \cdot (\mu + \rho) = |\kappa \cdot_o (\mu + \rho)| = |\kappa \cdot_o (\mu +_o \rho)|$ by Proposition 10.2.7. By applying now Proposition 8.5.17 and Lemma 10.2.1, we get

$$\kappa \cdot (\mu + \rho) = |(\kappa \cdot_o \mu) +_o (\kappa \cdot_o \rho)| = \kappa \cdot \mu + \kappa \cdot \rho$$

□

We use the following order in the class **Ord** $\times$ **Ord**: $\langle\alpha,\beta\rangle < \langle\gamma,\delta\rangle$ if either $\max(\{\alpha,\beta\}) < \max(\{\gamma,\delta\})$ or if those maxima are equal but $\langle\alpha,\beta\rangle < \langle\gamma,\delta\rangle$ in the lexicographic ordering. It is routine to check that this gives indeed a class of ordered pairs of the class **Ord** $\times$ **Ord**; and that this is a (strict) well-ordering, which is known as the *canonical well-ordering of* ***Ord*** $\times$ ***Ord***. It is important to note that in this ordering, the initial segment determined by any pair of the form $\langle 0,\alpha\rangle$ contains precisely all the elements of $\alpha\times\alpha$. Thus the restriction to $\langle 0,\alpha\rangle$ is a well-ordering of the set $\alpha\times\alpha$.

Lemma 10.2.9. *Let α be an infinite ordinal. Then $\alpha\times\alpha\simeq\alpha$.*

Proof. If the assertion is not true, there is a smallest infinite ordinal λ such that λ is not equipotent with $\lambda\times\lambda$. By Proposition 6.6.8, $\lambda\neq\omega$. If λ is not a cardinal, then $\lambda\simeq\mu$ for some infinite ordinal $\mu<\lambda$. But then $\mu\simeq\mu\times\mu$ and $\lambda\simeq\lambda\times\lambda$ by Proposition 10.2.7, which is absurd. So we conclude that λ must be a cardinal, hence a limit ordinal (Proposition 10.1.8).

We consider $\lambda\times\lambda$ with the well-ordering induced by the canonical well-ordering of **Ord** $\times$ **Ord**. By Proposition 8.3.9, there is an ordinal ρ and an isomorphism $g:\rho\to\lambda\times\lambda$. If $\rho<\lambda$ then we get an injective map $\lambda\times\lambda\to\rho\to\lambda$; since we have an obvious injective map $\lambda\to\lambda\times\lambda$, the CSB-theorem would give $\lambda\simeq\lambda\times\lambda$ and this is impossible by the choice of λ. The same reason gives $\rho\neq\lambda$ and thus we infer that $\lambda<\rho$; i.e. $\lambda\in\rho$.

Let us set $g(\lambda)=\langle\alpha,\beta\rangle$. Since $\alpha,\beta<\lambda$ and λ is limit, there is $\gamma\in\lambda$ such that $\alpha,\beta<\gamma$; and in the ordering of $\lambda\times\lambda$, $\langle\alpha,\beta\rangle<\langle 0,\gamma\rangle$. Thus the isomorphism g restricts to a morphism of well-ordered ordered sets $\lambda\to S_{\lambda\times\lambda}(\langle 0,\gamma\rangle)=\gamma\times\gamma$ (with the notation of Definition 8.1.2). Surely, γ is an infinite ordinal and thus $\gamma\times\gamma\simeq\gamma$, which gives $\lambda\leq|\gamma|<\lambda$ by Proposition 10.1.6, which is absurd. □

Proposition 10.2.10. *Let κ be an infinite cardinal. Then $\kappa\cdot\kappa=\kappa$.*

Proof. We know $\kappa\cdot\kappa=|\kappa\times\kappa|=\kappa$ by the lemma. □

As with sums, this settles the question of the product of two cardinals, one of them being infinite: $\mu\cdot\kappa=\sup(\{\mu,\kappa\})$. This shows that, while arithmetic may be extended to infinity through consideration of injective maps as corresponding to the relation $\leq$, this is not very illuminating in what concerns the basic operations of unions and cartesian products. An operation that will be considered later is by far more interesting.

10.3 The Axiom of Choice

10.3.1 Choice and cardinals of sets

Definition 10.3.1. Let A be a set such that $\emptyset\notin A$. A *choice function* for A is a function f with $\mathrm{D}(f)=A$ such that for each $a\in A$, we have $f(a)\in a$.

Existence of choice functions is not problematic for (usual) finite collections of non-empty sets: essentially, the function chooses an element belonging to each set of the collection, whence the name. The reason why one could accept this to be true also for infinite sets is the basic assumption that infinite and finite sets have analogous behavior; which is a justifiable assumption as long as no contradiction arise. We will later see that indeed the existence of choice functions for arbitrary sets does not produce, by itself, any contradiction. So, the last axiom of the theory stipulates that choice functions always exist.

Axiom 9 (of choice). If A is a set such that $\emptyset \notin A$, then there exists a choice function for A.

A ZF-universe where axiom 9 is satisfied will be called a ZFC-universe ("C" for choice).

Comment 10.3.2. Each of the nine axioms of the ZFC-theory is special in some sense, but the axiom of choice is more special than the others. It will be worth to make now a quick review of those axioms. The principle of extension can be seen as an axiom when it is restricted to sets; as such it seems to be forced by the very idea of a collection and does not normally raise objections. In their own ways, axioms 1 to 5 allow the mathematician to carry the basic operations on sets that are usual in mathematical reasoning; collectively, they state that one can form sets by setting some property, so that the sets that fulfill that property (and belong to some limited domain) form a set. Except axiom 4, these axioms specify quite precisely which are the objects that form the set whose existence is asserted; the axiom of the power set is a bit different because it simply gives the property that the elements of the postulated set have to satisfy; so, in principle, one would have to know all the objects of the universe to select which ones are the members of the new set. But in general, these axioms are relatively concrete about which are their elements.

Axioms 7 and 8 are quite different, and we have already mentioned that they are more conventional. They restrict the universe in a way that is perhaps not necessary, but which is convenient and does not diminish the power of set theory in its role as foundation for classical mathematics. Axiom 6, in its normal form, is a bold assertion of existence of some set with particular properties, but it seems necessary because mathematicians need to deal with infinity, and axiom 6 opens the way to the construction of the basic notions of mathematics.

Axiom 9, too, asserts the existence of some set corresponding to any given set, but this new set is not described through its elements nor through any property that these elements satisfy, there is no link whatever between its members and if some object is presented, one could have no idea whether it belongs to the new set or not. Thus it is not strange that many mathematicians have been reluctant to accept it as a principle that can legitimately be used in mathematical investigation, though this resistance has also had historical reasons. But there are many results in topology, analysis or algebra that can be proved by using the axiom and not without it; that is, the theory of ZFC-universes is richer, in many significant topics, than the more austere ZF-theory. More decisively, Gödel proved in 1938 that if the theory of ZF-universes is consistent (that is, it does not involve a contradiction; which amounts to say that it has some meaning), then the theory with the axiom of choice, the ZFC-theory is also consistent.

As a result, mathematicians may work either by assuming the ZF-axioms or the ZFC-axioms, and both possibilities are equally legitimate from the point of view of logic. Consequently, most of them develop their research in that stronger frame and the ZFC-theory is currently the standard set theory for use in each mathematical discipline.

This axiom is precisely what is needed to prove the following crucial property.

Proposition 10.3.3. *If V is a ZFC-universe, then any set has a cardinal.*

Proof. (Before giving the actual proof, let us explain the plan for that proof. You may skip this paragraph and go directly into the proof. If you find it too difficult to follow, it might be helpful to read this and get an idea of the basic lines of the proof and the difficulties of a too naive approach.

Our objective is to show that any set is equipotent to some ordinal. Given the set A, we construct a map H from ordinals to A; first we take $h(0) = a_0 \in A$ at our wish. Then we choose $h(1)$ as any element $a_1 \neq a_0$ looking for constructing an injective map. Then $h(2) = a_2$ with $a_2 \neq a_0, a_1$. In general, $h(\alpha)$ will be some $a_\alpha \in A$ with the property that a_α does not coincide with any of the previously chosen images; that is, $h(\alpha) \notin h[\alpha]$. Proceeding in this way, we will reach the situation where $h(\beta)$ cannot be found because $h[\beta] = A$, so there is no element left to choose. But then $h : \beta \to A$ will be a surjective map; and injective because it has been so constructed. Thus it is a bijection and $\beta \simeq A$, as we wanted to show.

There are several reasons why this is far from being a valid proof; let us list some of them.

(1) The most important problem is this: we are totally wrong in describing the process as "something we do". What we can do is limited to the finite so we could not go far with this type of reasoning. We need instead to use abstract arguments and replace the intuitive ideas of how we do the construction with the results of the theory. In particular, this repeated process of getting one element of A after another has to be seen through the abstract vision of the recursion theorems: we must specify a "rule" (by means of an adequate functional) that obtains $h(\alpha)$ from the previous values, i.e., from the set $h[\alpha]$.

(2) We have been implicitly assuming that we can *choose* at any point an element in $A \setminus h[\alpha]$ if this is not empty. This will require the axiom of choice applied to the set $\mathbb{P}(A) \setminus \{\emptyset\}$, but if a choice function f for this set is applied to choose $h(\alpha)$ from the set $h[\alpha]$, we would fail; what we need here is that $f(h[\alpha]) \in A \setminus h[\alpha]$. Hence a variation of that choice function will be necessary.

(3) We have supposed that at some point we exhaust A, so that for some ordinal β, $h[\beta] = A$, and we may stop then. But, in principle, it could happen that the process described has no end. Thus we must prove that this is not the case.

(4) Finally, a minor technical problem arises because of the use of recursion. What is obtained through Theorem 8.4.4 is a functional with domain **Ord** while what we need here is that the domain be limited to some ordinal β

which cannot be specified at the start. To deal with this, we need to extend the definition of h above to include the point where $h[\beta] = A$; in that case, we will define $h(\beta) = A$, so the codomain of h will be $A \cup \{A\} = s(A)$, instead of A. This explains some details in the proof that begins here).

By (iii) of Proposition 10.1.6, it will be enough to show that any set is equipotent to some ordinal. So, we start with a set A and we have to prove that there is a bijective map $h : \beta \to A$ for some ordinal β. First, we define the map $g : \mathbb{P}(A) \setminus \{A\} \to \mathbb{P}(A) \setminus \{\emptyset\}$ with $g(X) = A \setminus X$. By composing g with a choice function for $\mathbb{P}(A) \setminus \{\emptyset\}$ (which exists by axiom 9), we get $f : \mathbb{P}(A) \setminus \{A\} \to A$ such that $f(X) \in A \setminus X$. Finally, we extend this map to $m : \mathbb{P}(s(A)) \to s(A)$ by adding all the pairs $\langle Y, A \rangle$ when $A \in Y$ or $A = Y$. So, if $X \subsetneqq A$, then $m(X) = f(X) \in A \setminus X$.

We apply recursion (Theorem 8.4.4, simplified form) for the class $s(A)$ and the functional m and we obtain a functional H with $\mathrm{D}(H) = \mathbf{Ord}$ such that $H(\alpha) = m(H[\alpha])$ for any ordinal α, so that $\mathrm{cD}(H) \subseteq s(A)$. Let $C = \{\alpha \in \mathbf{Ord} | H[\alpha] \subsetneqq A\}$. We claim that the restriction functional $H|_C$ is injective and its codomain is a subset of A. This second part follows because for $\alpha \in C$, $H(\alpha) = m(H[\alpha]) = f(H[\alpha]) \in A$ as seen above. For the injectivity, take $\alpha \neq \beta \in C$; by the well-ordering of $\mathbf{Ord}$, let us suppose that $\alpha \in \beta$. Then $H(\alpha) \in H[\beta]$ but $H(\beta) = m(H[\beta]) \in A \setminus H[\beta]$, so $H(\alpha) \neq H(\beta)$.

It is impossible that $C = \mathbf{Ord}$. For in such case H is injective and then H^{-1} would be a functional, whose domain is the set $\mathrm{cD}(H|_C) \subseteq s(A)$ so that its codomain $\mathbf{Ord}$ would have to be a set by replacement, contradicting Proposition 8.3.7. By the well-ordering of $\mathbf{Ord}$, there is a smallest $\beta \notin C$. By the definition of H, $H(\beta) = m(H[\beta]) = A$, and thus H restricts to a map $h : \beta \to A$ which is bijective because $H|_C$ has been proved to be injective. □

By Definition 10.1.3, there is a unique cardinal κ such that $\kappa \simeq A$ and it is called *the cardinal* of A, and written $|A|$. Like this, the notions that roughly correspond to "having the same size" or "having smaller size" may be applied to all sets. In particular, item (vi) of Proposition 10.1.6 holds for arbitrary sets; hence $|A| \leq |B|$ if and only if there is an injective map $A \to B$. The following is a new characterization of that relationship.

Proposition 10.3.4. *Let A, B be non-empty sets in a ZFC-universe. $|A| \leq |B|$ if and only if there is a surjective map $g : B \to A$.*

Proof. This follows because, for $A \neq \emptyset$, there is an injective map $f : A \to B$ if and only if there is a surjective map $g : B \to A$. Indeed, assume the existence of f and let $X = \mathrm{cD}(f) \subseteq B$, so that we get a map $f^{-1} : X \to A$. By choosing $u \in A$ whatever, we complete f^{-1} to a map $g : B \to A$ by sending each element in $B \setminus X$ to u. This gives clearly a surjection. Conversely, assume the existence of the surjection $g : B \to A$. Then we may construct a map $h : A \to \mathbb{P}(B) \setminus \{\emptyset\}$ with $h(x) = g^{-1}[\{x\}]$. If f is a choice function for $\mathbb{P}(B) \setminus \{\emptyset\}$, the composition $f \circ h$ is a map $A \to B$. Since the sets $g^{-1}[\{x\}]$ are disjoint for different x, $f \circ h$ is injective. □

10.3.2 Equivalent forms of the axiom of choice

As we have seen, the axiom of choice allows us to assign to each set a unique cardinal. In turn, this property implies, in any ZF-universe, the statement of the axiom of choice. We say that these two statements are *equivalent* (of course, in a ZF-universe), because each of them implies the other. There are several other properties which are also equivalent to the axiom of choice. In a sense, we may understand these properties as forms of the axiom. We state and prove only a few.

Proposition 10.3.5. *The following three properties are equivalent for a ZF-universe V.*

(1) Axiom of choice.

(2) Every set is equipotent to some cardinal.

(3) For any set A, there exists a well-ordering of A (This assertion is known as the well-ordering principle*).*

Proof. (1) $\Rightarrow$ (2) is Proposition 10.3.3.

(2) $\Rightarrow$ (3) Take the set A and a bijection $f : A \simeq \alpha$ for some cardinal α. One may define a relation $<$ on A by setting $x < y \Leftrightarrow f(x) \in f(y)$. The bijection is therefore an order isomorphism, and since α is well-ordered by membership, A is well-ordered by the relation $<$.

(3) $\Rightarrow$ (1) Consider a set A such that $\emptyset \notin A$, and take $B = \cup A$ and a well-ordering of B. Each $a \in A$ is a non-empty subset of B and we define $g(a)$ as the minimum element of $a \subseteq B$ in the well-ordering of B. Then g is a function with domain A and $g(a) \in a$, so it is a choice function for A. □

We introduce the following notation: if A is a set, $\text{dj}(A) = \{\{a\} \times a | a \in A\}$. We call $\text{dj}(A)$ the *disjunction of A* and it is easy to check that $\text{dj}(A)$ is a set whose elements are pairwise disjoint. For example, $\text{dj}(\omega)$ is like a copy of ω, but instead of the element t it has $\{t\} \times t$; for instance, the elements of n are $0, 1, 2, \ldots, n-1$ and the elements of $\{n\} \times n \in \text{dj}(\omega)$ are $\langle n, 0\rangle, \langle n, 1\rangle, \langle n, 2\rangle, \ldots, \langle n, n-1\rangle$; any two elements of $\text{dj}(A)$ are disjoint because the first components of their elements are different. The next form of the axiom is basically the one given by Zermelo in 1908.

Proposition 10.3.6. *In a ZF-universe, the axiom of choice is equivalent to the following: given any set A whose elements are pairwise disjoint and* $\neq \emptyset$*, there is a set B which has exactly one element of each of the elements of A.*

Proof. We prove first Zermelo's axiom from the axiom of choice. Starting with A, there is a choice function for A, and its codomain is a set B which contains one element of each of the elements of A.

Conversely, let A be given such that $\emptyset \neq a$ for each $a \in A$. Then $\emptyset \notin \text{dj}(A)$, and the hypothesis applied to $\text{dj}(A)$ gives a set B such that for each $a \in A$

there is exactly one element $\langle a, x\rangle \in B$ and $x \in a$. B is then a choice function for A. □

There is another important result which is equivalent to the axiom of choice and is of extended use. It is *Zorn's lemma* and we give now its statement and proof. Let us mention that a subset S of an ordered set is called a *chain* if any two elements of S are comparable; i.e., if the restricted order is a total ordering of S.

Proposition 10.3.7. *(Zorn's lemma) In a ZFC-universe, let $\leq$ be an ordering in a non-empty set A. If every subset of A that is a chain admits some upper bound, then A has a maximal element.*

Proof. Suppose A is as stated but with no maximal element. Then every chain admits strict upper bounds, because if m is an upper bound of a chain S which has no strict upper bounds, then $m \in S$ is a maximal element of A since m has not strict upper bounds. We now seek a contradiction that will prove the result by RAA.

To this end we follow similar ideas to those in the proof of Proposition 10.3.3. We pick a particular element $a_0 \in A$ and, by using a choice function for $\mathbb{P}(A) \setminus \{\emptyset\}$, we obtain a map $m : \mathbb{P}(A) \to A$ such that $m(X)$ is a strict upper bound of X, if it exists (for instance, if X is a chain); and $m(X) = a_0$ otherwise. By recursion we obtain a functional H with $\mathrm{D}(H) = \mathbf{Ord}$, and $H(\alpha) = m(H[\alpha])$. When $H[\alpha]$ is a chain, $H(\alpha)$ is a strict upper bound of that chain.

We may easily see by ordinal induction that every $H[\alpha]$ is a chain. This is vacuously true for $\alpha = 0$. Then if it is true for α, then $H(\alpha)$ is an upper bound for $H[\alpha]$ hence $H[\alpha + 1] = H[\alpha] \cup \{H(\alpha)\}$ is a chain. If λ is limit and $H[\beta]$ is a chain for every $\beta \in \lambda$, then it is obvious that $H[\lambda]$ is a chain, the union of a chain (with respect to inclusion) of chains of A. This implies that H is strictly increasing so that it is injective and H^{-1} is a functional. Since $\mathrm{D}(H^{-1}) = \mathrm{cD}(H) \subseteq A$ is a set, $\mathrm{cD}(H^{-1}) = \mathbf{Ord}$ is a set, and by Proposition 8.3.7 we reach the contradiction we looked for. □

We show next the equivalence.

Proposition 10.3.8. *In any ZF-universe, the validity of Zorn's lemma implies the validity of the axiom of choice.*

Proof. Let A be a set such that $\emptyset \notin A$. Assuming Zorn's lemma, we shall prove that there exists a choice function for A. First, we define the class B of all ordered pairs $\langle x, f\rangle$ where $x \subseteq A$ and f is a choice function for x; B is a set because $B \subseteq \mathbb{P}(A) \times \mathbb{P}(A \times A)$ and is not empty because $\langle \emptyset, \emptyset\rangle \in B$. In B we consider the relation R given by $\langle x, f\rangle R\langle y, g\rangle \Leftrightarrow x \subseteq y \wedge f \subseteq g$. It is routine to verify that R is an ordering of B.

Let C be any chain of the ordered set B. We take $u = \cup\, \mathrm{D}(C) = \cup\{x|\exists f\, (\langle x, f\rangle \in C)\} \subseteq A$ and $h = \cup\, \mathrm{cD}(C) = \cup\{f|\exists x\, (\langle x, f\rangle \in C)\}$. We check that $\langle u, h\rangle \in B$ and it will follow immediately that it is an upper bound of C. Given $\langle a, t\rangle, \langle a, s\rangle \in h$ then $f(a) = t,\ g(a) = s$ for some f, g. But $f \subseteq g$ or $g \subseteq f$ because the ordering in C is total, hence $t = s$. Thus h is a function with domain u and is a choice function since $h(a) = f(a)$ for some $\langle x, f\rangle \in C$ so that $h(a) \in a$.

By Zorn's lemma, B has a maximal element $\langle v, g\rangle$. If $v \neq A$, let $a \in A \setminus v$ and $z \in a$ (since $a \neq \emptyset$). Then $g \cup \{\langle a, z\rangle\}$ is a choice function for $v \cup \{a\}$, which contradicts the maximality in B of $\langle v, g\rangle$. So, $v = A$ and g is a choice function for A. □

A final equivalence of the axiom of choice has to do with a kind of generalization of cartesian products. Let I be a set. A function f with domain I is also called an *I-family of sets* (or simply a family). Usually, I is called the *set of indices* of the family. If we write $f(i) = A_i$, we also denote this family as $\{A_i\}_{i\in I}$ or $\{A_i\}_I$. One is not to forget that the object denoted in this way is the function, i.e., the set of ordered pairs $\langle i, A_i\rangle$ and not the set $\{A_i|i \in I\}$ which denotes the codomain of f: if $A_i = A_j$ for some $i \neq j$, then A_i, A_j are the same element in the codomain, but they give different elements of the family.

The axiom of choice has an equivalent version for families. Given an I-family of non-empty sets, $\{A_i\}_{i\in I}$, an *I-choice function for* the family $\{A_i\}$ is a function g with $\mathrm{D}(f) = I$ and such that $g(i) \in A_i$ for every $i \in I$. To see that the axiom of choice is equivalent to the existence of I-choice functions, let us define a kind of disjunction for families: if $\{A_i\}_I$ is a family, then we write A_i^* for the set $\{i\} \times A_i$, and we may form the *disjoint I-family* $\{A_i^*\}_I$ in the obvious way. This is very much like the original family $\{A_i\}$, but its members are disjoint sets.

Proposition 10.3.9. *The axiom of choice is equivalent to the assertion that for every set I, every I-family of non-empty sets has an I-choice function.*

Proof. Let $\{A_i\}$ be an I-family with each $A_i \neq \emptyset$, and consider the codomain of its disjoint family $\{\{i\} \times A_i|i \in I\}$. By Proposition 10.3.6, there exists a set with one element $\langle i, a_i\rangle$ for each $i \in I$ with $a_i \in A_i$. This set is an I-choice function for the family $\{A_i\}$.

Conversely, let A be any set such that $x \in A \Rightarrow x \neq \emptyset$. The identity is an A-family $\{x\}_{x\in A}$: A is the set of indices and the members of the family are the elements of A. Our hypothesis means that there is an A-choice function g for this family, hence $g(x) \in x$, and g is a choice function for A. □

We now see the relationship with cartesian products.

Proposition 10.3.10. *Let $f = \{A_i\}_I$ be a family. The collection of all I-choice functions for the family is a set, that is called the* cartesian product *of the family $\{A_i\}_{i\in I}$. It is written $\times_{i\in I} A_i$ or $\times_I A_i$.*

Proof. This collection is clearly a class included in the set of all functions from I to $\cup_I A_i$. □

Remarks and Examples 10.3.11. Any set s can be written as a family: take s as the set of indices and the identity as the function, giving the family $\{x\}_{x \in s}$. When $I = \{0, 1\}$ and $f = \{\langle 0, a\rangle, \langle 1, b\rangle\}$ is a family, the above representation would be $\{a, b\}_{\{0,1\}}$. Then the elements of the cartesian product in the sense of Proposition 10.3.10 are all functions $\{\langle 0, x\rangle, \langle 1, y\rangle\}$ with $x \in a, y \in b$. Though this is not exactly $a \times b$, there is an obvious bijection between both sets, and this is how the cartesian product of two sets fits into this definition: the elements of $a \times b$ can be seen as I-choice functions with $I = \{a, b\}$. There exist also, in the general case, the canonical projections $p_j : \times_I A_i \to A_j$ which are maps that take each choice function g to the value $g(j) \in A_j$.

Thus, the definition of cartesian product in Proposition 10.3.10 provides a way for generalizing the usual cartesian product of two sets. One might think that it is not necessary for this generalization to use the concept of I-families: for defining $a \times b$ it could be enough to consider the set $X = \{a, b\}$ and the choice functions for X give essentially the elements of $a \times b$. But this misses the point: if we adopt this view, then to define $a \times a$, we would have to use the set $X = \{a, a\} = \{a\}$; and the choice functions would give simply the elements of a. This is where the concept of I-family is useful: we want to consider sets A_i, but possibly with repetitions, so the simple collection of the A_i (i.e., the codomain of the function I-family) is not what we need.

However, there is one construction for which the difference between $\mathrm{cD}(f)$ and the family f is unimportant: with the notation $\cup_{i \in I} A_i$ or $\cup_I A_i$ we refer to the union of the codomain of f, and considering A_i, A_j as two objects or only one when $A_i = A_j$ does not change this union.

An obvious consequence of Proposition 10.3.9 and the definition of cartesian products is: the axiom of choice is equivalent to the assertion that the cartesian product of any family of non-empty sets is not empty.

10.4 Finite and Infinite Sets

10.4.1 Finiteness criteria

From now on, we assume that our universe V is a ZFC-universe. By Definition 6.3.11 a set is finite when it is equipotent to an element of ω, i.e., to a finite cardinal. Thus a set is finite (in the sense of Definition 6.3.11) if its cardinal is a finite cardinal; and it is infinite if its cardinal is infinite. The following is another characterization, in the theory of ZFC-universes, of infinite sets.

Proposition 10.4.1. *A set A is infinite if and only if there is an injective map $f : \omega \to A$.*

Proof. If A is infinite, then $|A| = \mu \geq \omega$, because $|A| < \omega$ implies A is finite. Then $\omega \subseteq \mu$ and we obtain an injective map through composition $\omega \to \mu \simeq A$.

Conversely, if there is an injective map $\omega \to A$, then $\omega \leq |A|$ by Proposition 10.1.6. Hence A is infinite because A finite implies $|A| < \omega$. □

Still another characterization of finite and infinite sets is given by a criterion which goes back to Dedekind. It is the following.

Proposition 10.4.2. *Let A be a set. A is infinite if and only if there exists a proper subset $X \subsetneqq A$ and a bijection $X \simeq A$.*

Proof. In view of Proposition 10.3.3, it is enough to prove the property for cardinals. If κ is an infinite cardinal, then $\omega \leq \kappa$ and we get an injective function $\kappa \to \kappa$ by considering all pairs $\langle \beta, \beta \rangle$ (for $\omega \leq \beta \in \kappa$) and $\langle t, 2t \rangle$ (for $t \in \omega$). The codomain is a proper subset of κ which is equipotent with κ.

For the converse, we apply natural induction and show that, for $t \in \omega$, there is no bijection $t \to A$ when $A \subsetneqq t$. This is vacuously true for $t = 0$, so assume it is true for some t, but that $g : s(t) \to A$ is a bijective map for some $A \subsetneqq s(t)$. If $t \notin A$, then $A \subseteq t$ and $t \simeq A \setminus \{g(t)\} \subsetneqq t$ which contradicts the inductive hypothesis. If $t \in A$, then $A \setminus \{t\} \simeq s(t) \setminus \{g^{-1}(t)\} \simeq t$ by Lemma 6.6.1, again contradicting the inductive hypothesis because $A \setminus \{t\} \subsetneqq t$. □

10.4.2 The series of the alephs

Up to now, we have seen only one infinite cardinal, namely ω. The key to the proof of the existence of larger cardinals is the next theorem.

Proposition 10.4.3. *(Cantor's theorem) If A is a set, then $|A| < |\mathbb{P}(A)|$.*

Proof. Suppose, to the contrary, that $|\mathbb{P}(A)| \leq |A|$. By (vi) of Proposition 10.1.6 there is an injective map $f : \mathbb{P}(A) \to A$, but this contradicts Proposition 6.6.11. □

In particular, $|\mathbb{P}(\omega)| > \omega$ and this gives an uncountable infinite cardinal. When using cardinals, it is customary to write $\aleph_0$ instead of ω: they are the same, but by this writing we emphasize that we consider it as a cardinal. The above result shows that there are cardinals greater than $\aleph_0$. In fact, we may go much farther with Cantor's theorem, and construct an endless collection of bigger and bigger cardinals.

Proposition 10.4.4. *The collection of cardinals is a proper class, which we write* ***Card****.*

Proof. Cardinals are characterized inside the class of ordinals by the condition $\forall u \in \mathbf{Ord}\ ((u \in \mathbf{Ord} \wedge u \simeq x) \Rightarrow x \subseteq u)$. It follows that it is a class.

To see that it is proper, consider any set S of cardinals. Then $\cup S = \alpha$ is an ordinal, by Proposition 8.3.8 and is an upper bound for all elements of S, therefore $|\alpha|$ is also an upper bound of S by Proposition 10.1.6 (iv). According to Proposition 10.4.3, $|\mathbb{P}(|\alpha|)|$ is a strict upper bound of S, hence $S \neq \mathbf{Card}$. This entails that **Card** is a proper class. □

Let us denote the (proper) class of all infinite cardinals as $\mathbf{Card}(\infty)$; we will see now that they are order-isomorphic to the ordinals.

Proposition 10.4.5. *There exists a functional H with $D(H) = \mathbf{Ord}$ whose codomain is $\mathbf{Card}(\infty)$ and which is strictly increasing; i.e., $\alpha < \beta \Rightarrow H(\alpha) < H(\beta)$.*

Proof. We apply recursion (Theorem 8.4.4) to define H. For the functional F with domain $\mathbb{P}(\mathbf{Card}(\infty))$, we take $F(X)$ as the smallest (infinite) cardinal κ such that $\rho < \kappa$ for every $\rho \in X$. By the recursion theorem, there is H defined on all ordinals so that $H(\alpha)$ is the smallest infinite cardinal which is a strict upper bound for $H[\alpha]$.

If $\alpha < \beta$, then $H(\beta)$ is a strict upper bound for $H[\beta]$, hence $H(\alpha) < H(\beta)$. To see that $\mathrm{cD}(H) = \mathbf{Card}(\infty)$, we use RAA and suppose that there is a smallest infinite cardinal κ such that $\kappa \notin \mathrm{cD}(H)$. So, the set S of all infinite cardinals $< \kappa$ consists of cardinals $H(\beta)$. If $S' = H^{-1}[S]$, then S' is a transitive set of ordinals because H is strictly increasing, and thus S' is an ordinal μ (see, e.g., Exercise 4 of Chapter 8), and $H[\mu] = H[S'] = S$. Then $H(\mu) = F(H[\mu]) = F(S)$ is the smallest infinite cardinal which is a strict upper bound for the elements of S, that is, $H(\mu) = \kappa$ and $\kappa \in \mathrm{cD}(H)$, which is a contradiction. □

From the definition of H we see that $H(0) = \aleph_0$; extending this notation, $H(\alpha)$ will be written as $\aleph_\alpha$, so $\aleph_1$ is the smallest uncountable cardinal. By Proposition 10.4.5, the sequence of the alephs $\aleph_0, \aleph_1, \ldots$ is the well-ordered series of all the infinite cardinals; that is, every infinite cardinal is an aleph.

For any cardinal κ, there exists a next cardinal which is written κ^+: it is the smallest cardinal which is $> \kappa$. For a finite cardinal t, $t^+ = t + 1$, but the next cardinal of an infinite cardinal is an infinite cardinal. The cardinals of the form κ^+ are called *successor cardinals*, while cardinals that are not successor cardinals are called *limit cardinals*. These concepts are not to be confused with those of ordinals identified by the same terms. In fact, every infinite cardinal is a limit ordinal, as we know, but it may be a successor cardinal; that is, these are two different classifications, one for ordinals and one for cardinals, under the same names. It is plain that $\aleph_\alpha$ is a limit cardinal if and only if α is a limit ordinal. Thus $\aleph_0$ is a limit cardinal, each $\aleph_t$ is a successor cardinal (for $t \in \omega$) and $\aleph_\omega$ is the next limit cardinal.

10.5 Cardinal Exponentiation

10.5.1 Exponentiation and the continuum hypothesis

The third operation on cardinals is not dependent on operations with ordinals. Recall that if A and B are sets, then there is a set formed with all maps $B \to A$ (Proposition 5.2.8). We will write this set as ${}^{B}A$.

Definition 10.5.1. Let κ, μ be cardinals. We write κ^μ for $|{}^{\mu}\kappa|$.

Note that κ^2 means the cardinal of the set of all maps from $\{0,1\}$ to κ. But there is an easy bijection between this set and $\kappa \times \kappa$, hence $\kappa^2 = \kappa \cdot \kappa$.

Some basic properties of exponentiation are listed below.

Proposition 10.5.2. *Let A, B be sets, and κ, μ, ρ cardinals. Then:*
(i) $\kappa \neq 0 \Rightarrow \kappa^0 = 1, \quad 0^\kappa = 0.$
(ii) $\kappa^1 = \kappa, \quad 1^\kappa = 1.$
(iii) $|{}^{B}A| = |A|^{|B|}.$
(iv) $\kappa^{\mu+\rho} = \kappa^\mu \cdot \kappa^\rho.$
(v) $\kappa^\mu \cdot \rho^\mu = (\kappa \cdot \rho)^\mu.$
(vi) $(\kappa^\mu)^\rho = \kappa^{\mu \cdot \rho}.$
(vii) $\rho \leq \mu \Rightarrow \kappa^\rho \leq \kappa^\mu.$
(viii) $\kappa \leq \mu \Rightarrow \kappa^\rho \leq \mu^\rho.$

Proof. We sketch the proof of some of the items and leave the rest as exercises.

(iv) We may consider disjoint sets A, B and compare ${}^{A\cup B}C$ and ${}^{A}C \times {}^{B}C$. A bijection is easy to obtain: functions in the first set give a pair of restrictions.

(vi) Let $D = {}^{B}A$ and consider ${}^{C}D$. There is a canonical bijection with ${}^{(C\times B)}A$.

(viii) Since $\kappa \subseteq \mu$, maps $\rho \to \kappa$ can be seen as maps $\rho \to \mu$. □

As a consequence of these properties, it is possible to estimate the value of power cardinals when the basis is not greater than the exponent.

Proposition 10.5.3. *Let $2 \leq \kappa \leq \mu$ cardinals with μ infinite. Then $\kappa^\mu = 2^\mu$.*

Proof. We show that the result is true even for $\kappa \leq 2^\mu$. First, $2^\mu \leq \kappa^\mu \leq (2^\mu)^\mu$ follows by (viii) of Proposition 10.5.2. Now, $(2^\mu)^\mu = 2^{\mu\cdot\mu} = 2^\mu$ by (iv) of Proposition 10.5.2 and Proposition 10.2.10. It follows that $\kappa^\mu = 2^\mu$. □

From Cantor's theorem (Proposition 10.4.3) and the fact that there is a bijection between ${}^{\kappa}2$ and $\mathbb{P}(\kappa)$, it follows that $\kappa < 2^\kappa$ for every cardinal κ. In particular, $\aleph_0 < 2^{\aleph_0}$. The set of real numbers $\mathbb{R}$ is equipotent to $\mathbb{P}(\omega) = \mathbb{P}(\aleph_0)$ (see Exercise 15 of Chapter 6), i.e., $|\mathbb{R}| = 2^{\aleph_0}$. The cardinal of $\mathbb{R}$, which is in fact the same as the cardinal of the interval $(0, 1)$, is known as the cardinal of *the continuum*. Identifying the cardinal of the continuum as an exponential cardinal is not the same as identifying it as a concrete element in the series of the alephs. The first conjecture about which is the value of $2^{\aleph_0}$, the *power of the continuum* in the older terms, was that $2^{\aleph_0} = \aleph_1$, since we know that $\aleph_1$ is the first cardinal that is uncountable, i.e., $> \aleph_0$. This conjecture is known as the *continuum hypothesis* (CH, for short) and will be discussed later in this book. This was conjectured by Cantor himself and is the first of the problems proposed by Hilbert in his famous list of 23 problems, presented in 1900. Attempts at proving (or disproving) the CH have been numerous and have led to central developments of set theory.

The CH suggests the following generalization: for any infinite cardinal κ, the equation $\kappa^+ = 2^\kappa$ holds. This assertion is called the *generalized continuum hypothesis*, or GCH.

10.5.2 Infinite sums and products

Once that, by virtue of the axiom of choice, we have that every set has a cardinal, i.e. a formal version of its "size", the operations with cardinals reflect in fact operations with sets: sum, product and exponentiation give really the measure of the set operations union (taking disjoint copies of the sets), cartesian product and sets of maps. From this point of view, a natural extension of the operations presented so far consists in studying infinite unions or infinite cartesian products and evaluating their "size". With this goal in mind, recall our notations for I-families $\{A_i\}_I$, $\{A_i\}_{i \in I}$; recall also that the disjoint I-family $\{A_i^*\}$ of the I-family $\{A_i\}$ has $A_i^* = \{i\} \times A_i$. Note that there is an obvious surjective map $\cup_I A_i^* \to \cup_I A_i$ (induced by the projections onto the second component) and, consequently, $|\cup_I A_i| \leq |\cup_I A_i^*|$ by Proposition 10.3.4.

Definition 10.5.4. Let I be a non-empty set, and $\{\kappa_i\}_{i \in I}$ an I-family of cardinals. We define:

$$\sum_I \kappa_i = |\cup_{i \in I} \kappa_i^*|, \quad \prod_I \kappa_i = |\times_I \kappa_i|$$

Particular instances of these definitions give on the one hand the concepts of sum and product of two cardinals, respectively; on the other, product and exponentiation. Namely, if $I = 2$, the family consists of $\langle 0, \kappa_0 \rangle, \langle 1, \kappa_1 \rangle$ and the sum in the above definition gives $\kappa_0 + \kappa_1$ (see Proposition 10.2.3), while the product is $\kappa_0 \cdot \kappa_1$. When $I = \lambda$ is any non-zero cardinal and the I-family is constant (each $\kappa_i = \kappa$), then the sum above gives $\lambda \cdot \kappa$ and the product is κ^λ. This can be seen easily by constructing the corresponding bijections; for example, each element of $\times_I \kappa$ is a family with domain λ and values in κ, hence it is an element of ${}^\lambda\kappa$.

The following result follows immediately from the definitions, by using the cardinals of each of the sets below.

Proposition 10.5.5. *Let $\{A_i\}_I$ be an I-family. Then:*

$$\sum_I |A_i| = |\cup_{i\in I} A_i^*|, \quad \prod_I |A_i| = |\times_{i\in I} A_i|$$

Proposition 10.5.6. *Let $\{\kappa_i\}_I$ be an I-family of cardinals, λ another cardinal. Then*

(i) $(\sum_I \kappa_i) \cdot \lambda = \sum_I (\kappa_i \cdot \lambda)$.

(ii) $(\prod_I \kappa_i)^\lambda = \prod_I \kappa_i^\lambda$.

Proof. (i) The sets $\kappa_i^* \times \lambda$ are pairwise disjoint, and so the equality follows from the obvious bijection between $\cup_I \kappa_i^* \times \lambda$ and $(\cup_I \kappa_i^*) \times \lambda$.

(ii) Here too we have a natural bijection between ${}^\lambda(\prod_I \kappa_i)$ and $\prod_I ({}^\lambda \kappa_i)$. □

The couple of results that follow will find application later.

Proposition 10.5.7. *Let $\{\kappa_i\}, \{\mu_i\}$ be I-families of cardinals as above. If $\kappa_i \leq \mu_i$ for each $i \in I$, then $\sum_I \kappa_i \leq \sum_I \mu_i$ and $\prod_I \kappa_i \leq \prod_I \mu_i$.*

Proof. The injective maps $\kappa_i \to \mu_i$ give injections $\kappa_i^* \to \mu_i^*$ and these combine to give again an injective map from $\cup \kappa_i^*$ to $\cup \mu_i^*$ since these are disjoint unions. As for the second inequality, any element of the first cartesian product is an I-family f with $f(i) \in \kappa_i$; by composing with the inclusion $\kappa_i \subseteq \mu_i$, we see that f is also an element of the second cartesian product, and this gives the desired injective function. □

Proposition 10.5.8. *(J. König's theorem) Let $\{\kappa_i\}_I$ and $\{\mu_i\}_I$ be I-families of cardinals, with $I \neq \emptyset$. If $\kappa_i < \mu_i$ for each $i \in I$, then $\sum_I \kappa_i < \prod_I \mu_i$.*

Proof. We show first the relation $\sum_I \kappa_i \leq \prod_I \mu_i$, by giving an injective map $f : \cup_I \kappa_i^* \to \times_I \mu_i$. Given $\langle i, x_i \rangle \in \kappa_i^*$, we simply take for $f(\langle i, x_i \rangle)$ the choice function with pairs $\langle j, \kappa_j \rangle$ for $j \neq i$, and $\langle i, x_i \rangle$; this gives an element of $\times_I \mu_i$ since $\kappa_j \in \mu_j$ by the hypothesis, and $x_i \in \kappa_i \subseteq \mu_i$. It is easy to see that f is indeed injective.

To complete the proof, we show that every map $h : \cup_I \kappa_i^* \to \times_I \mu_i$ is non-surjective, and the conclusion follows by Proposition 10.3.4. For this we use a kind of argument similar to Cantor's diagonal argument (see, e.g., [23, p.99]): given the map h, each composition[1] $\kappa_j^* \hookrightarrow \cup \kappa_j^* \xrightarrow{h} \times_I \mu_i \xrightarrow{p_j} \mu_j$ (p_j being the canonical projection, so that $p_j(f) = f(j)$) is non-surjective, because $\kappa_j^* \simeq \kappa_j < \mu_j$. For each $j \in I$, let $u_j \in \mu_j$ be the smallest element of μ_j which

[1]The writing $A \hookrightarrow B$ is used sometimes, as it is here, to emphasize that the map is an inclusion.

does not belong to the image of that composition map, i.e., u_j is such that $h(\langle j, x_j \rangle)(j) \neq u_j$ for any $\langle j, x_j \rangle \in \kappa_j^*$. Then we get a choice function $g \in \times_I \mu_i$ by setting $g(j) = u_j$ for each $j \in I$. But g cannot equal $h(\langle i, x_i \rangle)$ for any $\langle i, x_i \rangle \in \kappa_i^*$, because $h(\langle i, x_i \rangle)(i) \neq u_i = g(i)$ by the choice of the u_j, and thus h is not surjective. □

10.6 Cofinalities

10.6.1 The cofinality of an ordinal

Besides the classifications of cardinals as finite, infinite, successor or limit, there is another, deeper, distinction to be made. It depends on the concept of the cofinality of an ordinal.

A subset S of an ordinal α is said to be a *cofinal subset* when for each $x \in \alpha$ there is $y \in S$ such that $x \leq y$. If α is an ordinal, a map $f : A \to \alpha$ is said *cofinal* when $\mathrm{cD}(f) \subseteq \alpha$ is a cofinal subset.

For a given ordinal α, the ordinals β for which there is a cofinal map $\beta \to \alpha$ form a non-empty class (obviously, 1_α is cofinal). Therefore there is a smallest member in that class.

Definition 10.6.1. Let α be an ordinal. The smallest ordinal β such that there exists a cofinal map $f : \beta \to \alpha$, is called the *cofinality* of α, written $\mathrm{cf}(\alpha)$.

Some immediately seen observations: (1) a successor ordinal $s(\gamma)$ has a maximum element γ, hence its cofinality is 1 (the map giving $0 \mapsto \gamma$ is cofinal); (2) $\mathrm{cf}(\alpha)$ is always a cardinal: if $f : \beta \to \alpha$ is a cofinal map and $g : \gamma \to \beta$ is a bijection with $\gamma \in \beta$, then $h = f \circ g$ is cofinal and this implies that β is not smallest; (3) $\mathrm{cf}(\alpha) \leq |\alpha|$, because $1 : \alpha \to \alpha$ is cofinal, hence $\mathrm{cf}(\alpha) \leq \alpha$ and the result now follows from item (2) and Proposition 10.1.6 (iii).

0 is the only ordinal with cofinality 0. Recall (Definition 8.2.1) that a map $f : A \to B$ between ordered sets is a *morphism* when $x < y \Rightarrow f(x) < f(y)$.

Lemma 10.6.2. *For any ordinal α, cf(α) is the smallest ordinal κ such that there exists a morphism $g : \kappa \to \alpha$ that is cofinal.*

Proof. It is enough to show that if $\kappa = \mathrm{cf}(\alpha)$, then there exists a morphism $g : \kappa \to \alpha$ that is cofinal. This is obvious when $\kappa = 1$, so we assume that α is a limit ordinal.

Let $f : \kappa \to \alpha$ a cofinal map. A functional F is defined on $\mathbb{P}(\alpha) \times \mathbf{Ord}$ as follows: $F(X, \rho)$ is the maximum between $f(\rho)$ (if $\rho \in \kappa$) and the smallest strict upper bound in α of the subset X (if it exists). When none of these elements exist, then $F(X, \rho) = 0$ and therefore $\text{cD}(F) \subseteq \alpha$. By Theorem 8.4.4, there exists a functional H defined on ordinals so that $H(\rho) = F(H[\rho], \rho)$. We claim that $g = H|_\kappa : \kappa \to \alpha$ is a morphism that is cofinal.

Let $x < y \in \kappa$. Then $H(x) \in H[y]$ and $H[y]$ is a subset of α that is not cofinal (if it is, then we have a cofinal map $y \to \alpha$ while $y < \kappa = \text{cf}(\alpha)$). Thus $H[y]$ has strict upper bounds, hence $g(y) > g(x)$. This shows that g is a morphism. On the other hand, for each $x \in \kappa$ we have $f(x) \leq g(x)$ by the definition of H, hence $\text{cD}(g)$ is cofinal in α because $\text{cD}(f)$ is cofinal in α. □

Proposition 10.6.3. *If $\kappa = cf(\alpha)$, then $cf(\kappa) = \kappa$.*

Proof. Let $\mu = \text{cf}(\kappa)$. By Lemma 10.6.2, there are cofinal morphisms $\mu \to \kappa$ and $\kappa \to \alpha$, so their composition is a cofinal morphism $\mu \to \alpha$ and $\kappa = \text{cf}(\alpha) \leq \mu$. Since $\mu \leq \kappa$ by observation (3) after Definition 10.6.1, the result follows. □

Lemma 10.6.4. *Let α be a nonzero limit ordinal and $S \subseteq \alpha$ a cofinal subset. Then $|\cup S| = \sum_{x \in S} |x| = |\alpha|$.*

Proof. We view S as an S-family and thus $|\cup S| = |\cup_{x \in S} x| \leq |\cup_{x \in S} x^*| = \sum_{x \in S} |x|$, according to Proposition 10.5.5. Thus, it will suffice to prove that $|\cup_{x \in S} x^*| \leq |\alpha| \leq |\cup S|$.

For the first inequality, there is an inclusion $\cup_{x \in S} x^* \subseteq \alpha \times \alpha$ because any element of the union is $\langle x, u \rangle$ with $u \in x$ and hence $\langle x, u \rangle \in \alpha \times \alpha$. Thus $|\cup_{x \in S} x^*| \leq |\alpha \times \alpha| = |\alpha| \cdot |\alpha| = |\alpha|$ by Proposition 10.2.10. On the other hand, if $\beta \in \alpha$, then there is $u \in S$ with $\beta \in u$ because S is cofinal and α is limit. Thus $\beta \in \cup S$ and $\alpha \subseteq \cup S$, from which it follows that $|\alpha| \leq |\cup S|$. □

10.6.2 More on cardinal arithmetic

Up to now, we know very few inequalities concerning exponentiation. The following two are noteworthy: (1) $\kappa < 2^\kappa$; and (2) $\kappa < \kappa^\kappa$ for infinite κ. In fact, (2) is the same as (1) if one takes into account Proposition 10.5.3. We next improve these two bounds.

Proposition 10.6.5. *For any infinite cardinal κ, we have $\kappa < \kappa^{cf(\kappa)}$.*

Proof. Let $\lambda = \text{cf}(\kappa)$ and $f : \lambda \to \kappa$ a map such that $\text{cD}(f)$ is a cofinal subset of κ. By Lemma 10.6.4 and Proposition 10.5.8,

$$\kappa = \sum_{x \in \lambda} |f(x)| < \prod_{x \in \lambda} \kappa = \kappa^\lambda = \kappa^{cf(\kappa)}$$

□

Corollary 10.6.6. *If κ is an infinite cardinal, then $\kappa < cf(2^\kappa)$.*

Proof. We use RAA and assume that $cf(2^\kappa) \leq \kappa$. Call $\lambda = 2^\kappa$. Then

$$2^\kappa = \lambda < \lambda^{cf(\lambda)} = \lambda^{cf(2^\kappa)} \leq \lambda^\kappa = (2^\kappa)^\kappa = 2^{\kappa^2} = 2^\kappa$$

by Propositions 10.6.5 and 10.5.2. □

An infinite cardinal κ is called *regular* if $\mathrm{cf}(\kappa) = \kappa$.

Thus cofinalities are regular cardinals by Proposition 10.6.3. An infinite cardinal which is not regular is *singular*.

Proposition 10.6.7. *$\aleph_0$ is a regular cardinal. Also, if κ is infinite, κ^+ is regular. On the other hand, if α is a nonzero limit ordinal, $cf(\aleph_\alpha) = cf(\alpha)$. In particular, $\aleph_\omega$ is singular.*

Proof. If $t \in \omega$, then a morphism $f : s(t) \to \aleph_0$ cannot be cofinal because $s(f(t))$ is a strict upper bound for $\mathrm{cD}(f)$; thus $\aleph_0 = \mathrm{cf}(\aleph_0)$ and $\aleph_0$ is regular.

Next, let κ be infinite and assume that $\rho = \mathrm{cf}(\kappa^+)$ and $\rho \leq \kappa$. Let $f : \rho \to \kappa^+$ a cofinal map so that $\sum_{x \in \rho} |f(x)| = \kappa^+$ by Lemma 10.6.4. Since each $|f(x)| < \kappa^+$, we have $|f(x)| \leq \kappa$. According to Proposition 10.5.7, $\sum_{x \in \rho} |f(x)| \leq \sum_{x \in \rho} \kappa = \rho \cdot \kappa = \kappa$, and thus we arrive to a contradiction.

For the next assertion, observe that if κ is a limit cardinal, then the set of cardinals $< \kappa$ is cofinal in κ. This is because if $\beta \in \kappa$, then $|\beta|^+$ is a cardinal $< \kappa$ and $\beta < |\beta|^+$. Thus the functional H of Proposition 10.4.5 gives a cofinal morphism $\alpha \to \aleph_\alpha$. Moreover, the same property implies that there is a cofinal monotone map $g : \aleph_\alpha \to \aleph_\alpha$ with $g(\beta) = |\beta|^+$. Since H^{-1} is defined on all cardinals of $\aleph_\alpha$, there is a monotone cofinal map $H^{-1} \circ g : \aleph_\alpha \to \alpha$. It is then easy to infer that α and $\aleph_\alpha$ have the same cofinality: see Exercise 30 of this chapter. □

It was shown in Proposition 6.6.9 that in 6-universes and under a certain additional condition, the countable union of countable sets is countable. In ZFC-universes, this property holds without any restriction, and in fact, it can be generalized to any infinite cardinal λ.

Proposition 10.6.8. *Let λ be an infinite cardinal. If $\{A_i\}_I$ is an I-family such that $|A_i| \leq \lambda$, and $|I| \leq \lambda$, then $|\cup_I A_i| \leq \lambda$.*

Proof. By Propositions 10.5.7 and 10.2.10, and the observations before Proposition 10.5.5, $|\cup_I A_i| \leq |\cup_I A_i^*| = \sum_I |A_i| \leq \sum_I \lambda \leq \lambda^2 = \lambda$. □

A similar property with strict inequalities characterizes regular cardinals.

Proposition 10.6.9. *Let κ be an infinite cardinal. Then κ is regular if and only if the union of any I-family with $|I| < \kappa$ and whose elements have cardinal $< \kappa$ has cardinal $< \kappa$.*

Proof. Let κ be singular and $\mathrm{cf}(\kappa) = \lambda < \kappa$. Let $f : \lambda \to \kappa$ be a cofinal morphism, and take the λ-family $\{A_\beta\}_{\beta\in\lambda}$ with $A_\beta = f(\beta)$. Then $\cup_{\beta\in\lambda} A_\beta = \kappa$ because of the cofinality of f. On the other hand, the set of indices has cardinal $|\lambda| < \kappa$; and each set A_β has cardinal $|f(\beta)| < \kappa$ as $f(\beta) \in \kappa$. Thus κ does not satisfy the condition of the statement.

Suppose now that $\kappa = |\cup_I A_i|$ for $|I| < \kappa$ and each $|A_i| = \kappa_i < \kappa$. Let $\lambda = \sup_I(\kappa_i) \leq \kappa$. Thus $\kappa = |\cup_I A_i| \leq |\cup_I A_i^*| = \sum_I \kappa_i = \sum_I |A_i| \leq \sum_I |\lambda| = |I| \cdot |\lambda|$, and therefore $|\lambda| = \kappa$ since $|I| < \kappa$. Then the map $f : I \to \kappa$ with $f(i) = \kappa_i$ is cofinal, hence $\mathrm{cf}(\kappa) < \kappa$ and κ is singular. □

Proposition 10.5.3 identifies powers κ^μ when $\kappa \leq \mu$. We want to know more in the other direction, i.e., when the exponent is smaller than the base. In universes where GCH holds, computing exponentiation is easy.

Proposition 10.6.10. *Assume GCH, and let κ, μ be infinite cardinals. Then:*
(i) If $\mu < \mathrm{cf}(\kappa)$, then $\kappa^\mu = \kappa$.
(ii) If $\mathrm{cf}(\kappa) \leq \mu < \kappa$, then $\kappa^\mu = \kappa^+$.

Proof. (i) Since $\mu < \mathrm{cf}(\kappa)$, every map $f : \mu \to \kappa$ has a bounded codomain, hence $\mathrm{cD}(f) \subseteq \alpha < \kappa$. Thus ${}^\mu\kappa$ is the union of the sets ${}^\mu\alpha$ for $\alpha \in \kappa$. It follows that $\kappa^\mu \leq \sum_{\alpha\in\kappa} |\alpha|^\mu$ by the definition of infinite sums.

Now, $|\alpha| < 2^{|\alpha|}$ implies that $|\alpha|^\mu \leq (2^{|\alpha|})^\mu = 2^{|\alpha|\cdot\mu} = (|\alpha| \cdot \mu)^+$. But $|\alpha|, \mu < \kappa$ and it follows that $|\alpha| \cdot \mu < \kappa$ and $|\alpha|^\mu \leq (|\alpha| \cdot \mu)^+ \leq \kappa$. Therefore $\kappa^\mu \leq \sum_{\alpha\in\kappa} \kappa = \kappa \cdot \kappa = \kappa$. The other inequality is obvious.

(ii) $\kappa < 2^\kappa$ implies that $\kappa^\mu \leq (2^\kappa)^\mu = 2^\kappa = \kappa^+$. For the other direction, there exists some cofinal subset $A \subseteq \kappa$ such that $|A| \leq \mu$, so that $\kappa^{|A|} \leq \kappa^\mu$. Since A is cofinal in κ, $\kappa = \cup A$ and $\kappa \leq \sum_{\alpha\in A} |\alpha| < \Pi_{\alpha\in A}\, \kappa = \kappa^{|A|} \leq \kappa^\mu$ where König's theorem has been applied to get the strict inequality. Thus $\kappa < \kappa^\mu$, and $\kappa^+ \leq \kappa^\mu$, completing the proof of the equality. □

Of course, when $\kappa \leq \mu$, $\kappa^\mu = 2^\mu = \mu^+$. We turn now to the general situation, without GCH. The next result is known as *the Hausdorff formula.*

Lemma 10.6.11. *Let $\mu \leq \kappa$ be infinite cardinals. Then $(\kappa^+)^\mu = \kappa^\mu \cdot \kappa^+$.*

Proof. By Proposition 10.6.7 κ^+ is regular, hence $\mu < \mathrm{cf}(\kappa^+)$. As in the proof of Proposition 10.6.10, we see that $(\kappa^+)^\mu$ is the cardinal of the union of the sets ${}^\mu\alpha$ for $\alpha < \kappa^+$. If $A = \{\beta \in \kappa^+ | \kappa \leq \beta\}$, then every set ${}^\mu\alpha$ is included in ${}^\mu\beta$ for some $\beta \in A$. Thus $(\kappa^+)^\mu = |\cup_{\alpha\in A}({}^\mu\alpha)| \leq \sum_{\alpha\in A} |\alpha|^\mu = \sum_{\alpha\in A} \kappa^\mu$. Since $\kappa^+ = A \cup \kappa$ and $A \cap \kappa = \emptyset$, $|A| = \kappa^+$ and $(\kappa^+)^\mu \leq \kappa^\mu \cdot \kappa^+$ by the known property of constant infinite sums (see, e.g., Exercise 23). The converse inequality is obvious, because $\kappa^\mu, \kappa^+ \leq (\kappa^+)^\mu$ and Proposition 10.2.10 and its consequence can be applied. □

We summarize the results about the values of exponentials without GCH that can be derived from what we have found.

Proposition 10.6.12. *Let κ, μ be infinite cardinals. Then:*

(1) If $\kappa \leq \mu$, then $\kappa^\mu = 2^\mu$.

Let $\mu < \kappa$. Then:

(2) If κ is a limit cardinal, then:

(2.1) If there is a cardinal ρ such that $\rho < \kappa \leq \rho^\mu$, then $\kappa^\mu = \rho^\mu$.

(2.2) If $\rho < \kappa \Rightarrow \rho^\mu < \kappa$ and $\mu <$ cf(κ), then $\kappa^\mu = \kappa$.

(2.3) If $\rho < \kappa \Rightarrow \rho^\mu < \kappa$ and cf(κ) $\leq \mu$, then $\kappa^\mu = \kappa^{cf(\kappa)}$.

(3) If $\kappa = \rho^+$ is a successor cardinal, then:

(3.1) If $\kappa \leq \rho^\mu$, then $\kappa^\mu = \rho^\mu$.

(3.2) If $\rho^\mu < \kappa$, then $\kappa^\mu = \kappa$.

Proof. (1) This is Proposition 10.5.3.

(2.1) This is direct: $\kappa^\mu \leq (\rho^\mu)^\mu = \rho^\mu$. The converse inequality is obvious.

(2.2) Since $\mathrm{cf}(\kappa) > \mu$, ${}^\mu\kappa$ is the union of the sets ${}^\mu\alpha$, with α ranging over the cardinals $< \kappa$, as maps $\mu \to \kappa$ cannot be cofinal. Thus $\kappa^\mu \leq \sum_{\alpha<\kappa} |\alpha|^\mu \leq \sum_{\alpha<\kappa} \kappa = \kappa$ because $\rho < \kappa \Rightarrow \rho^\mu < \kappa$. Therefore $\kappa^\mu \leq \kappa$ and the equality follows immediately.

(2.3) Let $\mathrm{cf}(\kappa) = \lambda \leq \mu$. Let $\psi : \lambda \to \kappa$ be a cofinal morphism and write $\kappa_\alpha = \psi(\alpha)$; then $\kappa = \cup_{\alpha\in\lambda}\kappa_\alpha \leq \sum_{\alpha\in\lambda}|\kappa_\alpha| \leq \prod_{\alpha\in\lambda}|\kappa_\alpha|$; this inequality can be obtained with an easy variation of the first part of the proof of Proposition 10.5.8. Then

$$\kappa^\mu \leq (\prod_{\alpha\in\lambda}|\kappa_\alpha|)^\mu \leq \prod_{\alpha\in\lambda}|\kappa_\alpha|^\mu \leq \prod_{\alpha\in\lambda}\kappa$$

by the hypothesis and Proposition 10.5.6. This gives $\kappa^\mu \leq \kappa^\lambda = \kappa^{cf(\kappa)}$, and the equality follows because $\mathrm{cf}(\kappa) \leq \mu$.

(3.1) This follows exactly as (2.1).

(3.2) By Lemma 10.6.11, $\kappa^\mu = \rho^\mu \cdot \kappa = \kappa$.

□

10.7 Exercises

In the exercises that follow, the universe is assumed to be a ZF-universe.

1. By using the results of Chapter 8, show that the ordinals $\omega \cdot 2, \omega^2$ or ω^{ω^ω} are not cardinals.
2. Prove Lemma 10.2.1, by using the definition of ordinal sum.
3. Prove the following properties of addition for arbitrary cardinals μ, ρ, σ:

   (i) $\mu \leq \rho \Leftrightarrow \exists\sigma \ (\rho = \mu + \sigma)$

   (ii) $\mu \leq \rho \Rightarrow ((\mu + \sigma \leq \rho + \sigma) \wedge (\sigma + \mu \leq \sigma + \rho))$

4. For cardinals μ, ρ, σ, prove that $\mu \leq \rho \Rightarrow \mu \cdot \sigma \leq \rho \cdot \sigma$. Does this property hold when $\sigma \neq 0$ and $<$ is substituted for $\leq$?
5. Prove Hartogs' theorem (in ZF-universes): for every set X, there is an ordinal α such that there is no injective map $\alpha \to X$ (Hint: Suppose there is X such that every ordinal has an injective map to X. Consider the set of all relations $\langle R, y \rangle$ where R is a well-ordering of the subset y of X, and define a map from this set to ordinals.)
6. Prove that in a ZF-universe the class of cardinals has no maximum. Deduce, as in the proof of Proposition 10.4.4, that the class of cardinals is a proper class.
7. (Assuming property NNS) Compare the restriction to $\omega \times \omega$ of the canonical well-ordering of **Card** $\times$ **Card** and the well-ordering of $\omega \times \omega$ that is obtained from Exercise 14 of Chapter 6. Make drawings to visually explain each of these orderings.
8. Recall that ${}^{A}B$ is the set of all maps $A \to B$. Suppose A is a non-empty set and B contains at least two elements. Find an injective map $A \to {}^{A}B$, but prove that there is no surjective map $F : A \to {}^{A}B$ (Hint: Use RAA, suppose that F exists, and try to construct a map $g : A \to B$ which cannot be $F(a)$ for any $a \in A$).
9. Recall that, in an ordered set A, a *chain* is a subset S of A with the property that any two elements of S are comparable. The Hausdorff maximal principle (HMP) is the assertion that, given a chain S in a non-empty ordered set, there exists a chain $S' \supseteq S$ which is a maximal chain (C is a maximal chain when there is no chain C' of A such that $C \subsetneq C'$). Prove that Zorn's lemma implies the Hausdorff maximal principle.
10. Let A be a well-ordered set. By using the symmetric difference of sets $\triangle$ defined in Exercise 23 of Chapter 2, construct a total ordering of the set $\mathbb{P}(A)$.
11. A selector for a set A is a set S with the property that, for each $x \in A$, the set $x \cap S$ has exactly one element. Show that the axiom of choice is equivalent to the property that every partition of a set has a selector.
12. Prove that the axiom of choice is equivalent to the property that for every non-empty set A and map $f : A \to B$, there exists a map $g : B \to A$ such that $f = f \circ g \circ f$ (Hint: For the sufficiency of the condition, consider a partition P of the set A and the canonical projection map $f : A \to P$ (where $f(x) = \overline{x}$) to obtain a selector for P, and then apply the preceding exercise).
13. Prove that the Hausdorff maximal principle implies the existence of a choice function for any set A with $\emptyset \notin A$ (Hint: Consider the set of all choice functions defined on subsets of A, and order this set by inclusion).

14. (Some notions of topology are necessary in this exercise) A set x is said to have the *finite intersection property or FIP* when for every finite non-empty subset S of x, $\cap S \neq \emptyset$. A topological space X is compact when any set of closed subsets of X with the FIP has a non-empty intersection. The following result is the Tychonoff's theorem: if $\{X_i\}$ is a family of compact topological spaces, then its product $\times_I X_i$ is compact (with the product topology). Prove that this property implies the axiom of choice (Hint: Consider a family $\{A_i\}$ of non-empty sets, add a point $p_i \notin A_i$ to each of them, and give a structure of compact topological space to each set $X_i = A_i \cup \{p_i\}$ by taking A_i and $\{p_i\}$ as open sets, besides X_i and $\emptyset$. In the product $\times_I X_i$, show that $\times_I A_i$ is an intersection of a set of closed subsets with the FIP, and apply Tychonoff's theorem.).

15. Prove that the axiom of choice implies that every vector space has a maximal set of linearly independent vectors; and deduce that every vector space has a basis.

16. Prove that the axiom of choice implies the following result. Let A be a set, $c_0 \in A$ and R a relation in A, and assume that A has no maximal element for R. Then there exists a function f with domain ω such that $f(0) = c_0$ and for each $t \in \omega$, $f(t)Rf(t+1)$. This property is known as the *principle of dependent choices*, DC.

17. Assume that the principle DC of the previous exercise holds. Prove that then the following holds: let A be a set and R a relation, $R \subseteq \omega \times A$. Suppose that for each $x \in \omega$, $\{y \in A | xRy\} \neq \emptyset$. Then there exists a map $f : \omega \to A$ with the property that, for every $x \in \omega$, $xRf(x)$ (Hint: Apply DC to some relation in the set $\omega \times A$). This property is known as the *countable principle of choice*, AC(ω).

18. Prove that the principle AC(ω) is equivalent to the following assertion: for every denumerable set s whose elements are pairwise disjoint and non-empty, there is a set containing one element from each of the elements of s.

The universe in the exercises that follow is assumed to be a ZFC-universe.

19. Complete the proof of Proposition 10.5.2.

20. By using Exercise 15 of Chapter 6, justify that the cardinal of the set of all maps $\omega \to \omega$ is the cardinal of $\mathbb{R}$.

21. Prove that if X is an infinite set and S is the set of all the finite subsets of X, then $|S| = |X|$.

22. Suppose V is a ZF-universe. Show that one half of Proposition 10.4.1 and one half of Proposition 10.4.2 are valid, and give a proof in each

case. Moreover, prove that both properties are valid in a ZF-universe where the principle of dependent choices holds (see Exercise 16).

23. Let λ be a cardinal and $\{\kappa_i\}_{i\in\lambda}$ a constant λ-family of cardinals with $\kappa_i = \kappa$. Justify that, as said in the text, $\sum_{i\in\lambda}\kappa_i = \lambda\cdot\kappa$, and $\prod_{i\in\lambda}\kappa_i = \kappa^\lambda$. Deduce that if $\kappa < \mu$ are cardinals, then $\kappa\cdot\lambda < \mu^\lambda$ for any cardinal λ.

24. Prove that for any I-family of cardinals $\{\mu_i\}$, $\kappa^{\sum_I \mu_i} = \prod_I \kappa^{\mu_i}$, for a cardinal κ.

25. Prove $\prod_{t\in\omega, t\neq 0} t = 2^{\aleph_0}$, $\prod_{t\in\omega}\aleph_t = \aleph_\omega^{\aleph_0}$ (For the second part, you may find useful to define for each $g : \aleph_0 \to \aleph_\omega$ an auxiliary function $\varphi : \aleph_0 \to \aleph_0$ with $\varphi(t) =$ the smallest $x \in \omega$ such that: $g(t) \in \aleph_x$ and $x \neq \varphi(z)$ for every $z < t$).

26. Prove that $\aleph_\omega < 2^{\aleph_0} \Rightarrow \aleph_\omega^{\aleph_0} = 2^{\aleph_0}$.

27. In the last sentence of the first part of the proof of J.König's theorem (Proposition 10.5.8), it is asserted that "it is easy to see that f is indeed injective". Give a detailed proof of the fact that f is injective.

28. Let S be a subset of an ordinal α. Show that S is cofinal in α if and only if S is strictly unbounded (i.e., every element of α is strictly smaller than some element of S) or α has a maximum u and $u \in S$.

29. Let λ, κ be infinite cardinals with $\lambda < \kappa$ and κ regular. Prove that, if μ is the smallest cardinal with the property $\alpha \in \kappa \Rightarrow \alpha^\lambda \leq \mu$, then $\kappa^\lambda = \kappa\cdot\mu$.

30. Show that if α, β are ordinals and there exist monotone cofinal maps $\alpha \to \beta$ and $\beta \to \alpha$, then α and β have the same cofinality.

31. Prove that if F is a normal functional (see Exercise 12 of Chapter 8) with domain **Ord** and ordinal values, then $\mathrm{cf}(F(\alpha)) = \mathrm{cf}(\alpha)$ for any limit ordinal α.

32. Prove that the functional H from ordinals to cardinals which assigns $\aleph_\alpha$ to α is a normal functional. Deduce that $\alpha \leq \aleph_\alpha$, but that for each ordinal α there is some $\rho \geq \alpha$ with $\rho = \aleph_\rho$. Prove also that if $\aleph_\alpha$ is an uncountable limit cardinal which is regular, then $\aleph_\alpha = \alpha$.

Part II

Independence Results

11

Countable Universes

11.1 The Metatheory of Sets

It is natural to think that the goal of studying any theory in Mathematics is to find the properties that the objects under study satisfy, and to find proofs that they do satisfy those properties. In our case, we study ZF(C)-universes[1] and thus in the first part of this book we aimed at studying properties that these universes satisfy. For a change, we will be concerned in this second part with properties that ZF(C)-universes do not satisfy; or, more precisely, properties that cannot be proven to hold in every universe, but such that their negation cannot be proven neither.

It is not that one has a morbid interest in looking for elusive properties, those such that there is no proof for it nor for its contrary. The fact is that results of this type imposed themselves by appearing in the development of the theory and providing the answers, perhaps unexpected, to central questions to which most mathematicians had hoped to find a univocal solution.

The most notable of these properties was the continuum hypothesis (CH) which we have already presented in Section 10.5.1. It is such a natural and seemingly concrete question that it was felt at the beginning (by Cantor who first studied it and probably by the rest of mathematicians during perhaps the first third of the 20th century) that it should have a definite solution, whatever difficult it could be to find and prove it. But the true solution came in two steps and was not as presumed. In 1938, the Austrian (but US citizen since 1948) mathematician Kurt Gödel (1906-1978) showed that if V is a ZF-universe, then there is a certain class L of V such that L is a ZF-universe and CH holds in L. Consequently, it is impossible to obtain a proof that the continuum hypothesis is false in every ZF-universe (unless the axioms for ZF-universes are inconsistent, see next paragraph). In 1963, the American Paul Cohen (1934-2007) gave the complementary result that CH cannot be proven to hold in every ZFC-universe. Leaving aside the fact that his new and powerful methods gave rise to an extremely important area of research in set theory, and considering only the concrete consequence referring to CH,

[1]By writing ZF(C)-universes we refer in just one movement both to ZF-universes and to ZFC-universes, following respectively the non-parenthetical version or the parenthetical one.

DOI: 10.1201/9781003449911-11

this was a result of a new type, and fits well our explanations of the first paragraph: a result that states not a property of universes, but the fact that the property cannot be proved.

In order to make the explanation more precise we must introduce two classical terms referring to axiomatic systems. These terms specify properties that one considers to be desirable for the axiomatic systems to have. The first and more crucial of these health rules for a collection of axioms is that of the *consistency* of the axioms. A collection[2] of axioms is called *inconsistent* when a contradiction can be deduced from these axioms; i.e., that a certain assertion P and its contrary $\neg(P)$ can both be proven. Two important consequences of the fact that an axiom system is inconsistent are: (1) There cannot exist a universe of objects, whatever they be, in which the axioms hold; and non-existence here has a strong meaning: we cannot even imagine a universe which could satisfy the axioms as long as we cannot imagine as existing some structure for which a certain property is true and false simultaneously. (2) Any assertion will be a consequence of inconsistent axioms: indeed, if P is an assertion whatever, we may prove P by RAA: assume $\neg(P)$, obtain a contradiction $Q \wedge \neg(Q)$ (it exists by assumption), and deduce P by RAA. On the other hand, a collection of axioms is *consistent* if it is not inconsistent.

The second rule is that of the *independence* of the axioms. The property that each particular axiom is *independent* of the other axioms means that this axiom cannot be deduced from the rest. The interest of the concept is easily understood: if, for instance, when presenting the axiom system for ZF-universes it turned out that axiom 8 follows from axioms 1 to 7 (i.e., if every ZF$^-$-universe is a ZF-universe), then axiom 8 would be useless, and there would be no point in introducing that axiom in the system. In principle one prefers to avoid such situations of dependence. However, finding an independent system of axioms is not always possible; fortunately it is not a compelling requirement for the system to work: having more axioms than necessary can be inelegant, but it is still correct. For example, the axioms for ZF-universes are not independent: if A, B are classes in a universe such that the A and B-instances of the axiom of replacement hold, then the $(A \cup B)$-instance of the axiom is a consequence of them (and the rest of axioms), hence it is not an independent axiom. However, this lack of independence of the axioms has not undesirable consequences. On the contrary, being inconsistent is of course deadly: we would be building on the void if studying a theory with inconsistent axioms.

These notions of consistency and independence find their place into what is known as the *metatheory*; in our case, the metatheory of sets. In considering set theory, we can identify three different levels; and, old-fashioned as this terminology is, the distinction between actual and potential infinity may help in understanding these three levels. The universe of the theory is the first one;

[2]This is an instance where we use "collection" in its general meaning, not in the particular one of collection of objects of the universe V.

we view it as really existing, hence objective and *actually* infinite, with objects that may be infinite too in that actual sense. A second level is what can be called the *theory*. It is formed with the arguments we may construct and the results we may obtain about the objects of the universe. The concept of class belongs to this second level, being a concept created by us in order to assist in the investigation of the universe. Consisting of concepts, arguments or results, this level has an infinity of the potential type. The number of results[3] we may effectively obtain and present is necessarily finite in practice, though there is no limit to the number of results that generations of mathematicians (or machines) can produce.

The third level is that of the metatheory. The metatheory imagines the theory as the **actual** infinite collection of all its results and proofs, assuming it has been completed. And the metatheory studies the theory from this point of view, asking for instance if such or such assertion appears or not in the presumably infinite list of propositions and theorems that form the totality of the theory. Like this, consistency or independence are genuine metatheoretical questions.

It is easy to understand why a proof of consistency of a system of axioms is too frequently difficult to find: we are asked to show that a contradiction (i.e., a conjunction $P \wedge (\neg(P))$) cannot be proved; and this is a property that, indirectly, concerns all the propositions of the theory, because we are saying that none of them is a contradiction. However, something less ambitious is usually within reach. An assertion P is said to be *consistent with* $\mathcal{C}$ (or consistent relative to $\mathcal{C}$) when the following holds: if $\mathcal{C}$ is consistent, then the system obtained by adding P to the axioms of $\mathcal{C}$ (it is usual to write $\mathcal{C} + P$ to refer to the result of this addition) is consistent as well. This definition of relative consistency can be extended to collections of assertions, say $\mathcal{C}'$: $\mathcal{C}'$ is consistent with $\mathcal{C}$ if the collection formed with all the assertions of $\mathcal{C}$ and $\mathcal{C}'$ ($\mathcal{C} + \mathcal{C}'$) is consistent, provided $\mathcal{C}$ is consistent. Equivalently: either $\mathcal{C} + \mathcal{C}'$ is consistent, or $\mathcal{C}$ is inconsistent.

The traditional concept of independence mentioned above can be reduced to relative consistency. Because saying that the assertion P is independent from the system of axioms $\mathcal{C}$ means that $\mathcal{C}$ is consistent (otherwise, P, as any other assertion, could be deduced from $\mathcal{C}$) and moreover, $\mathcal{C} + \neg(P)$ is consistent, that is $\neg(P)$ is consistent relative to $\mathcal{C}$. This is because, by RAA, $\mathcal{C} + \neg(P)$ inconsistent would mean precisely that there is a proof of P from $\mathcal{C}$. This reduction of the old meaning of "independent" is one of the reasons for the shifting that this meaning has experienced. Most authors nowadays use the term "independent" in a different sense. Namely, they (and, henceforth, we

[3]We understand "result" in the sense that has been already explained: it should be an assertion which refers to a particular class (possibly the class of all sets) or, at most, to finitely many classes; and which has been shown to hold through a finite proof, hence using finitely many axioms. Like this, a result which appears as stating a property that **all** classes satisfy, is to be seen as an unattainable infinity of results, having a proof for each particular class.

too) call an assertion P *independent* of a system of axioms $\mathcal{C}$ when both $\mathcal{C}+P$ and $\mathcal{C}+\neg(P)$ are consistent relative to $\mathcal{C}$. If $\mathcal{C}$ is consistent, this means that neither $\neg(P)$ nor P can be proved from $\mathcal{C}$. Of course, when $\mathcal{C}$ is inconsistent, both assertions are provable from $\mathcal{C}$. Put otherwise, the axioms $\mathcal{C}$ cannot tell between P and $\neg(P)$: either both are deducible from $\mathcal{C}$ or else none of them is deducible from $\mathcal{C}$. An older and highly expressive word for calling what we now know as an "independent assertion" was *undecidable*, a term coined by Gödel (always relative to the axiom collection $\mathcal{C}$).

Implicitly, we have already encountered proofs of relative consistency in the first part of this book. For instance, we have not justified that the axioms for ZF$^-$-universes (axioms 1 to 7) are consistent. But we know that, if they are, then the axioms for ZF-universes are consistent; that is, axiom 8 is consistent relative to the axioms for ZF$^-$-universes. To show this we use Proposition 9.2.12 and Corollary 9.2.13: these results say that if U is a ZF$^-$-universe, then there is a related ZF$^-$-universe V in which axiom 8 is satisfied, and the proof of the existence and properties of V is entirely set-theoretic, it is developed inside the theory. The only metatheoretic ingredient is then the following derivation: if there is a proof of the negation of axiom 8 from the axioms of ZF$^-$-universes, then every ZF$^-$-universe satisfies the negation of axiom 8, and in particular, V satisfies both the axiom and its negation. This shows that a contradiction is obtained from the axioms for ZF$^-$-universes so that these axioms are inconsistent. In this same way one may prove that the axiom of extension is consistent with the axioms for 6-universes: by Proposition 7.2.5 and Corollary 7.2.7, from any 6-universe we get another 6-universe which satisfies axiom 7; the rest is the metatheoretical straightforward derivation shown above in the similar situation.

However, there is something we have overlooked in the above description of the ideal proof of relative consistency, something that was already hinted at just after the proof of Proposition 9.2.7. In fact, Proposition 9.2.12 that we have used above in showing relative consistency of the regularity axiom has not been proven *inside the theory* of ZF$^-$-universes, since that proof would require to justify every instance of the axiom of replacement and thus it cannot be given in finitely many steps, as a proof should. It is more accurate to say that Proposition 9.2.12 is part of the metatheory: we know that there is a proof of each instance of the axiom of replacement, hence we conclude that V must be a ZF$^-$-universe, but this conclusion is made through a consideration *about* the theory. This difference may have not much importance in these cases, as it merely uses a metatheoretical argument instead of an argument in the theory. But it will play a fundamental part in some of the developments to come, hence it is worth to stop a bit here and carefully consider some aspect of this issue.

Though we cannot give a proof in the theory of the result that V is a ZF$^-$-universe, we do have a proof that V satisfies all axioms of ZF$^-$-universes other than replacement; and that, moreover, given any particular class C, it also satisfies the C-instance of replacement. This of course can be extended to finitely many classes, so there is indeed a proof, for any class sequence $\mathcal{C}$, that V satisfies all the axioms for ZF^--universes other than replacement, as well as all the C_i-instances of replacement for the classes C_i in the sequence $\mathcal{C}$. Now we may recover our proof that axiom 8 is consistent relative to the axioms for ZF^--universes: imagine there is a proof of the negation of axiom 8 for ZF^--universes; such a proof, being finite, uses only finitely many axioms, thus finitely many instances of replacement, but we have a proof that V satisfies all those instances of replacement, hence the negation of axiom 8 has to be valid in V. The argument is now completed as in the previous version: we know that axiom 8 is valid in V by Corollary 9.2.13, and we thus get a contradiction from the axioms for ZF^--universes.

We present in this part of the book two independence results. That is, for an assertion P, we prove that both P and its negation $\neg(P)$ are consistent with the system of axioms. These two independence results refer to the continuum hypothesis CH, which is independent (or undecidable) from the axioms of ZFC-universes; and to the axiom of choice (AC, for short) which is independent from the axioms of ZF-universes. Gödel proved that AC and CH are consistent relative to ZF and ZFC respectively (equivalently, $\neg$(AC) and $\neg$(CH) are not provable in ZF and ZFC); Cohen showed later that the negations of AC and of CH are consistent with ZF and ZFC, respectively.

11.2 Extensions and Reliability

The principal, but impossible, aim for the rest of this chapter is to get from a ZF(C)-universe, a denumerable set which is a ZF(C)-universe itself. But we succeed at a more modest quest: for any finite collection of the ZFC-axioms, there is a denumerable transitive set which satisfies those axioms. Except otherwise stated, V will be a ZFC-universe throughout this chapter. For the rest of the book, we accept property NNS. So, "finite" has the intuitive usual meaning and this will make our arguments easier to follow, though it is not strictly necessary for getting the main results.

In our search for a denumerable set which is a ZFC-universe, we proceed by successive steps. The first one is simple: we just want, starting with any set X, find an apt set (Definition 7.3.4 introduced the concept of an apt set) $F(X)$ including X, and including also the set ω and all its elements. In order to explain the idea for the proof we present an example. Suppose that our starting set is $X = \emptyset$. We are going to add elements to X until we get a set $F(X)$ with the desired properties. To begin, we add all the elements of $s(\omega)$ because, as mentioned, they must belong to $F(X)$. $s(\omega)$ is transitive, but it is not an apt set because it is not closed for pairs: $n, m \in s(\omega)$, but $\{n, m\}$, $\langle 0, \omega\rangle$ are not natural numbers. Our first step here is to add to $s(\omega)$ all the ordered pairs $\langle x, y\rangle$ with $x, y \in s(\omega)$ (the unordered pairs, like $\{0, \omega\}$ will be obtained in a second step). But still $s(\omega) \cup (s(\omega) \times s(\omega))$ is not an apt set. Among others, we have now new elements like $\langle \omega, n \rangle, \langle m, k \rangle$, while the ordered pair $\langle \langle \omega, n \rangle, \langle m, k \rangle \rangle$ formed with these two objects is not an element of our set. So, we need to repeat the step of adding all ordered pairs of objects in our set, and all the elements of arbitrary pairs: in this way, $\{0, \omega\}$, being an element of $\langle 0, \omega \rangle$ will now be added. And then the same construction has to be repeated again and again; only after infinitely many such steps (which we of course must obtain by using natural recursion) will we get an apt set.

Lemma 11.2.1. *There exists a functional class F with $D(F) = V$ and the following property: given $X \in V$, $F(X)$ is the smallest apt set such that $X \cup s(\omega) \subseteq F(X)$. Moreover, $|F(X)|$ is the smallest infinite cardinal that is $\geq |X|$.*

Proof. Firstly, we prove that there exists a functional class G with domain V and such that for every $Z \in V$ the following holds: $Z \subseteq G(Z)$, $|Z| = |G(Z)|$ (if Z is infinite) and if $a, b, c \in Z$ with $c = \{x, y\}$, then $\langle a, b \rangle, x, y \in G(Z)$. To this end, we consider the class $\mathcal{L}$ whose elements are all pairs $\{a, b\}$ (with $a \neq b$ or not). Then we define

$$G(Z) = (\cup(\mathcal{L} \cap Z)) \cup Z \cup (Z \times Z)$$

This is obviously a set and G is a functional. By this definition, every ordered pair of elements of Z and every component of a (unordered) pair belonging to Z are elements of $G(Z)$. On the other hand, each of the three members of the union that gives $G(Z)$ has cardinal $\leq |Z|$ if Z is infinite (by Propositions 10.2.9 and 10.3.3). Therefore $|Z| = |G(Z)|$ in this case. Of course, if Z is finite, then $G(Z)$ is finite and $|Z| \leq |G(Z)|$.

Next we prove the existence of $F(X)$ for each $X \in V$, by using natural recursion: $F_0(X) = X \cup s(\omega)$, $F_{n+1}(X) = G(F_n(X))$. Finally, $F(X) = \cup_{n \in \omega} F_n(X)$. This gives a functional F such that $F(X)$ is a countable union of sets having cardinal $|X|$ in the infinite case. Thus $|F(X)| = |X|$ or $|F(X)| = \aleph_0$ if X is finite.

From the description of $G(Z)$, it is plain that every apt set $Y \supseteq X \cup s(\omega)$ satisfies $F_1(X) = G(F_0(X)) \subseteq Y$, because Y is closed under taking ordered pairs and components of unordered pairs. By the same reason $F_n(X) \subseteq Y \Rightarrow F_{n+1}(X) \subseteq Y$; therefore $F(X) \subseteq Y$. To see that $F(X)$ is apt, suppose first that $\{a, b\} \in F(X)$; then $\{a, b\} \in F_n(X)$ for some n, hence $a, b \in F_{n+1}(X) \subseteq F(X)$. Similarly, if $a, b \in F(X)$, then $a, b \in F_n(X)$ for some n and $\langle a, b \rangle \in F_{n+1}(X)$, from which it follows that $\{a, b\} \in F_{n+2}(X) \subseteq F(X)$. □

The second step is an important improvement of the first lemma, finding some $C \supseteq X$ which is not only apt, but also extensional and making a given class to be C-reliable. Recall Definitions 7.3.1 and 7.3.12 where the concepts of extensional set and reliable class appeared for the first time. To present the idea of the proof that follows, let us choose the same example given before Lemma 11.2.1. So, let $X = \emptyset$ and take $A = \mathcal{M}$, the membership class. We want to extend X now to an extensional apt set C including $s(\omega)$ and such that A is C-reliable, so we start with $s(\omega)$. A is not $s(\omega)$-reliable: $\omega \in s(\omega) \cap \mathrm{D}(A)$, because, for instance $\langle \omega, \omega+\omega \rangle \in A$, but $\mathrm{D}(A \cap s(\omega))$ does not contain ω, as there is no ordered pair $\langle \omega, x \rangle \in s(\omega)$ with $\omega \in x$. So, we have to add to $s(\omega)$ at least one element x (for instance $\omega+1$) with the property $\omega \in x$. In general, we have to do this with every element of the starting set.

$s(\omega)$ happens to be extensional. But if we start adding elements to satisfy the other properties, the bigger set may cease being extensional. Imagine we have added to $s(\omega)$ the elements $\omega+1$ and $\omega+\omega$, but none of the ordinals $\omega+n$ with $1 < n < \omega$. Now, $\omega+1, \omega+\omega$ have the same visible elements, but they are not the same, so the set we have constructed is not extensional. To remedy this, we must include other elements from $(\omega+\omega) \setminus (\omega+1)$, for instance, $\omega+2$. In general, if $u \nsubseteq v$, it will be necessary to add elements of $u \setminus v$ to our set. On the other hand, if we complete an updating of our set with the above additions, we must also apply the functional F of the preceding lemma to obtain an apt set. As was the case with Lemma 11.2.1, these additions have to be repeated infinitely many times to reach our objective.

A technical point is worth mentioning. If we could effectively choose just one element from each of the sets $u \setminus v$ or $(X_n \cap \mathrm{D}(A)) \setminus \mathrm{D}(X_n \cap A)$ in each of the steps we have described, then the condition about the cardinality of the new apt extensional set given in the statement would be fulfilled. However, this is only possible after we have limited the domain where these new objects have to be taken, so that this domain is a set; in principle, $\mathrm{D}(A)$ is a class whatever, so we have no clear rule as to how to choose elements in there. It is for this reason that the proof that follows has two parts; and the first part serves only to provide us with that necessary limitation of the domain.

Lemma 11.2.2. *Let A be a V-class, X a set. There exists an apt and extensional set $C \supseteq X \cup s(\omega)$, such that A is C-reliable and $|C|$ is the smallest infinite cardinal which is $\geq |X|$.*

Proof. We first show the existence of a set Y satisfying exactly the same conditions as those in the statement for C except possibly that of the cardinality. We then use this set Y to obtain the set C as in the statement.

Let us define the function f with domain $\mathrm{D}(f) = X \cap \mathrm{D}(A)$: we set $f(x) = \{y | \langle x, y \rangle \in A \wedge (\langle x, z \rangle \in A \Rightarrow \mathrm{rk}(y) \leq \mathrm{rk}(z))\}$. Next, we define the function g with $\mathrm{D}(g) = \{\langle u, v \rangle \in X \times X | u \nsubseteq v\}$ and $g(\langle u, v \rangle) = u \setminus v$.

Note that $f(x), g(\langle u, v \rangle) \neq \emptyset$ for elements $x, \langle u, v \rangle$ in the respective domains.

Let us now set $\hat{X} = X \cup (\bigcup(\mathrm{cD}(f))) \cup (\bigcup(\mathrm{cD}(g)))$ so that $\hat{X}$ is a set and $X \subseteq \hat{X}$. Then we define $H(X) = F(\hat{X})$, F being the functional constructed in Lemma 11.2.1. $H(X)$ is an apt set that satisfies the properties: (1) $\mathrm{D}(A) \cap X \subseteq \mathrm{D}(A \cap H(X))$, and (2) $u \nsubseteq v$ for $u, v \in X$ implies $u \cap H(X) \nsubseteq v$. (2) holds because $u \setminus v \subseteq \hat{X} \subseteq H(X)$; for (1), note that if $x \in \mathrm{D}(A) \cap X$, then there is $y \in \hat{X}$ such that $\langle x, y \rangle \in A$. Since $x, y \in H(X)$ and $H(X)$ is apt, $\langle x, y \rangle \in H(X) \cap A$. Moreover $X \cup s(\omega) \subseteq H(X)$ and H is a functional class defined for every set X.

By natural recursion, we define $Y_0 = X$ and $Y_{n+1} = H(Y_n)$. As we have seen, (1) $\mathrm{D}(A) \cap Y_n \subseteq \mathrm{D}(A \cap Y_{n+1})$; and (2) if $u, v \in Y_n$, then $u \nsubseteq v \Rightarrow u \cap Y_{n+1} \nsubseteq v$. Next, we take $Y = \cup_{n \in \omega} Y_n$ and thus Y is an apt set because each Y_n is apt. Also A is Y-reliable: the property $\mathrm{D}(A) \cap Y \subseteq \mathrm{D}(A \cap Y)$ follows from (1), and the reverse inclusion is simple because Y is semi-transitive. Finally, Y is extensional: if $a, b \in Y$ with $a \nsubseteq b$ then $a, b \in Y_n$ (for some n) so that $a \cap Y_{n+1} \nsubseteq b$, hence $a \cap Y \nsubseteq b$. It follows immediately that $a \cap Y = b \cap Y \Rightarrow a = b$ and Y is extensional.

We now need to find C with the same properties but with the additional condition on $|C|$. For this we will use a fixed well-ordering of the set Y, and we parallel the previous construction from the start with sets C_n replacing the Y_n. As before, we take $C_0 = X$, and show how to recursively construct the sequence of the sets C_n with $C_n \subseteq C_{n+1}$ so that: each C_n satisfies the condition on the cardinal; $C_n \subseteq Y_n$; and for $n \geq 1$, C_n is apt and satisfies also (1) and (2).

Given C_n with $C_n \subseteq Y_n$, we define functions f_1, g_1 with $\mathrm{D}(f_1) = \mathrm{D}(A) \cap C_n$ and $\mathrm{D}(g_1) = \{\langle u, v\rangle \in C_n \times C_n | u \nsubseteq v\}$. Specifically, for $x \in \mathrm{D}(A) \cap C_n \subseteq \mathrm{D}(A) \cap Y_n \subseteq \mathrm{D}(A \cap Y_{n+1})$, $f_1(x)$ is the minimum (in the well-ordering of Y) of the set of those y such that $\langle x, y\rangle \in Y_{n+1} \cap A$. Similarly, for $u \nsubseteq v$ and $u, v \in C_n \subseteq Y_n$, $g_1(\langle u, v\rangle)$ is the minimum $y \in Y_{n+1}$ such that $y \in u \setminus v$. Then we let $C'_n = C_n \cup (\mathrm{cD}(f_1)) \cup (\mathrm{cD}(g_1))$ and $C_{n+1} = F(C'_n)$. Like this, C_{n+1} is apt and satisfies properties (1) and (2) above – with C_n, C_{n+1} substituted for Y_n, Y_{n+1}. Also, C'_n is the union of three sets each of them with cardinal $\leq |C_n|$ in the infinite case (the finite case can only happen for $n = 0$ and is trivial), hence it satisfies the condition on the cardinal provided C_n does. And then $C_{n+1} = F(C'_n)$ has the same cardinal by Lemma 11.2.1. Finally, $C'_n \subseteq Y_{n+1}$ by construction; and $C_{n+1} \subseteq Y_{n+1}$ again by Lemma 11.2.1 because Y_{n+1} is apt. Therefore all the conditions claimed for C_{n+1} are fulfilled.

The rest is as above: we define $C = \cup_{n \in \omega} C_n$ and still C, as Y, is apt and extensional with A being C-reliable, but now, C is the countable union of sets with the smallest infinite cardinality $\geq |X|$, hence $|C| = |X|$ (when X is infinite). When X is finite, then C is denumerable which proves the condition of the statement. □

We may easily extend this property to finitely many classes A_i.

Proposition 11.2.3. *Let $A_1, \ldots, A_k$ be finitely many V-classes, X a set. There exists an apt and extensional set $C \supseteq X \cup s(\omega)$, such that $A_1, \ldots, A_k$ are C-reliable and $|C|$ is the smallest infinite cardinal which is $\geq |X|$.*

Proof. The proof is basically that of Lemma 11.2.2, the only difference being that, in the construction of $\hat{X}$ and $H(X)$, we need to define functions $f_1, \ldots, f_k$, one for each class A_i, instead of the function f of the lemma. Each of these functions is defined in the same way, with $f_i(x) = \{y | \langle x, y\rangle \in A_i \wedge (\langle x, z\rangle \in A_i \Rightarrow \mathrm{rk}(y) \leq \mathrm{rk}(z))\}$. Then $\hat{X} = (\cup_{i=1}^{k} (\cup \mathrm{cD}(f_i))) \cup (\cup \mathrm{cD}(g)) \cup X$ and we obtain that $\mathrm{D}(A_i) \cap X \subseteq \mathrm{D}(A_i \cap H(X))$ for each index i. The rest of the proof is as in the lemma, and it is left as an exercise. □

Corollary 11.2.4. *Let V be a ZFC-universe, $X \in V$ a countable set, $\mathcal{C}$ a finite collection of V-classes. Then there exists a denumerable apt and extensional set $M \in V$ such that $X \cup s(\omega) \subseteq M$ and each $C \in \mathcal{C}$ is M-reliable.*

The next result is another corollary to Proposition 11.2.3. But it has a special interest, so we emphasize it as an independent statement.

Proposition 11.2.5. *Let V be a ZFC-universe, $X \in V$ a countable set. Let $B_1, \ldots, B_n$ be a class sequence of V. Then there exists a denumerable apt and extensional set $M \in V$ such that $X \cup s(\omega) \subseteq M$, the class sequence $B_1, \ldots, B_n$ is M-absolute and each B_i is M-reliable.*

Proof. Let $x_1, \ldots, x_r$ be the terms of the class sequence $B_1, \ldots, B_n$ which are introduced by rule 1, i.e., as sets; and let $X' = X \cup \{x_1, \ldots, x_r\}$. By Corollary 11.2.4, there is a denumerable apt and extensional set $M \in V$ with $X' \cup s(\omega) \subseteq M$ and such that all the classes B_i are M-reliable. Thus $B_1, \ldots, B_n$ is a class sequence over M, and by Proposition 7.3.13, the sequence $B_1, \ldots, B_n$ is M-absolute. □

11.3 ZF(C)*-Universes

Definition 11.3.1. A universe U will be called a *ZF(C)*-universe* if U satisfies all the axioms for ZF(C)-universes other than the replacement axioms.

From now on, we will have to deal with more than one universe, typically a given ZFC-universe V and an apt and extensional V-class B. As we know from discussions in Chapter 7, constructions on sets, like the union or the power set of a set, may differ when the construction is carried out in B or in V, even though their definition is the same. We already saw that the way to handle this issue is to consider the class sequence that gives the construction in study, and take the relativization of that sequence. This is especially effective for neat classes (Definition 3.1.3): a neat class sequence is completely determined by the labels of its terms. Thus a given sequence of labels specifies the construction from membership and equality classes of a neat V-class C by means of the operations listed in Definition 3.1.2; and the same sequence of labels specifies the construction of the relativized class C^B exactly with the same operations in the same order, but in the (apt and extensional) universe B. Therefore C and C^B represent the same construction, each one in its own universe. We must bear this in mind already in the observations that follow.

The axioms of the ZFC-theory may be seen as assertions about some relations between neat classes (assuming that the universe is a 2-universe satisfying the extension axiom, so that speaking of classes is meaningful and every object is a set). For example, let us write in this form axioms $3, 4, 8, 9$ and the instances of the axioms of replacement.

(1) Unions. Let $C_1 = \{\langle x, z\rangle | \forall y\ (y \in z \Leftrightarrow (\exists u\ (u \in x \wedge y \in u)))\}$. In an abbreviated form, $C_1 = \{\langle x, z\rangle | z = \cup x\}$. The axiom of unions (for a universe V) is the equation $\mathrm{D}(C_1) = V$; that is, for every set, there is an object which is its union. For the supposed class B above, the axiom holds when $\mathrm{D}(C_1)^B = B$.

(2) Power set. Take $C_2 = \{\langle x, z\rangle | z = \mathbb{P}(x)\}$ (with the usual definition for the power set). Axiom 4 is the assertion $\mathrm{D}(C_2) = V$.

(4) Regularity. Let $C_3 = \{x | \exists y\ (y \in x \wedge (y \cap x = \emptyset))\}$. Axiom 8 is the equation $C_3 = V \setminus \{\emptyset\}$.

(5) Choice. $C_4 = \{\langle x, f\rangle | f$ is a function, $\mathrm{D}(f) = x \wedge (\forall u \in x\ (f(u) \in u))$; and $C_5 = \{x | \emptyset \notin x\}$.

Axiom 9 is the assertion that $C_5 \subseteq \mathrm{D}(C_4)$.

(6) We finally consider the C-instance of the axiom of replacement, assuming that C is a neat class. Then we define the following two classes.

$$C_6(C) = C_6 = \{\langle u, s\rangle | \forall x \in s\ (\langle u, x, y\rangle, \langle u, x, y'\rangle \in C \Rightarrow y = y')\}$$

$$C_7(C) = C_7 = \{\langle u, s\rangle | \exists z (\forall y (y \in z \Leftrightarrow (\exists x \in s\ (\langle u, x, y\rangle \in C))))\}$$

The C-instance of the replacement axiom is the assertion that $C_6 \subseteq C_7$.

Proposition 11.3.2. *Let V be a ZFC-universe, $X \in V$ a countable set and $\mathcal{C}$ a finite collection of neat V-classes. There exists a denumerable apt and extensional set $B \in V$ such that $B \supseteq X \cup s(\omega)$ and B is a ZFC*-universe, which moreover satisfies each of the C-instances of the axiom of replacement for $C \in \mathcal{C}$.*

Proof. The classes C_1, C_2, C_4, C_5 and the $C_6(C), C_7(C)$ for each $C \in \mathcal{C}$ are finitely many neat classes that can be inserted as members of a neat class sequence (for instance, one may take neat class sequences leading to each of them and glue all these sequences together). By Proposition 11.2.5, there is a denumerable apt and extensional class B with $X \cup s(\omega) \subseteq B$ and such that all the classes in the sequence are B-absolute.

Since B is an apt extensional set, axioms $1, 2, 7$ hold in B by Propositions 7.3.2 and 7.3.5. By the construction, $\omega \in B$ and $\omega \subseteq B$. This entails that ω is an inductive set of B, because for each $n \in \omega$, all elements of n are B-visible and thus $s(n) \in \omega$ is the same in B. Regularity can also be seen directly: let $x \in B$ with $\emptyset \neq x$. By extension, x contains some visible elements, hence $x \cap B$ is not empty; by regularity in V, there is $y \in x \cap B$ such that $y \cap x \cap B = \emptyset$. This entails that $y \cap x = \emptyset$ when we view that intersection in B, and this proves the axiom. We prove now the rest of the axioms.

Axiom 3. The class C_1 is B-absolute and B-reliable. Hence $\mathrm{D}(C_1)^B = \mathrm{D}(C_1^B) = \mathrm{D}(C_1 \cap B) = \mathrm{D}(C_1) \cap B = B$ because $\mathrm{D}(C_1) = V$ as V is a ZFC-universe. Thus $\mathrm{D}(C_1)^B = B$ and this proves the axiom for the universe B.

Axiom 4. $\mathrm{D}(C_2)^B = \mathrm{D}(C_2) \cap B = B$, as in the preceding item.

Axiom 9. $C_5^B = C_5 \cap B \subseteq \mathrm{D}(C_4) \cap B = \mathrm{D}(C_4 \cap B) = \mathrm{D}(C_4^B)$ by the same reasons above. But this is the condition for satisfaction of axiom 9, as seen above.

C-instance of axiom 5. $C_6^B = C_6 \cap B \subseteq C_7 \cap B = C_7^B$; so $C_6^B \subseteq C_7^B$, which is the condition for the axiom. $\square$

The result of Proposition 11.3.2 may be obtained also for a countable **transitive** set M. But we need first the following lemma.

Lemma 11.3.3. *Let V be a ZFC-universe, and $B \in V$ an apt and extensional non-empty set. The membership relation in B is extensional and well-founded, so let π denote the isomorphism $\pi : B \to M$ of Theorem 9.3.9, where $M = \pi[B]$. The following properties hold.*

(i) If $\langle x, y\rangle \in B$, then $\pi(\langle x, y\rangle) = \langle \pi(x), \pi(y)\rangle \in M$. Moreover, any ordered pair of M is of that form.

(ii) If A is a neat class of V, then $D(\pi[A^B]) = \pi[D(A^B)]$.

(iii) If A is a neat class of V, then $\pi[A^B] = A^M$.

Proof. (i) Recall that $\pi(x) = \pi[x \cap B]$ for any $x \in B$. Then, if $x, y \in B$, then $\{x, y\} \in B$ by the hypothesis; and $\pi(\{x, y\}) = \pi[\{x, y\}] = \{\pi(x), \pi(y)\}$. This immediately implies the property for the ordered pairs.

(ii) Let $u \in \mathrm{D}(\pi[A^B])$ so that $\langle u, v\rangle \in \pi[A^B]$ and $u = \pi(x), v = \pi(y)$ with $\langle u, v\rangle = \pi(\langle x, y\rangle)$ by (i). Thus $x \in \mathrm{D}(A^B)$ and $u \in \pi[\mathrm{D}(A^B)]$. The converse is equally straightforward.

(iii) We prove this by finite induction: we consider a neat class sequence $A_1, A_2, \ldots, A_k = A$ leading to A and show that (iii) holds for each item in the sequence, under the assumption that it holds for the previous items.

Let A_i be the membership or equality relation; by (i) and the fact that π is an isomorphism for the membership relation, we see that $\pi[A_i^B] = A_i^M$ in these cases. Application of rule 3 for the formation of classes keeps the property true, because clearly $\pi[A_j^B \setminus A_k^B] = \pi[A_j^B] \setminus \pi[A_k^B] = A_j^M \setminus A_k^M$, by the isomorphism and the induction hypothesis. Same arguments and item (i) justify that $\pi[A_j^B \times A_k^B] = \pi[A_j^B] \times \pi[A_k^B] = A_j^M \times A_k^M$ if we accept the property for A_j, A_k. So the property still holds by application of rule 4.

Suppose now that $A_i = \mathrm{D}(A_j)$ and $A_j^M = \pi[A_j^B]$. Thus $A_i^M = \mathrm{D}(A_j^M) = \mathrm{D}(\pi[A_j^B]) = \pi[D(A_j^B)] = \pi[A_i^B]$, by (ii).

Finally, (i) entails that $\pi(\langle x, y, z\rangle) = \langle \pi(x), \pi(y), \pi(z)\rangle$, and the same for pseudo-triples. Therefore, if A_i is obtained from A_j by application of some of the rules 6 or 7 and $\pi[A_j^B] = A_j^M$, it follows that $\pi[A_i^B] = A_i^M$. This finishes the proof. $\square$

Theorem 11.3.4. *Let V be a ZFC-universe. Given any finite collection $\mathcal{C}$ of neat V-classes, there exists a denumerable transitive set M with $s(\omega) \subseteq M$; and which is a ZFC*-universe such that the C-instance of replacement for each $C \in \mathcal{C}$ is satisfied in M.*

Proof. By Proposition 11.3.2, there exists a denumerable apt and extensional set $B \supseteq s(\omega)$ which is a ZFC*-universe in which the C-instances of replacement hold for any $C \in \mathcal{C}$. By Mostowski's theorem (Theorem 9.3.9), we get the transitive denumerable set M through the isomorphism π. Since $\omega \subseteq B$, it follows inductively that $\pi(x) = x$ for any $x \in \omega$, and therefore $\pi(\omega) = \omega \in M$.

We check that the axioms hold in M. Extension and regularity follow from the transitivity of M. The axiom of pairs is an immediate consequence of the proof of Lemma 11.3.3. Infinity also holds because $\omega \in M$ and $\omega \subseteq M$ and we may apply Proposition 7.1.11.

For the remaining axioms, recall the classes C_i defined before Proposition 11.3.2 and which determine the validity of the axioms. These axioms are valid in the universe B, and from this we may deduce the same results for M:

Unions. $C_1^M = \pi[C_1^B] = \pi[B] = M$, applying (iii) of Lemma 11.3.3.

Power set. $C_2^M = \pi[C_2^B] = \pi[B] = M$, by the same reasons.

Choice. $C_5^M = \pi[C_5^B] \subseteq \pi[C_4^B] = C_4^M$.

C-instance of replacement. $C_6(C)^M = \pi[C_6(C)^B] \subseteq \pi[C_7(C)^B] = C_7(C)^M$. □

11.4 Reflection Theorems

The results in the two final sections of this chapter will not be used in the rest of the book.

Theorem 11.3.4 gives transitive countable ZFC*-universes which will be of great interest. But if we drop the countable condition, it is easy to find a huge collection of transitive ZFC*-universes.

Proposition 11.4.1. *Let λ be a limit ordinal with $\omega < \lambda$. Then V_λ is a transitive set which is a ZFC*-universe.*

Proof. V_λ is transitive by Proposition 9.2.2. Thus V_λ satisfies axiom 7 by Proposition 7.3.3, and also axiom 8. We check the rest of axioms except replacement.

Axioms 1 and 2 are obvious. For unions and power set, let $x \in V_\lambda$ so that $x \in V_\mu$ for some $\mu \in \lambda$, as λ is limit. Then $\cup x \subseteq V_\mu$ by transitivity, hence $\cup x \in V_{\mu+1} \subseteq V_\lambda$ and axiom 3 holds by Proposition 7.1.9. Also, $\mathbb{P}(x) \subseteq V_{\mu+1}$ as every subset of x is included in V_μ. Therefore $\mathbb{P}(x) \in V_{\mu+2} \subseteq V_\lambda$ and $\mathbb{P}(x) = \mathbb{P}(x) \cap V_\lambda$; axiom 4 holds again by Proposition 7.1.9.

Since $\omega \subseteq V_\omega$ by Proposition 9.2.3, $\omega \in V_{\omega+1} \subseteq V_\lambda$ and axiom 6 is fulfilled by Proposition 7.1.11. We finally check the axiom of choice: if $\emptyset \notin x \in V_\lambda$, then $x \in V_\mu$ with $\mu \in \lambda$, and there is a choice function f with $\mathrm{D}(f) = x$ and $\mathrm{cD}(f) \subseteq \cup x$. But $x, \cup x \subseteq V_\mu$ and it follows that $f \subseteq V_{\mu+1}$ and thus $f \in V_\lambda$. But f is a choice function for x in V_λ. □

This result suggests the question whether V_λ can be shown to be a ZFC-universe for some special λ. In this direction, a variation on the arguments given for Lemma 11.2.2 will show how we can obtain ZFC*-universes V_λ satisfying any finite collection of instances of the replacement axioms.

Theorem 11.4.2. *Let V be a ZFC-universe and $\mathcal{C}$ a finite collection of V-classes. Given any ordinal α, there exists an ordinal $\rho > \alpha$ such that every class of $\mathcal{C}$ is V_ρ-absolute.*

Proof. As in the proof of Proposition 11.3.2, we may consider that the classes in $\mathcal{C}$ appear in some class sequence $A_1, \ldots, A_n$ and, since there are only finitely many elements of V introduced by rule 1 in the sequence, we may also assume that the sequence is a class sequence over V_α – if necessary, by replacing the original α with a greater one.

Assume w.l.o.g., that $A_1, \ldots, A_k$ are those terms of the sequence such that $\mathrm{D}(A_i)$ is obtained in that sequence by application of rule 5 of Definition 3.1.2. We claim that for each ordinal $\beta \geq \alpha$ there exists an ordinal $\gamma \geq \beta$ such that $\mathrm{D}(A_i) \cap V_\beta \subseteq \mathrm{D}(A_i \cap V_\gamma)$ for $i = 1, \ldots, k$. For each of the classes A_i define a function which to $x \in \mathrm{D}(A_i) \cap V_\beta$ assigns the smallest ordinal $\mu \geq \beta$ such that $x \in \mathrm{D}(A_i \cap V_\mu)$. The (finite) union of the codomains of these functions is a set of ordinals, which is bounded because it is a set. If we take γ as the smallest upper bound of that set, the correspondence $\beta \mapsto \gamma$ will be a functional F between ordinals $\geq \alpha$ satisfying $\mathrm{D}(A_i) \cap V_\beta \subseteq \mathrm{D}(A_i \cap V_{F(\beta)})$ for each index $i \leq k$.

By natural recursion, we define the function g with $g(0) = \alpha$ and $g(n+1) = F(g(n))$. Again $\mathrm{cD}(g)$ is a set of ordinals so $\rho = \bigcup_{n \in \omega} g(n)$ is an ordinal $\geq \alpha$. The relations $\mathrm{D}(A_i) \cap V_{g(n)} \subseteq \mathrm{D}(A_i \cap V_{g(n+1)})$ entail that $\mathrm{D}(A_i) \cap V_\rho \subseteq \mathrm{D}(A_i \cap V_\rho)$; and in fact we have equality because V_ρ is transitive. Then each A_i (for $i = 1, \ldots, k$) is V_ρ-reliable and every A_i for $i = 1, \ldots, n$ is V_ρ-absolute by Proposition 7.3.13. Since we can always replace α with a larger ordinal, the inequality $\alpha \leq \rho$ can be made strict. □

Both this theorem and the next one are simplified versions of the results which are known in the literature as the *reflection theorems.*

Theorem 11.4.3. *Let V be a ZFC-universe and $\mathcal{C}$ a finite collection of V-classes. Given any ordinal α, there exists a limit ordinal $\lambda > \alpha$ such that V_λ is a ZFC*-universe which satisfies the C-instances of the axiom of replacement for each $C \in \mathcal{C}$.*

Proof. Note that if in the proof of Theorem 11.4.2 we define the function g with $g(0) = \max(\alpha, \omega)$ and $g(n+1) = F(g(n)+1)$, then the ordinal $\rho = \bigcup g(n)$ would be a limit ordinal, as it is the union of a strictly increasing sequence of ordinals.

Now, for each $C \in \mathcal{C}$, take the classes $C_6(C), C_7(C)$ as defined before Proposition 11.3.2, and add them to $\mathcal{C}$. By applying Theorem 11.4.2, we get a limit ordinal $\lambda > \alpha, \omega$ such that all those classes are V_λ-absolute. With the same proof of Proposition 11.3.2, each C-instance of the axiom of replacement is satisfied in V_λ. On the other hand, V_λ is a ZFC*-universe by Proposition 11.4.1. □

We give next an application of this result to the metatheory. But a previous comment on proofs will be necessary, in order to make precise a general observation we have already made. We have indeed remarked that any proof of a property about the objects of the ZF(C)-universe V uses only finitely many instances of the axiom of replacement, say the C_i-instances for $C_1, \ldots, C_n$; and these may be seen as forming a class sequence if we want. There is however a difficulty: the argument of that possible proof has to be given for any ZF(C)-universe, while the classes $C_1, \ldots, C_n$ are collections of objects of the particular universe V, and the class C_i of the universe V is generally different from the class C_i of another universe. Can we speak like that of classes in different universes as if they were the same?

The fact is that we can. Proposition 5.1.10 says that we may assume that the class sequence $C_1, \ldots, C_n$ is a neat class sequence. And we have remarked before Proposition 11.3.2 that neat class sequences represent constructions that can be made in any ZF(C)-universe in the same way. Thus the neat class C of the universe V defines also a well determined class in any other universe; according to Remark 3.2.9, the class is described by a formula which has no parameters, i.e., no elements of any particular universe, and hence it can be seen as "the same" class in every universe – see also item (1) of Remarks 7.3.9.

Theorem 11.4.4. *Assume that the axioms for ZFC-universes are consistent; and that $\mathcal{C}$ is a collection of neat classes with the property that every neat instance of the axiom of replacement is provable from the C-instances for $C \in \mathcal{C}$, along with the rest of axioms. Then $\mathcal{C}$ is not finite.*

Proof. Admit that there exists such a collection $\mathcal{C}$ that is finite. Consequently, there is a formula $P(x)$ stating that x is a limit ordinal and V_x satisfies the C-instances of the replacement axiom for $C \in \mathcal{C}$. By Corollary 3.2.7 the ordinals satisfying $P(x)$ form a class; and this class is not empty by Theorem 11.4.3. Therefore there exists a smallest limit ordinal λ with that property. By the hypothesis and Proposition 11.4.1, V_λ is a ZFC-universe, let us call $V_\lambda = W$.

Since W is a ZFC-universe, we may apply 11.4.3 to W and obtain a limit ordinal $\beta \in W$ such that W_β satisfies the C-instances of replacement for every $C \in \mathcal{C}$. By Propositions 9.2.2 and 9.2.17, the (limit) ordinals of W are (limit) ordinals of V, hence β is a limit ordinal of V. Moreover, one may use induction to show that $W_\alpha = V_\alpha$ for each $\alpha \in \lambda$, because W is closed under V-subsets; hence $\mathbb{P}(V_\alpha) = \mathbb{P}(W_\alpha) = W_{\alpha+1}$. We conclude that $V_\beta = W_\beta$ satisfies the C-instances of replacement for $C \in \mathcal{C}$ and this contradicts the choice of λ. □

This shows that no finite subcollection of the axioms for ZFC-universes can give all the results of the ZFC-theory, if this is consistent.

11.5 Inaccessible Cardinals

Definition 11.5.1. A cardinal κ is said to be a *strong limit cardinal* when the implication $\lambda < \kappa \Rightarrow 2^\lambda < \kappa$ holds.

We note that a strong limit cardinal is a limit cardinal. Because if $\kappa = \rho^+$ then $\rho < 2^\rho$ implies $\kappa \leq 2^\rho$ while $\rho < \kappa$. On the other hand, if the universe V satisfies the generalized continuum hypothesis GCH, then limits are strong limits: if κ is limit and $\lambda < \kappa$, then $2^\lambda = \lambda^+ \leq \kappa$; and $2^\lambda < \kappa$ because κ is limit.

By Proposition 10.6.7, if α is a nonzero limit ordinal, then $\mathrm{cf}(\aleph_\alpha) = \mathrm{cf}(\alpha)$ and thus if $\aleph_\alpha$ is moreover regular, then $\aleph_\alpha = \alpha$ (see Exercise 32 in Chapter 10). This gives an indication that regular limit cardinals are probably very rare: in fact, while $\aleph_0$ is regular (Proposition 10.6.7), there is no other infinite limit cardinal which is known to be regular (and for a good reason, as we shall see). This justifies the interest of the following definition.

Definition 11.5.2. A cardinal is called *weakly inaccessible* when it is an uncountable limit cardinal which is regular. It is *inaccessible* when it is uncountable, regular and a strong limit.

Since strong limits are limit cardinals, an inaccessible cardinal is weakly inaccessible. Under the GCH, weakly inaccessible cardinals are inaccessible. The next result generalizes the fact that the elements of V_ω are finite sets.

Proposition 11.5.3. *Let κ be a regular and strong limit cardinal. For $x \in V$, $x \in V_\kappa \Leftrightarrow x \subseteq V_\kappa \wedge |x| < \kappa$.*

Proof. We show inductively that $\rho < \kappa \Rightarrow |V_\rho| < \kappa$, and this gives the implication to the right because $x \in V_\kappa \Rightarrow x \subseteq V_\rho$ for some $\rho < \kappa$.

Assuming $|V_\rho| < \kappa$, $|V_{\rho+1}| = |\mathbb{P}(V_\rho)| = 2^{|V_\rho|} < \kappa$ by Definition 11.5.1. If ρ is a limit ordinal and $|V_\mu| < \kappa$ for each $\mu \in \rho$, then $|V_\rho| = |\cup_{\mu\in\rho} V_\mu| < \kappa$ by Proposition 10.6.9, since $|\rho| < \kappa$ and κ is regular.

For the converse implication, we consider the rank functional restricted to V_κ. Since $x \subseteq V_\kappa$, this rank function gives a map $g : x \to \kappa$. Also, $|x| < \kappa$ and κ regular imply that $\mathrm{cD}(g)$ cannot be cofinal in κ. Consequently there exists $\alpha \in \kappa$ with $\mathrm{cD}(g) \subseteq \alpha$ and $x \subseteq V_\alpha$. But then $x \in V_{\alpha+1}$ and $x \in V_\kappa$. □

Proposition 11.5.4. *If κ is an inaccessible cardinal, then V_κ is a ZFC-universe.*

Proof. We know from Proposition 11.4.1 that V_κ is a ZFC*-universe. We now prove that if C is a V_κ-class of triples, the C-instance of the replacement axiom is satisfied in V_κ. By Proposition 7.3.6, C is a class of the universe V included in V_κ, and it satisfies the replacement axiom in V. Thus given $u, s \in V_\kappa$ with $C|_{\{u\}\times s}$ functional, we know that $\mathrm{cD}(C|_{\{u\}\times s}) = y \in V$. This induces a surjection $s \to y$ and $|y| \leq |s|$ by Proposition 10.3.4; since $|s| < \kappa$ by Proposition 11.5.3, $|y| < \kappa$. So, $y \subseteq V_\kappa$ implies finally that $y \in V_\kappa$ by the same proposition. □

We will obtain as a consequence another metatheoretic result. But the essence of this consequence is the next proposition which, basically, tells something about the universe.

Corollary 11.5.5. *Let V be a ZFC-universe. Then there is some ZFC-universe $V_1 \subseteq V$ such that V_1 does not contain inaccessible cardinals (of the universe V_1).*

Proof. Suppose to the contrary that every ZFC-universe U included in V has the property that it contains some inaccessible cardinal of U. In particular, this is true for V and hence there is a smallest inaccessible cardinal $\kappa \in V$. By Proposition 11.5.4, V_κ is a ZFC-universe and by our assumption, there is some cardinal $\alpha \in V_\kappa$ which is inaccessible in the universe V_κ. By Proposition 9.2.17, α is an ordinal of V with $\alpha < \kappa$. We shall show that it is also a cardinal of V; and, as such, it is a regular uncountable strong limit, i.e., inaccessible. This obviously yields the contradiction with the choice of κ.

So, let α be an inaccessible, hence regular cardinal of V_κ and assume there is in V a cofinal map $h : \beta \to \alpha$ for some ordinal $\beta \in V$ and $\beta < \alpha$. Then $\beta \in V_\kappa$ and every element of the function h belongs to V_κ hence $h \subseteq V_\kappa$. Moreover $|h| = |\beta| < \kappa$ so that $h \in V_\kappa$ by Proposition 11.5.3. Thus h is a cofinal function $\beta \to \alpha$ in V_κ which is impossible because α is inaccessible in V_κ. It follows that α is a regular cardinal of V. Similarly, α is uncountable in V.

We note finally that for any cardinal $\rho < \alpha$, the set of maps $\rho \to 2$ is the same in V or in V_κ. This is because any such map (in V) belongs to V_κ by the same argument given above for h. In V_κ, $2^\rho < \alpha$ and this gives an injective map, also in V, ${}^\rho 2 \to \alpha$, hence $2^\rho \leq \alpha$ in V. Finally, a possible bijection ${}^\rho 2 \to \alpha$ in V would be a bijection also in V_κ, by the same argument above. Thus $2^\rho < \alpha$ also in V, which proves that α is a strong limit in V. □

The metatheoretic interpretation of that result follows now.

Proposition 11.5.6. *If the axioms of ZFC-universes are consistent, then the following assertion is not provable from the axioms: in every ZFC-universe there is an inaccessible cardinal.*

Proof. Suppose that there is a proof of the assertion from the ZFC axioms. Given a ZFC-universe V, take $V_1 \subseteq V$ a ZFC-universe without inaccessible cardinals, which exists by Corollary 11.5.5. But V_1 contains inaccessible cardinals by the presumed proof, thus we get a contradiction from the ZFC-axioms, which are therefore inconsistent. □

11.6 Exercises

1. Give axioms 6 and 7 in the form of class relations as done before Proposition 11.2.5 for the other axioms.
2. Prove that, if $\omega < \alpha$, then the axiom of infinity holds in V_α.
3. Find which are the axioms of ZFC that are satisfied in V_ω.
4. Prove that the negation of the axiom of infinity is consistent with the rest of the ZFC-axioms.
5. Following the proof of Lemma 11.2.1, show that the lemma still holds true if V is assumed to be a ZF-universe and X a countable set, with $F(X)$ being denumerable in this case (Hint: Proposition 6.6.9 will have to be used).
6. With reference to Lemma 11.2.2, justify that the function f defined in its proof is indeed a set.
7. Prove that Lemma 11.2.2 is valid for ZF-universes if one drops the condition about the cardinality of the set C.
8. Give and prove the adequate version of Proposition 11.2.5 for ZF-universes with X an arbitrary set.
9. Check that the reflection theorem (Theorem 11.4.2) is valid for ZF-universes.

Except stated otherwise, V will now be a ZFC-universe.

10. Complete the proof of Proposition 11.2.3.
11. With reference to the proof of Theorem 11.3.4, check directly that π preserves unions. That is, $\pi((\cup x)^B) = (\cup\pi(x))^M$.
12. Prove that if t is a transitive set and $\alpha = \mathrm{rk}(t)$, then every ordinal $< \alpha$ equals $\mathrm{rk}(u)$ for some $u \in t$ (Hint: Use RAA, choose the smallest $\beta < \alpha$ with $\beta \neq \mathrm{rk}(u)$ for every $u \in t$; prove that there is $x \in t$ such that $\mathrm{rk}(x) > \beta$ and then find a contradiction).
13. Let t be an infinite transitive set. Prove that $|\mathrm{rk}(t)| \leq |t|$.
14. $\mathrm{tc}(x)$ denotes the transitive closure of the set x. For each infinite cardinal κ, we define H_κ as the class of all sets x such that $\kappa > |\mathrm{tc}(x)|$. Prove that $H_\kappa \subseteq V_\kappa$ and that, consequently, each H_κ is a set.
15. Describe the set $H_{\aleph_0}$. Prove that any H_κ is transitive, the set of ordinals belonging to H_κ is κ, and H_κ is closed under subsets.
16. Prove that every set of $H_{\aleph_1}$ is countable in the transitive 3-universe $H_{\aleph_1}$.

17. Prove that if κ is an uncountable regular cardinal, H_κ satisfies all the axioms of ZFC except perhaps the power set axiom.

18. Let λ be a non-zero limit ordinal. Show that each of the instances of the principle of separation (Proposition 5.1.11) is satisfied in V_λ (Hint: Observe that for $x \in V_\lambda$, $\mathbb{P}(x)$ is the same in V or V_λ; and every V_λ-class is a V-class).

19. Prove that the property that every well-ordered set is isomorphic to an ordinal is not true in the ZFC*-universe $V_{\omega \cdot 2}$ (Hint: Use the set $\omega \times 2$ to find the counterexample and mind which are the ordinals in such a universe as $V_{\omega \cdot 2}$).

20. Deduce from Theorem 11.4.4 that, if the ZFC-theory is consistent, it is impossible to have proofs for all the results of the theory from a finite collection of theorems – one says that the theory is not finitely axiomatizable.

21. Find the flaw in the following argument "proving" that ZFC is inconsistent: let $\mathcal{C}$ be the finite collection of neat classes such that the C-instances (for $C \in \mathcal{C}$) of the replacement axioms, along with the rest of axioms except replacement, are enough for the proof of Theorem 11.4.3. By Theorem 11.4.4, there is a smallest λ with V_λ a ZFC*-universe where C-instances of replacement hold when $C \in \mathcal{C}$. Since V_λ satisfies the necessary properties, we may apply Theorem 11.4.3 to the universe V_λ to get an ordinal $\alpha < \lambda$ such that V_α is a ZFC*-universe with the same property as V_λ, a contradiction.

22. Use ordinal recursion to define the cardinals $\beth_\alpha$ for each ordinal α so that $\beth_0 = \aleph_0$, $\beth_{s(\alpha)} = 2^{\beth_\alpha}$ and $\beth_\lambda = \sup(\{\beth_\alpha | \alpha \in \lambda\})$ for λ limit. Justify that the functional with all pairs $\langle \alpha, \beth_\alpha \rangle$ is normal (see Exercise 12 of Chapter 8); and that the GCH holds if and only if $\beth_\alpha = \aleph_\alpha$ for each ordinal α.

23. Compare (for each ordinal α) $\beth_\alpha$ and $|V_\alpha|$. Determine the smallest ordinal α such that $|V_\alpha| = \beth_\alpha$ and justify that this equality holds true from that value of α onwards. Deduce from the preceding exercise that there are cardinals κ such that $\kappa = \beth_\kappa$; and that if κ satisfies that equation, then κ is a strong limit cardinal.

24. Let κ be an inaccessible cardinal. Prove that the classes of pairs $\langle \alpha, \mathrm{cf}(\alpha) \rangle$ and $\langle \alpha, \beth_\alpha \rangle$ (α ranging over ordinals) are V_κ-absolute.

25. Prove that if κ is an uncountable regular cardinal, then $H_\kappa = V_\kappa$ if and only if κ is inaccessible.

26. Again κ is an uncountable regular cardinal. Show that H_κ is a ZFC-universe if and only if $H_\kappa = V_\kappa$. Deduce that the axiom of the power set is not provable from the rest of the axioms for ZFC-universes, if these axioms are consistent.

12

The Constructible Universe of a ZF-Universe

At face value, this chapter is entirely similar to Chapters 7 and 9. In Chapter 7, we added the axiom of extension to the effect of restricting the assumed universe $\mathcal{U}$ to the class of pure sets; in Chapter 9 another limitation of the universe was obtained by means of the axiom of regularity, which reduces the given universe U to the von Neumann universe of well-founded sets. In both cases, a justification for accepting this reduction was that it preserved a fundamental objective of set theory, to give account of classical mathematics. Indeed, the set ω, which we may reasonably see as a version of the natural numbers, is a pure and well-founded set, so those reductions do not change the most basic construction of mathematics, as seen inside set theory. Also, if x belongs to the restricted universe, say V, then $\cup x$ or $\mathbb{P}(x)$ is the same in V or in $\mathcal{U}$, because subsets or elements of well-founded pure sets are also well-founded pure sets. In the present chapter, we will do apparently the same thing: the universe V suffers a new reduction, to the class of the so called *constructible sets* and this may be done through the introduction of a new axiom, the axiom of constructibility. It turns out that ω is constructible, and therefore the basic sets of natural, integer or rational numbers are the same if we consider only the constructible universe.

However, there is an important difference between these axioms. Reducing to the von Neumann universe of sets does not change the set of real numbers (or complex) neither: because the real numbers come from subsets of $\mathbb{Q}$, and all subsets of $\mathbb{Q}$ (and of ω) in the bigger universe $\mathcal{U}$ are elements of the von Neumann universe V. But the axiom of constructibility may change the set of real numbers, because we cannot assure that all subsets of $\mathbb{Q}$ or ω inside the universe V are constructible. This explains why there is a substantial difference, which makes that the constructible universe cannot be considered as the right universe (if there is such thing) but just one possible universe among others; in exchange, this universe is useful to obtain some important results in the metatheory.

DOI: 10.1201/9781003449911-12

12.1 X-Constructible Sets

Unless stated otherwise, V will be a ZF-universe throughout this chapter. The few properties on cardinality that will be used do not require the axiom of choice and hence will be valid in V. We recall that we are assuming property NNS, hence ω is the set of the "natural numbers" $0, s(0), s^2(0), \ldots$; and "finite" has the intuitive meaning.

Definition 12.1.1. Let X be a non-empty set. A finite sequence of sets $c_0, c_1, \ldots, c_n$ together with corresponding labels $e(k), u(k)$ (where each $e(k)$ is a positive integer ≤ 8), will be called an *X-constructible sequence* when the following rules are fulfilled.

(i) If $e(k) = 1$, then $c_k \in X$ and $u(k) = (0, 0)$.

(ii) If $e(k) = 2$, then $u(k) = (i, j)$ with $i, j < k$ and $c_k = \{c_i, c_j\}$.

(iii) If $e(k) = 3$, then $u(k) = (i, j)$ with $i, j < k$ and $c_k = c_i \setminus c_j$.

(iv) If $e(k) = 4$, then $u(k) = (i, j)$ with $i, j < k$ and $c_k = c_i \times c_j$.

(v) If $e(k) = 5$, then $u(k) = (i, i)$ for some $i < k$ and $c_k = \mathcal{M} \cap c_i$ (as always, $\mathcal{M}$ is the membership relation).

(vi) If $e(k) = 6$, then $u(k) = (i, i)$ for some $i < k$ and $c_k = \mathrm{D}(c_i)$.

(vii) If $e(k) = 7$, then $u(k) = (i, i)$ for some $i < k$, and c_k is the set of all triples $\langle a_1, a_2, a_3\rangle$ such that $\langle a_2, a_1, a_3\rangle \in c_i$.

(viii) If $e(k) = 8$, then $u(k) = (i, i)$ for some $i < k$, and c_k is the set of all pairs $\langle a_1, \langle a_2, a_3\rangle\rangle$ such that $\langle a_1, a_2, a_3\rangle \in c_i$.

A set c is called *X-constructible* if it appears in some X-constructible sequence. For each step k in the sequence, if $e(k) = j$ we say that c_k has been obtained by rule j.

A usual term for X-constructible in the literature is *X-definable*, which makes a reference to the notion of definability in a formal language. Rules 2 to 8 may be viewed as functionals which produce an object of V from either an ordered pair or an object of V. These are called *Gödel operations*

Proposition 12.1.2. *Let X be a non-empty set. There is a set* $\mathrm{cns}(X)$ *formed with all those objects $y \in V$ such that y is X-constructible.*

Proof. We start by defining functionals $G_2, \ldots, G_8$ which represent the Gödel operations. Thus $\mathrm{D}(G_i) = V \times V$ for $i = 2, 3, 4$ and $\mathrm{D}(G_i) = V$ for $i = 5, 6, 7, 8$. For example, $G_2(\langle u, v\rangle) = \{u, v\}$ or $G_4(\langle u, v\rangle) = u \times v$. Similarly, $G_6(u) = \mathrm{D}(u)$ and $D_7(u) = \sigma[u \cap T]$ if σ stands for the permutation of triples of rule 7 and T denotes the class of triples.

Next, F will be a functional defined on all sets: $F(A) = A \cup (\cup_{i=2}^{4} G_i[A \times A]) \cup (\cup_{i=5}^{8} G_i[A])$.

By Definition 12.1.1, if all the elements of A are X-constructible, then $F(A)$ consists of X-constructible elements. Now, we define by natural recursion the function H with $H(0) = X$, and $H(n+1) = F(H(n))$. By induction, each

$H(n)$ consists of X-constructible elements. On the other hand, it follows easily, also by induction, that all the elements in an X-constructible sequence of length n belong to $H(n)$. Consequently, $\cup_{n\in\omega}H(n)$ contains precisely all the X-constructible sets; that is, it is $\text{cns}(X)$, which is therefore a set. □

The reader will have noticed that Definition 12.1.1 is reminiscent of the definition of class sequences. This suggests that some properties of X-constructible sets might be obtained in the way used for obtaining the corresponding properties for classes. We see next that this is indeed the case. First, we define, for any given set X, the set $\mathcal{C}(X) = \text{cns}(s(X)) \cap \mathbb{P}(X)$, $s(X)$ meaning the successor set of X, $s(X) = X \cup \{X\}$. It is obvious that, if $X \subseteq Y$, then $\text{cns}(X) \subseteq \text{cns}(Y)$.

Proposition 12.1.3. *Let $X \in V$. If $x, y \in \mathcal{C}(X)$, then $\mathcal{M} \cap x, x \setminus y, x \cap y, x \cup y$ belong to $\mathcal{C}(X)$.*

Proof. $\mathcal{M} \cap x, x \setminus y$ are $s(X)$-constructible by rules 3 and 5. The proof of Proposition 3.1.5 can be easily adapted to show that $x \cap y, x \cup y$ belong to $\text{cns}(s(X))$, since $X \setminus x$ and $X \setminus y$ are $s(X)$-constructible. Finally, it is immediate to see that the four sets are included in X, hence they belong to $\mathcal{C}(X)$. □

Recall that X is an apt set (Definition 7.3.4) if it contains $\emptyset$, is semi-transitive and is closed under the formation of pairs.

Proposition 12.1.4. *Let X be an apt set. If R is a relation and $R \in \mathcal{C}(X)$, then $R^{-1} \in \mathcal{C}(X)$. Also, if $a, b \in \mathcal{C}(X)$, then $D(a)$, $cD(a), a \times b \in \mathcal{C}(X)$.*

Proof. That R^{-1}, $\text{D}(a)$, $\text{cD}(a), \{a, b\}, a \times b$ belong to $\text{cns}(s(X))$ follows easily from rules $2, 4, 5, 6, 7$ as in Proposition 3.1.8 and Corollary 3.1.9. By the hypotheses, $R \subseteq X \times X$ hence $R^{-1} \subseteq X$. Obviously, $\text{D}(a)$, $\text{cD}(a), a \times b \subseteq X$ if $a, b \subseteq X$. □

Proposition 12.1.5. *Let X be an apt set. Then $\mathcal{C}(X)$ is closed under permutations of triples and under conversion of pseudo-triples into triples.*

Proof. Due to rules 7 and 8, this is similar to Propositions 3.1.14 and 3.1.15. □

Proposition 12.1.6. *Let X be a transitive apt set. If $a \in \mathcal{C}(X)$, then $\mathcal{I} \cap a \in \mathcal{C}(X)$ ($\mathcal{I}$ denotes the inclusion relation). Also $\mathcal{E} \cap a \in \mathcal{C}(X)$.*

Proof. Clearly, it will be enough to show that $\mathcal{I} \cap a, \mathcal{E} \cap a$ are $s(X)$-constructible. By the hypothesis on X and rules 1 and 4, $X \times X \in \mathcal{C}(X)$, and it follows by Proposition 12.1.3 that $a' = a \cap (X \times X)$ belongs to $\mathcal{C}(X)$. Since $\mathcal{I} \cap a = \mathcal{I} \cap a'$, we may assume that the elements of a are ordered pairs.

The set of triples $\langle x, y, z \rangle \in a \times X$ such that $z \in x$, is $s(X)$-constructible: it is the intersection of $a \times X$ and the adequate permutation of $[(X \times X) \cap \mathcal{M}] \times X$. Similarly, the set of triples $\langle x, y, z \rangle \in a \times X$ such that $z \notin y$ is $s(X)$-constructible. Taking the domain of the intersection of these two sets, we deduce by the transitivity of X that the set Y of ordered pairs $\langle u, v\rangle \in a$ with $u \not\subseteq v$ is $s(X)$-constructible. Finally, $\mathcal{I} \cap a = a \setminus Y$ and thus it is $s(X)$-constructible.

For $\mathcal{E} \cap a$, we have $\mathcal{E} \cap a = \mathcal{I} \cap a \cap (\mathcal{I} \cap a^{-1})^{-1}$. By the first part of this proof and Propositions 12.1.4 and 12.1.3, $\mathcal{E} \cap a$ is $s(X)$-constructible. □

Proposition 12.1.7. *Let X be transitive. If $a \in \mathcal{C}(X)$, then $\cup a \in \mathcal{C}(X)$.*

Proof. $\cup a \subseteq X$ follows by transitivity. Then $Y = (X \times a) \cap \mathcal{M}$ is $s(X)$-constructible by rules 4 and 5. Its domain is then $s(X)$-constructible by rule 6 and equals $\cup a$. □

12.2 The Constructible Hierarchy

Starting with the ZF-universe V, we use recursion on ordinals to construct a class $L \subseteq V$, which we are going to view as a universe. The construction parallels that of the von Neumann universe from a ZF$^-$-universe U and is the following:

$$L_0 = \emptyset, \quad L_{\alpha+1} = \mathcal{C}(L_\alpha), \quad L_\lambda = \cup_{\beta\in\lambda} L_\beta \quad (\text{for } \lambda \text{ limit})$$

Finally, $L = \cup_{\alpha\in Ord} L_\alpha$. It is called the class of *constructible sets* of the universe V. This allows us to define the c-rank (c for constructible) of any $x \in L$: $\operatorname{crk}(x)$ is the smallest ordinal α such that $x \in L_{\alpha+1}$. The ordinals α are, of course, the ordinals of V.

Example 12.2.1. By definition, $L_0 = \emptyset$. Then $L_1 \subseteq \mathbb{P}(\emptyset) = \{\emptyset\}$. The only element of this set is $\emptyset$ and it is obviously $\{\emptyset\}$-constructible, hence $L_1 = \{\emptyset\}$. As for $L_2 = \mathcal{C}(L_1)$, it must be a subset of $\mathbb{P}(L_1) = \{\emptyset, \{\emptyset\}\}$. These two objects belong to $\operatorname{cns}(s(L_1))$ because both are elements of $s(L_1)$, hence they are obtained by rule 1 of Definition 12.1.1. Therefore $L_2 = \mathbb{P}(L_1)$.

Next, $\mathbb{P}(L_2)$ has four elements, $\emptyset, \{\emptyset\}, \{\{\emptyset\}\}, L_2$. Besides, $s(L_2) = \{\emptyset, \{\emptyset\}, L_2\}$. And $\{\{\emptyset\}\}$ is $s(L_2)$-constructible as it can be obtained by rule 2 from $\{\emptyset\}$. Thus $\mathcal{C}(L_2) = L_3 = \mathbb{P}(L_2)$. The reader will see in the exercises that, likewise, $L_{n+1} = \mathbb{P}(L_n)$, but $L_{\omega+1} \neq \mathbb{P}(L_\omega)$.

We give now several properties of L and the L_α.

Proposition 12.2.2. *For each ordinal α, we have that: (1) $\beta < \alpha \Rightarrow L_\beta \subseteq L_\alpha$; and (2) L_α is transitive. Therefore L is transitive and apt.*

Proof. We prove both assertions simultaneously by ordinal induction. The property is obvious for 0. Let now $\alpha + 1$ a successor ordinal and suppose α satisfies (1) and (2). To prove (1) for $\alpha+1$ it is enough to see that $L_\alpha \subseteq L_{\alpha+1}$. Clearly, $x \in L_\alpha$ implies $x \subseteq L_\alpha$ by transitivity, and x is $s(L_\alpha)$-constructible by rule 1; therefore $x \in L_{\alpha+1}$. For (2), $u \in x \in L_{\alpha+1} \Rightarrow u \in L_{\alpha+1}$ because $x \subseteq L_\alpha \subseteq L_{\alpha+1}$.

For λ limit, we have clearly that L_λ being a union of transitive sets is also transitive. Moreover, L_λ is closed under the formation of pairs and hence L_λ is apt. This shows that L is transitive and apt. □

Proposition 12.2.3. *For each ordinal α, $\alpha \in L$ and crk(α) $= \alpha$.*

Proof. Note that, for any $x \in L$, the smallest ordinal ρ such that $x \in L_\rho$ is a successor ordinal; because if λ is a limit ordinal and $x \in L_\lambda = \cup_{\beta\in\lambda} L_\beta$, then $x \in L_\beta$ for some $\beta < \lambda$. This entails that, for any x, $\mathrm{crk}(x) = \mu$ means that $x \in L_{\mu+1}$ and $x \notin L_\mu$. We thus separate the proof that $\mathrm{crk}(\alpha) = \alpha$ into two: (1) $\alpha \notin L_\alpha$; (2) $\alpha \in L_{\alpha+1}$.

(1) We use induction and assume that (1) holds for every $\rho < \alpha$. Now, if α is a non-zero limit ordinal and $\alpha \in L_\alpha$, then $\alpha \in L_\beta$ for some $\beta < \alpha$; and Proposition 12.2.2 implies that $\beta \in L_\beta$, contradicting the inductive hypothesis. On the other hand, if $\alpha = \beta + 1$ and $\alpha \in L_\alpha$, then $\alpha \subseteq L_\beta$ by the definition of $L_{\beta+1}$. Hence $\beta \in L_\beta$, again a contradiction. Of course, $0 \notin L_0 = \emptyset$.

(2) We use again induction. Clearly $0 \in L_1 = \mathcal{C}(\emptyset)$. If $\alpha = \beta + 1$ and we assume $\beta \in L_{\beta+1} = L_\alpha$, then $\beta \subseteq L_\alpha$. Since $\beta, \{\beta\}$ are $s(L_\alpha)$-constructible by rules 1 and 2, $\alpha = \beta \cup \{\beta\}$ belongs to $\mathcal{C}(L_\alpha)$ by Proposition 12.1.3. Thus $\alpha \in L_{\alpha+1}$.

Let now λ be limit; by the inductive hypothesis $\lambda \subseteq L_\lambda$. By (1), every ordinal of L_λ is $< \lambda$, and thus $\lambda = L_\lambda \cap \mathbf{Ord}$. By the same methods of the proof of Proposition 12.1.6, one may see that the following sets belong to $\mathrm{cns}(s(L_\lambda))$:

$Y_1 = \{\langle x, y, z\rangle \in L_\lambda^3 | y \in x \wedge z \in y \wedge z \notin x\}, \quad X_1 = \mathrm{D}(\mathrm{D}(Y_1))$.
$Y_2 = \mathcal{M}^{-1} \cap (L_\lambda \times X_1), \quad X_2 = \mathrm{D}(Y_2)$.
$Y_3 = \{\langle x, y, z\rangle \in L_\lambda^3 | y \in x \wedge z \in x \wedge z \notin y \wedge y \notin z\}, \quad X_3 = \mathrm{D}(\mathrm{D}(Y_3))$.

Then the intersection X of the three sets $L_\lambda \setminus X_i$ for $i = 1, 2, 3$ is the set of all $x \in L_\lambda$ which are transitive and such that every element of x is transitive and any two elements of x are comparable as to membership. It follows from Proposition 9.2.16 that X is the set of ordinals of V that belong to L_λ, so $X = \lambda$; and X belongs to $\mathcal{C}(L_\lambda)$ because $X \subseteq L_\lambda$. Thus $\lambda \in L_{\lambda+1}$. □

L is moreover a *big class*, in the following sense:

Proposition 12.2.4. *If x is an object of V and $x \subseteq L$, then there exists $y \in L$ such that $x \subseteq y$.*

Proof. Let $x \in V$ and $x \subseteq L$. The restriction to x of the functional that assigns $\mathrm{crk}(u)$ to any $u \in L$ is a function whose codomain is a set of ordinals by the replacement axiom for that functional. Therefore, there is an ordinal μ which is an upper bound for all $\mathrm{crk}(u)$ with $u \in x$. This entails that every $u \in x$ belongs to $L_{\mu+1}$ hence $x \subseteq L_{\mu+1}$. □

The next observation synthesizes many of the properties seen so far by giving them in terms of the class L. We leave the proof as an exercise.

Corollary 12.2.5. *The operations given by rules* 2 *to* 8 *of Definition 12.1.1, when applied to elements of L produce elements of L. The same happens with the operations implicitly defined in Propositions 12.1.3, 12.1.4, 12.1.5, 12.1.6 and 12.1.7.*

12.3 The Constructible Universe

We want to prove that L is a ZFC-universe when V is a ZF-universe.

Some axioms are trivially checked. For instance $\emptyset \in L$ from Proposition 12.2.3. Extension and regularity follow from the transitivity of L. If $a, b \in L$, then $\{a, b\} \in L$ by Corollary 12.2.5. Proposition 12.2.3 shows that $\omega \in L$ and by Proposition 7.1.11 it is inductive, hence L satisfies the axiom of infinity (in fact, L satisfies property NNS by our assumption that V satisfies NNS). Let us look at the remaining axioms.

Proposition 12.3.1. *L satisfies the axioms of unions and power set.*

Proof. Let $X \in L$. By Proposition 12.1.7, $\cup X \in L$ and thus the axiom of unions holds in L by Proposition 7.1.9. By this same proposition, the axiom of the power set holds in L if $\mathbb{P}(X) \cap L \in L$. By Proposition 12.2.4, there is $y \in L$ such that $\mathbb{P}(X) \cap L \subseteq y$ and hence $\mathbb{P}(X) \cap L \subseteq L_\alpha$ for some ordinal α with $X \in L_\alpha$. We may choose some $\beta > \alpha$ such that $L_\alpha \times \{X\}$ belongs to L_β and L_β is apt. Then $\mathcal{I} \cap (L_\alpha \times \{X\})$ belongs to $L_{\beta+1}$ by Proposition 12.1.6. But the domain of this set is $\mathbb{P}(X) \cap L$ and this completes the proof. □

Proposition 12.3.2. *L satisfies the axioms of replacement.*

Proof. Let F be an L-class of triples. We show first that the F-instance of the simple principle of separation, as stated in Proposition 5.1.14, implies the F-instance of the replacement axiom. To this end, assume that Proposition 5.1.14 holds in L for the L-class F, and take $u, s \in L$ such that $F|_{\{u\}\times s}$ is functional. Then $\mathrm{cD}(F|_{\{u\}\times s}) = x$ is an element of V (by replacement in V applied to the V-class F) with the property $x \subseteq L$. It follows that $x \subseteq L_\alpha$ for some α by Proposition 12.2.4 and thus $F|_{\{u\}\times s} = F \cap ((\{u\} \times s) \times L_\alpha)$. By the F-instance of separation, this is a set of L and hence its codomain x belongs to L by Proposition 12.1.4 and Corollary 12.2.5. Thus $x \in L$ and the F-instance of replacement holds in L.

We prove that separation holds by finite induction, showing that for any L-class sequence $C_1, \ldots, C_n$, one has that $C_i \cap s \in L$ for any set $s \in L$.

(i) Let $C_i = c \in L$, a set; then $c \cap s \in L$ by Corollary 12.2.5.

(ii) Suppose $C_i = \mathcal{M} \cap L$. Then $s \in L \Rightarrow C_i \cap s = \mathcal{M} \cap s \in L$ by the transitivity of L and Corollary 12.2.5.

If $C_i = \mathcal{E} \cap L$, then $C_i \cap s = \mathcal{E} \cap s$ and it belongs to L by Proposition 12.1.6 and Corollary 12.2.5.

(iii) Let $C_i = Y \setminus Z$ for classes Y, Z for which the property holds. Then $C_i \cap s = (Y \cap s) \setminus (Z \cap s)$ and $Y \cap s, Z \cap s \in L$ by the induction hypothesis. Thus they belong to some L_α and $C_i \cap s \in L_{\alpha+1}$ by Corollary 12.2.5.

(iv) Let $C_i = Y \times Z$ for classes Y, Z for which the principle of separation holds. Take $Y' = \mathrm{D}(s) \cap Y$ and $Z' = \mathrm{cD}(s) \cap Z$. Then $C_i \cap s = (Y' \times Z') \cap s$. But $Y', Z' \in L$ by the inductive hypothesis and Corollary 12.2.5, and so does $Y' \times Z'$. Therefore $C_i \cap s \in L$ by Corollary 12.2.5.

(v) Let $C_i = \mathrm{D}(Y)$ for Y an L-class which satisfies separation. Define the functional F consisting of all pairs $\langle x, \alpha \rangle$ with $x \in C_i$ and α the smallest ordinal such that there exists $y \in L$ with $\mathrm{crk}(y) = \alpha$ and $\langle x, y \rangle \in Y$; thus $\mathrm{D}(Y) = \mathrm{D}(F)$. By applying simplified replacement (Proposition 5.1.4) in V to the functional F and the set $s \in L$, we obtain a function with domain $\mathrm{D}(Y) \cap s$ whose codomain is a set (in V) of ordinals. If α is an upper bound of this set, which exists by Proposition 8.3.8, then for each $x \in \mathrm{D}(Y) \cap s$ there exists $y \in L_{\alpha+1}$ such that $\langle x, y \rangle \in Y$. By the inductive hypothesis $Y \cap (s \times L_{\alpha+1})$ belongs to L and its domain is $\mathrm{D}(Y) \cap s$, hence $C_i \cap s \in L$.

(vi) Finally, let C_i be obtained from an L-class Y by some of the rules 6, 7 of Definition 3.1.2, so we may write $C_i = \sigma[Y]$ for the transformation σ of triples which gives rule 6 or 7. Let $x = \sigma^{-1}[s]$ so that $\sigma[x]$ is the set of triples of s; by Proposition 12.1.5 and Corollary 12.2.5, $x \in L$. Now, $C_i \cap s = \sigma[Y] \cap s = \sigma[Y] \cap \sigma[x] = \sigma[Y \cap x]$. But $Y \cap x \in L$ by the inductive assumption, and thus $C_i \cap s$ belongs to L by the adequate rule, 7 or 8, of Definition 12.1.1. □

Corollary 12.3.3. *The class L of constructible sets of the ZF-universe V is a ZF-universe. The ordinals in the universe L are the same ordinals of V.*

Proof. That the axioms hold in L has just been shown. Every ordinal of V belongs to L by Proposition 12.2.3. Besides, the class of ordinals is L-absolute by Proposition 9.2.17, since L is transitive and a ZF-universe. Thus every ordinal of L is an ordinal of V and vice versa. □

In Chapter 7, we saw that (Proposition 7.2.6) if U is the class of pure sets of the 6-universe $\mathcal{U}$, then U is a 6-universe whose class of pure sets is precisely U. The similar result about the class of well-founded sets in a ZF$^-$-universe was seen in Proposition 9.2.12. We show now the analogous result for the class L of constructible sets in a ZF-universe.

Proposition 12.3.4. *Let L be the class of constructible sets of a ZF-universe V. Then L is also the class of constructible sets of the universe L, i.e., $L = L^L$.*

Proof. We have seen in Corollary 12.3.3 that L and V have the same ordinals. We are now going to prove by ordinal induction that for each ordinal α,

$L_\alpha = L_\alpha^L$, where the second set identifies the αth-step of the constructible hierarchy obtained when taking L as the universe. The case $\alpha = 0$ is trivial.

It is an easy exercise to check that the Gödel operations, when applied to elements of a set $X \in L$, give the same result regardless of the universe where the operation is considered, L or V. Therefore $\text{cns}(X)^L = \text{cns}(X)$ and hence $\mathcal{C}(X)^L = \mathbb{P}(X) \cap L \cap \text{cns}(s(X)) = \mathcal{C}(X) \cap L$ – note that $\mathbb{P}(X) \cap L$ is the power set of X in the universe L, as seen in Proposition 12.3.1. By the inductive hypothesis, $L_\alpha^L = L_\alpha$, and then $L_{\alpha+1}^L = \mathcal{C}(L_\alpha) \cap L = L_{\alpha+1} \cap L = L_{\alpha+1}$. By transitivity, the union of a set or class of elements of L is the same viewed in V or in L, and hence $L_\lambda^L = L_\lambda$ holds also by induction for limit λ. The same reason gives now $L^L = L$. □

As with the aforementioned examples (pure sets, well-founded sets), we may introduce a new "axiom" (but this is not an axiom of the ZFC-theory) which reduces the universe to the constructible sets.

Axiom L (of constructibility) Every set of the universe is constructible.

Briefly, axiom L asserts $L = V$. As in those previous occasions, we give a name to the ZF-universes that satisfy the axiom of constructibility: they will be identified as *ZFL-universes*. And we add here a metamathematical consequence of the previous results: the statement of the axiom L is consistent with the axioms for ZF-universes. Because if the negation of axiom L is deduced from those axioms, then we reach a contradiction: given a ZF-universe V, the class $L \subseteq V$ has to satisfy axiom L, by Proposition 12.3.4; on the other hand, L must satisfy also the negation of axiom L by the assumption, because L is a ZF-universe by Corollary 12.3.3. Hence the ZF-theory would be inconsistent.

12.4 Constructibility Implies Choice

In this section, we are going to prove the title's result: the statement of the axiom of choice holds in any ZFL-universe V. By Proposition 10.3.5, it will suffice to show that every set of V admits a well-ordering. Indeed, we prove something stronger. But first, a lemma.

Lemma 12.4.1. *Let V be a ZFL-universe and let C be the class of all ordered pairs $\langle X, W\rangle$ where X is a non-empty set and W is a (strict) well-ordering of X. There is a functional class K with $D(K) = C$ and $cD(K) \subseteq C$, such that $K(\langle X, W\rangle) = \langle cns(X), W'\rangle$ and $W \subseteq W'$.*

Proof. Recall the functional F given in the proof of Proposition 12.1.2: $F(A) = A \cup (\cup_{i=2}^4 G_i[A \times A]) \cup (\cup_{i=5}^8 G_i[A])$, with the G_i denoting the Gödel operations. We are going to show that there is a functional K_0 with domain C giving $K_0(\langle A, W\rangle) = \langle F(A), W_0\rangle \in C$ and $W \subseteq W_0$.

Let us denote the strict well-ordering W of A as $<$, and let us write $A_2, A_3, \ldots, A_8$ for the sets of all objects obtained from elements of A by the corresponding Gödel functional $G_2, G_3, \ldots, G_8$ respectively; that is, $A_2 = \{\{a,b\}|a,b \in A\}$, $A_3 = \{a \setminus b|a,b \in A\}$, $A_5 = \{a \cap \mathcal{M}|a \in A\}$ and so on. Then we set $B_1 = A$ and $B_{i+1} = B_i \cup A_{i+1}$ for $i = 1, \ldots, 7$, so that $B_8 = F(A)$. We define now a well-ordering of B_{i+1} which extends the order of B_i, with the ordering in B_1 being the given order $<$. By repetition, this will give a well-ordering of $B_8 = F(A)$. In each case, the order of B_{i+1} consists of the union of the previous order in B_i, the order we are to define in $B_{i+1} \setminus B_i$ plus all the ordered pairs $\langle x, y\rangle$ with $x \in B_i, y \in B_{i+1} \setminus B_i$. So, it will be enough to define a well-ordering in $B_{i+1} \setminus B_i$: and it is easy to infer from this that this gives a well-ordering of B_{i+1}: for the subsets $S \subseteq B_{i+1}$ which contain elements of B_i, their minimum is the minimum of $S \cap B_i$; the rest of subsets are included in $B_{i+1} \setminus B_i$ and they have a minimum.

Let, for example, $u, v \in B_2 \setminus B_1$, so that $u, v \in A_2 \setminus A$. We let $u < v$ if and only if there exist $a, b \in A$ with $a \leq b$ and $u = \{a, b\}$ so that whenever $v = \{a', b'\}$ with $a', b' \in A$ and $a' \leq b'$, we have $\langle a, b\rangle < \langle a', b'\rangle$ in the lexicographic order of $A \times A$. Or, if $u, v \in B_3 \setminus B_2$, $u < v$ if and only if there exist $a, b \in A$ with $u = a \setminus b$ so that whenever $v = a' \setminus b'$ with $a', b' \in A$, we have $\langle a, b\rangle < \langle a', b'\rangle$ in the lexicographic order. The order for B_4 is defined analogously.

Next, if $u, v \in B_5 \setminus B_4$, then $u < v$ if and only if there is $a \in A$ such that $u = \mathcal{M} \cap a$ and whenever $v = \mathcal{M} \cap b$ for $b \in A$ we have $a < b$. The definitions for B_6, B_7, B_8 are analogous to this. For instance, if $u, v \in B_7 \setminus B_6$, then $u < v$ if and only if there is $a \in A$ such that $u = \sigma[a]$ (here σ is the transformation which changes the order of the triples in a); and whenever $v = \sigma[b]$ for $b \in A$, we have $a < b$. It is straightforward to see that this gives indeed a well-ordering of $F(A)$ by successive extensions of the ordering of $A = B_1$ to B_2, B_3, etc. Moreover this gives a functional K_0; that is, for every ordered pair $\langle A, W\rangle \in C$, $h(\langle A, W\rangle) = \langle F(A), W_0\rangle$ where W_0 is the well-ordering of $F(A)$ just described. Since this description can be given through some formula, K_0 is indeed a functional class.

As in Proposition 12.1.2, $H(0) = X$ and $H(n+1) = F(H(n))$, giving $\text{cns}(X) = \cup_{n\in\omega} H(n)$. By using K_0, we recursively define $\langle H(n+1), W_{n+1}\rangle$ from $\langle H(n), W_n\rangle$ with $H(n+1) = F(H(n))$ and $\langle F(H(n)), W_{n+1}\rangle = K_0(\langle H(n), W_n\rangle)$. In this way we obtain extended well-orderings W_n for $H(n)$. The union of all these well-orderings is then a well-ordering of $\text{cns}(X)$. This gives the functional K of the statement. □

Theorem 12.4.2. *Any ZFL-universe has a well-ordering.*

Comment 12.4.3. It would be inaccurate to understand this statement as saying that there is some object of the universe which is a well-ordering of V, because a well-ordering of V is necessarily a proper class. A correct wording would be to describe a particular class of a ZFL-universe V, and assert (and later prove) that that class is a well order relation in V. Of course this would be too long, but this is the idea: that it is possible to construct a specific class that is a well order relation in V.

Proof. We know $V = L$ and thus L_α will denote the set of all elements of V with c-rank $< \alpha$. For the proof, we define by ordinal recursion a new functional G, which to each α assigns a well-ordering of L_α so that $\beta < \alpha \Rightarrow G(\beta) \subseteq G(\alpha)$. To this end, assume that $G(\alpha)$ is a well-ordering of L_α, and call $G_1(\alpha)$ to the ordering of $s(L_\alpha)$ which extends $G(\alpha)$ by setting L_α as the maximum in $s(L_\alpha)$. Then we define $G(\alpha+1)$ as the relation of $L_{\alpha+1}$ which is the union of $G_1(\alpha)$, the set of all ordered pairs $\langle x, y\rangle$ where $x \in L_\alpha, y \in L_{\alpha+1} \setminus L_\alpha$, and the restriction to $L_{\alpha+1} \setminus L_\alpha$ of $K(\langle s(L_\alpha), G_1(\alpha)\rangle)$, K being the functional of Lemma 12.4.1. For limit λ, we simply take $G(L_\lambda) = \cup_{\alpha\in\lambda} G(L_\alpha)$. It is easy (though perhaps cumbersome) to verify that this gives a well-ordering of each set L_α, and that the union of all these orderings is a well-ordering of $L = V$. □

Corollary 12.4.4. *Any ZFL-universe is a ZFC-universe.*

Proof. Let W be a well-ordering of V according to Theorem 12.4.2. If $c \in V$, then the restriction $W \cap (c \times c)$ of W is a well-ordering of c. Thus every set has a well-ordering, and the axiom of choice holds by Proposition 10.3.5. □

Proposition 12.4.5. *Let V be a ZF-universe, L its constructible class. For each infinite ordinal α, L_α has a cardinal and $|\alpha| = |L_\alpha|$.*

Proof. We first prove the equality when V is a ZFC-universe. Note that for each $n \in \omega$, $L_n \subseteq V_n$ is a finite set; hence L_ω is a denumerable union of different finite sets (as $L_n \in L_{n+1}$) and thus $|L_\omega| = \aleph_0 = |\omega|$.

We now use ordinal induction. Suppose that $|\alpha| = |L_\alpha|$ for α infinite. Following the notation in the proof of Proposition 12.1.2 that introduces the functionals F and H, $|F(A)| = |A|$ for infinite A, and thus $\mathrm{cns}(L_\alpha) = \cup_{n\in\omega} H(n)$ (with $H(0) = L_\alpha$) is a denumerable union of the sets $H(n)$, each of which is in turn a finite union of sets with cardinal $\leq |L_\alpha|$ or $\leq |L_\alpha \times L_\alpha|$, which is again $|L_\alpha|$ according to Proposition 10.2.9. This entails that the cardinal of $\mathcal{C}(L_\alpha) = L_{\alpha+1}$ is $\leq |L_\alpha|$. In the opposite direction, $L_\alpha \subseteq L_{\alpha+1}$ and so $|L_{\alpha+1}| = |L_\alpha| = |\alpha| = |\alpha + 1|$.

Let now λ be a limit ordinal such that the property holds for each $\alpha < \lambda$. Since $L_\lambda = \cup_{\alpha<\lambda} L_\alpha$, we have $|L_\lambda| \leq \sum_{\alpha\in\lambda} |L_\alpha| = \sum_{\alpha\in\lambda} |\alpha| \leq \sum_{\alpha\in\lambda} |\lambda| = |\lambda|$ (see also Exercise 23 of Chapter 10). Therefore $|L_\lambda| \leq |\lambda|$. The equality follows because $\lambda \subseteq L_\lambda$ by Proposition 12.2.3. This completes the induction.

Suppose now that V is just a ZF-universe, $L \subseteq V$ the class of constructible sets, which is a ZFC-universe by Corollaries 12.3.3 and 12.4.4. By Proposition 12.2.3, α is an ordinal in L. Moreover, $L^L_\alpha = L_\alpha$, as seen in the proof of Proposition 12.3.4. It follows that there is a bijection in L (and hence in V) between α and L_α. By Proposition 10.1.6, α has a cardinal $|\alpha|$, and the bijection between $|\alpha|$ and L_α entails that L_α has a cardinal equal to $|\alpha|$. □

12.5 The Continuum Hypothesis

12.5.1 Overview

We prove in this section that the continuum hypothesis (CH) holds in any ZFL-universe V. The proof requires a series of lemmas; hence having an overview of the ideas in the proof will be helpful in understanding the part that each lemma plays in the whole.

The proof uses Proposition 12.4.5, because the goal is to show that $\mathbb{P}(\omega) \subseteq L_{\aleph_1}$. By the proposition, $|L_{\aleph_1}| = \aleph_1$ and thus $2^{\aleph_0} = |\mathbb{P}(\omega)| \leq \aleph_1$, proving CH. To prove that $X \in \mathbb{P}(\omega) \Rightarrow X \in L_{\aleph_1}$, a crucial use is made of Proposition 11.2.5: it allows us to obtain an apt and extensional set $A \supseteq X$ with the property that all the neat classes in a well chosen finite list are A-reliable. The Mostowski's collapse π (Theorem 9.3.9) is then applied to this set A (where membership is obviously well-founded) to get that $\pi(X) = X$ and $\pi[A] \subseteq L_{\aleph_1}$, so $X \subseteq L_{\aleph_1}$.

The main technical issue in this scheme is the behavior of the collapse isomorphism π in two respects: the easy one is its application to ordinals, and we had already hinted at this in Example 9.3.10. The second is more involved in its realization, though the idea is quite simple: the isomorphism π preserves the constructible sequences as far as providing the decisive result that, in the situation we consider, $\pi[L_\alpha \cap A] = L_\beta$ for some β. This is the main ingredient in obtaining the above mentioned equation $\pi[A] \subseteq L_{\aleph_1}$.

12.5.2 Mostowski's isomorphism and constructibility

Throughout this section, V is a ZFL-universe, so $V = L$. Recall from Definition 7.3.12 that for A an apt and extensional class, a class C is A-reliable when $\mathrm{D}(C) \cap A = \mathrm{D}(C \cap A)$. The fact that A is extensional entails that membership is an extensional and well-founded relation in A, so that we may apply Mostowski's theorem 9.3.9. We use the notation $\pi : A \to M$ for the $\in$-isomorphism of the Mostowski's collapse, so that $\pi(x) = \pi[x \cap A]$ for each $x \in A$. We will keep this notation for the rest of the section.

Lemma 12.5.1. *Let A be an extensional and apt class, $C_1 = \{\langle u, v, w\rangle | \langle v, w\rangle \in u\}$ and assume that C_1 is A-reliable. As in the proof of Proposition 12.1.2, $G_2, \ldots, G_8$ denote the functionals giving the Gödel operations. Then for $x, y \in A$, $\pi[G_i(\langle x, y\rangle) \cap A] = G_i(\langle \pi(x), \pi(y)\rangle)$ for $i = 2, 3, 4$; and $\pi[G_i(x) \cap A] = G_i(\pi(x))$ for $i = 5, 6, 7, 8$.*

Proof. The property for G_2 was given in the proof of Lemma 11.3.3 (i), where it was also shown that $\pi(\langle x, y\rangle) = \langle \pi(x), \pi(y)\rangle$. For G_3, $\pi[(x \setminus y) \cap A] = \{\pi(u) | u \in x \cap A \wedge u \notin y\} = \pi[x \cap A] \setminus \pi[y \cap A] = \pi(x) \setminus \pi(y)$. Similarly, $\pi[(x \times y) \cap A] = \{\langle \pi(u), \pi(v)\rangle | u, v \in A, u \in x, v \in y\} = \pi(x) \times \pi(y)$.

If σ is the transformation of triples of rule 7, $\pi[G_7(x) \cap A] = \{\pi(\sigma(u)) | u \in x \cap T \wedge u \in A\}$, where T is the class of triples, and the elements of a triple belong to A if and only if each of the components belongs to A, by the apt hypothesis. Now, $\pi(\sigma(u)) = \sigma(\pi(u))$ by the reason stated at the beginning concerning ordered pairs, and $\pi[G_7(x) \cap A] = \{\sigma(\pi(u)) | u \in x \cap T \cap A\} = G_7[\pi(x \cap A)] = G_7(\pi(x))$. The proof for rule 8 is similar. Also, $\pi[x \cap \mathcal{M} \cap A] = \{\pi(u) | u \in x \cap A \wedge u \in \mathcal{M}\} = \{\pi(u) \in \mathcal{M} | u \in x \cap A\}$ because A is apt and π is an isomorphism for membership. Thus $\pi[(x \cap \mathcal{M}) \cap A] = \pi(x) \cap \mathcal{M}$.

As for rule 6, we show first that for $u \in A$, $\mathrm{D}(u) \cap A = \mathrm{D}(u \cap A)$. The inclusion $\mathrm{D}(u \cap A) \subseteq \mathrm{D}(u) \cap A$ is valid for every extensional apt class A. For the converse inclusion, let $x \in \mathrm{D}(u) \cap A$ so that $\langle x, y \rangle \in u$ for some y. Then $\langle u, x, y \rangle \in C_1$ and $\langle u, x \rangle \in \mathrm{D}(C_1) \cap A = \mathrm{D}(C_1 \cap A)$. That is, there is $z \in A$ with $\langle u, x, z \rangle \in C_1$ and thus $\langle x, z \rangle \in u \cap A$ and $x \in \mathrm{D}(u \cap A)$. Therefore $\pi[\mathrm{D}(u) \cap A] = \pi[\mathrm{D}(u \cap A)] = \mathrm{D}(\pi[u \cap A])$. This last equality follows again from the fact that $\pi(\langle x, y \rangle) = \langle \pi(x), \pi(y) \rangle$ for $x, y \in A$. Thus we get $\pi[\mathrm{D}(u) \cap A] = \mathrm{D}(\pi(u))$. □

Consider the following seven classes $C_2, \ldots, C_8$ related to the Gödel operations of Definition 12.1.1: for the binary operations of rules 2, 3, 4, C_i will be the class of all triples $\langle a, b, z \rangle$ where z is the result of the corresponding operation carried on a, b. For instance, $C_4 = \{\langle a, b, z \rangle | z = a \times b\}$. For the unary operations of rules 5, 6, 7, 8, C_i will be the class of ordered pairs $\langle a, z \rangle$ with z being the result of the operation on a. For instance $C_6 = \{\langle a, z \rangle | z = \mathrm{D}(a)\}$; or $C_7 = \{\langle u, z \rangle | z = \{\langle a_1, a_2, a_3 \rangle | \langle a_2, a_1, a_3 \rangle \in u\}$. Note that all classes C_2 to C_8 are functional.

Lemma 12.5.2. *In the ZFL-universe V, let A be an extensional and apt class, X a non-empty set. If all the classes $C_1, \ldots, C_8$ are A-reliable and $x_0, x_1, \ldots, x_n$ is an $(X \cap A)$-constructible sequence, then each $x_i \in A$ and the sequence $\pi(x_0), \ldots, \pi(x_n)$ is a $\pi[X \cap A]$-constructible sequence which has exactly the same labels than the given sequence.*

Proof. The hypotheses of A-reliability imply that any Gödel operation applied to elements of A gives elements of A. For instance, if $a, b \in A$, then $\langle a, b, a \times b \rangle \in C_4$ and $\langle a, b \rangle \in \mathrm{D}(C_4) \cap A$. But then $\langle a, b, a \times b \rangle \in A$ because C_4 is functional, and hence $a \times b \in A$. As another example, $a \in A \Rightarrow \langle a, \mathrm{D}(a) \rangle \in C_6$. Since $a \in \mathrm{D}(C_6) \cap A = \mathrm{D}(C_6 \cap A)$ and C_6 is functional, $\mathrm{D}(a) \in A$.

As a consequence, every term x_i of an $(X \cap A)$-constructible sequence belongs to A. Also, the elements $\pi(x_i)$ form a sequence which is $\pi[X \cap A]$-constructible with each $\pi(x_i)$ being introduced by the same rule that worked for x_i. This follows immediately by Lemma 12.5.1 bearing in mind that if $x_i \in A$ is introduced in the original sequence as $G_j(u, v)$, then $\pi(x) = G_j(\pi(u), \pi(v))$. Thus the new sequence will be a $\pi[X \cap A]$-constructible sequence with the same labels than the original one. □

Let $X, u \in V$. If $u_1, \ldots, u_s = u$ are the terms of an X-constructible sequence, we call the s-tuple $\langle u_1, \ldots, u_s \rangle$ a *X-genealogy of u*. Let C_0 be the class of all triples $\langle X, u, g\rangle$ such that g is a X-genealogy of u. In particular, if $\langle X, u, g\rangle \in C_0$, then u is X-constructible. Note that, since A is apt, a genealogy g belongs to A if and only if each term of g belongs to A. In the next lemma, we impose conditions on A that guarantee that $\mathrm{cns}(X \cap A) = \mathrm{cns}(X) \cap A$.

Lemma 12.5.3. *Let A be an extensional and apt class and let $\emptyset \neq X \in A$. Suppose that all classes C_0 to C_8 are A-reliable. Then $cns(\pi(X)) = \pi[cns(X) \cap A]$.*

Proof. Let $y \in \mathrm{cns}(\pi(X)) = \mathrm{cns}(\pi[X \cap A])$ so that there is a $\pi[X \cap A]$-constructible sequence $y_1, \dots, y_k$ such that $y_k = y$. The lines introduced by rule 1 (w.l.o.g., let us suppose that these are $y_1, \dots, y_r$) are $y_i = \pi(x_i)$ with $x_i \in X \cap A$. Let us now consider the sequence $x_1, \dots, x_r, x_{r+1}, \dots x_k$ where the $x_{r+1}, \dots, x_k$ are obtained from $x_1, \dots, x_r$ through exactly the same Gödel operations as the corresponding y_i. Thus this will be a $(X \cap A)$-constructible sequence and Lemma 12.5.2 shows that $\pi(x_1), \dots, \pi(x_k)$ is a $\pi[X \cap A]$-constructible sequence with the same labels. But this sequence has the same labels as the sequence of the y_i and moreover $y_i = \pi(x_i)$ for $i = 1, \dots, r$; therefore, this last sequence coincides with $y_1, \dots, y_k$ by an easy inductive argument using Lemma 12.5.1; and hence $y_k = y = \pi(x_k)$ and $y \in \pi[\mathrm{cns}(X \cap A)] \subseteq \pi[\mathrm{cns}(X) \cap A]$, again by Lemma 12.5.2 because each term of the constructible sequence belongs to A. Thus $\mathrm{cns}(\pi[X \cap A]) \subseteq \pi[\mathrm{cns}(X) \cap A]$.

For the converse inclusion, let $u \in \mathrm{cns}(X) \cap A$. By hypothesis, there is some X-genealogy g of u and thus $\langle X, u, g\rangle \in C_0$ and $\langle X, u\rangle \in \mathrm{D}(C_0)$. Since $X, u \in A$, we infer from the A-reliability of C_0 that there is $g' \in A$ so that $\langle X, u, g'\rangle \in C_0$, so u is X-constructible with a sequence that consists of elements of A and thus u is $(X \cap A)$-constructible. If we apply π to each term of this sequence, we obtain by Lemma 12.5.2 a $\pi[X \cap A]$-constructible sequence giving $\pi(u)$. Therefore $\pi[\mathrm{cns}(X) \cap A] \subseteq \mathrm{cns}(\pi[X \cap A])$. □

With just one additional hypothesis, Lemma 12.5.3 holds also with $\mathcal{C}(X)$ substituted for $\mathrm{cns}(X)$.

Lemma 12.5.4. *Let A be an extensional apt class satisfying the hypotheses of Lemma 12.5.3 and such that the inclusion relation $\mathcal{I}$ is A-absolute. Then if $X, s(X) \in A$, $\mathcal{C}(\pi(X)) = \pi[\mathcal{C}(X) \cap A]$.*

Proof. Note that $\pi(s(X)) = \pi[(X \cup \{X\}) \cap A] = \pi[X \cap A] \cup \{\pi(X)\} = \pi(X) \cup \{\pi(X)\} = s(\pi(X))$. Also, for $u, v \in A$, $u \subseteq v \Leftrightarrow u \cap A \subseteq v \cap A$ because inclusion is A-absolute by hypothesis. Since π is an isomorphism for membership, $u \subseteq v \Leftrightarrow \pi(u) \subseteq \pi(v)$ (for $u, v \in A$). Therefore $\pi[\mathbb{P}(X) \cap A] = \mathbb{P}(\pi(X)) \cap \pi[A]$.

Then $\mathcal{C}(\pi(X)) = \mathbb{P}(\pi(X)) \cap \mathrm{cns}(s(\pi(X))) = \mathbb{P}(\pi(X)) \cap \mathrm{cns}(\pi(s(X))) = \mathbb{P}(\pi(X)) \cap \pi[\mathrm{cns}(s(X)) \cap A] = \mathbb{P}(\pi(X)) \cap \pi[A] \cap \pi[\mathrm{cns}(s(X)) \cap A] = \pi[\mathbb{P}(X) \cap A] \cap \pi[\mathrm{cns}(s(X)) \cap A] = \pi[\mathbb{P}(X) \cap \mathrm{cns}(s(X)) \cap A] = \pi[\mathcal{C}(X) \cap A]$. □

12.5.3 Mostowski's isomorphism and ordinals

Informally, we have seen in Example 9.3.10 how the $\in$-isomorphism $\pi : A \to M$ deals with ordinals: basically, it assigns to each ordinal $\alpha \in A$ the ordinal that measures the well-ordered set $\alpha \cap A$. We now extend this functional to all ordinals.

Lemma 12.5.5. *Let A be an extensional apt class and let us define for each ordinal α, $\mu(\alpha) = \pi[\alpha \cap A]$. Then every $\mu(\alpha)$ is an ordinal and μ is a monotone functional. Moreover, $\mu(\alpha) = \pi(\alpha)$ for every $\alpha \in A$.*

Proof. It is clear that $\alpha \in A \Rightarrow \mu(\alpha) = \pi(\alpha)$ because $\pi(\alpha) = \pi[\alpha \cap A]$. We show that every $\mu(\alpha)$ is an ordinal by using RAA; if some $\mu(\alpha)$ is not an ordinal, then there is a smallest ordinal τ such that $\mu(\tau)$ is not an ordinal, by the well-ordering of the ordinals. Then $\mu(\tau) = \pi[\tau \cap A]$ is a set of ordinals by the choice of τ. Let $x \in \mu(\tau)$ so that $x = \pi(\gamma)$ for some $\gamma \in \tau \cap A$; if $y \in x = \pi[\gamma \cap A]$, then $y = \pi(\rho)$ for some ordinal $\rho \in \gamma \cap A$. Hence $\rho \in \tau \cap A$ and hence $y \in \mu(\tau)$. This shows that $\mu(\tau)$ is a transitive set of ordinals and therefore an ordinal (see, e.g., Exercise 4 of Chapter 8), a contradiction that proves our claim.

Obviously, if $\alpha \subseteq \beta$ are ordinals, then $\alpha \cap A \subseteq \beta \cap A$ hence $\mu(\alpha) \subseteq \mu(\beta)$, and μ is a monotone functional. $\square$

Let us introduce some other classes. $B_1 = \{\langle x, r\rangle | r = \mathrm{crk}(x)\}$; $B_2 = \{\langle x, y\rangle | y = \mathrm{s}(x)\}$; $B_3 = \{\langle \alpha, Y\rangle | \alpha \in \mathbf{Ord} \wedge Y = L_\alpha\}$.

Lemma 12.5.6. *Let A be an extensional and apt class such that the classes B_2, B_2^{-1} are A-reliable. Then for every ordinal $\alpha \in A$, $\mu(\alpha + 1) = \mu(\alpha) + 1$; while $\mu(\alpha + 1) = \mu(\alpha)$ if $\alpha \notin A$. Also, $\mu(\lambda)$ is a limit ordinal whenever λ is a limit ordinal; and in this case $\mu(\lambda) = \cup_{\beta\in\lambda}\mu(\beta)$.*

Proof. If $\alpha \in A$ then $\alpha \in \mathrm{D}(B_2) \cap A$, so $\alpha \in \mathrm{D}(B_2 \cap A)$ because B_2 is A-reliable. Since B_2 is functional, we deduce that $\alpha + 1 \in A$. As in the first sentence of the proof of Lemma 12.5.4, we see that $\pi(\alpha + 1) = \pi(\alpha) + 1$ and this gives $\mu(\alpha + 1) = \mu(\alpha) + 1$ by Lemma 12.5.5. On the other hand, if $\alpha \notin A$, then $\alpha + 1 \notin A$ because B_2^{-1} is A-reliable and functional. Thus $\pi[(\alpha + 1) \cap A] = \pi[\alpha \cap A]$ and therefore $\mu(\alpha + 1) = \mu(\alpha)$.

We next consider the case of a limit ordinal λ and use RAA. If $\mu(\lambda) = \rho+1$ for some ordinal ρ, then $\rho + 1 = \pi[\lambda \cap A]$ and thus there is $\beta \in \lambda \cap A$ with $\pi(\beta) = \rho$. Since λ is limit, $\beta + 1 \in \lambda \cap A$ and $\pi(\beta + 1) \in \mu(\lambda) = \rho + 1$, but $\pi(\beta + 1) = \rho + 1$ by the first part of this proof. Then $\rho + 1 \in \rho + 1$, a contradiction. This shows that $\mu(\lambda)$ is a limit ordinal. Finally, $\mu(\lambda) = \mu[\lambda \cap A] = \mu[(\cup_{\beta\in\lambda}\beta) \cap A] = \mu[\cup_{\beta\in\lambda}(\beta \cap A)] = \cup_{\beta\in\lambda}\mu[\beta \cap A] = \cup_{\beta\in\lambda}\mu(\beta)$. $\square$

We obtain now the main result in this series. The set L_α below is the α-th term of the constructible hierarchy of $V = L$.

Lemma 12.5.7. *Let A be an extensional apt class such that the inclusion class $\mathcal{I}$ is A-absolute and the classes $C_0, \ldots, C_8, B_1, B_2, B_2^{-1}, B_3, B_3^{-1}$ introduced in this section are all A-reliable. Then for every ordinal α, the following equation holds: $\pi[L_\alpha \cap A] = L_{\mu(\alpha)}$.*

Proof. We prove the result by ordinal induction. The case $\alpha = 0$ is trivial because $\mu(0) = 0$ and $L_0 = \emptyset$. Suppose that $\pi[L_\alpha \cap A] = L_{\mu(\alpha)}$ and $\alpha \in A$; then $\alpha + 1 \in A$ because the class B_2 is A-reliable and $\alpha \in \mathrm{D}(B_2)$. Thus L_α, $\mathrm{s}(L_\alpha)$ and $L_{\alpha+1} \in A$ because B_2, B_3 are A-reliable. By Lemma 12.5.4, $\pi[L_{\alpha+1} \cap A] = \pi[\mathcal{C}(L_\alpha) \cap A] = \mathcal{C}(\pi(L_\alpha)) = \mathcal{C}(\pi[L_\alpha \cap A]) = \mathcal{C}(L_{\mu(\alpha)}) = L_{\mu(\alpha)+1}$. But $\mu(\alpha+1) = \mu(\alpha) + 1$ by Lemma 12.5.6. That is, $\pi[L_{\alpha+1} \cap A] = L_{\mu(\alpha+1)}$.

Suppose now that $\alpha \notin A$ and $\pi[L_\alpha \cap A] = L_{\mu(\alpha)}$. Since B_2^{-1} is A-reliable, $\alpha + 1 \notin A$ and $\mu(\alpha+1) = \mu(\alpha)$ by Lemma 12.5.6. B_3^{-1} is A-reliable and therefore $L_\alpha, L_{\alpha+1} \notin A$. From the fact that B_1 is A-reliable, it follows that there is no element $x \in A$ with $\mathrm{crk}(x) = \alpha$ and hence $L_{\alpha+1} \cap A = L_\alpha \cap A$. Consequently, $\pi[L_{\alpha+1} \cap A] = \pi[L_\alpha \cap A] = L_{\mu(\alpha)} = L_{\mu(\alpha+1)}$.

It remains to see the case when α is a limit ordinal. Then $\mu(\alpha)$ is a limit ordinal by Lemma 12.5.6 and hence $L_{\mu(\alpha)} = \cup_{x \in \mu(\alpha)} L_x$. Since, by the same lemma, $\mu(\alpha) = \cup_{\beta \in \alpha} \mu(\beta)$, $L_{\mu(\alpha)} = \cup_{\beta \in \alpha} L_{\mu(\beta)} = \cup_{\beta \in \alpha} \pi[L_\beta \cap A] = \pi[\cup_{\beta \in \alpha}(L_\beta \cap A)] = \pi[(\cup_{\beta \in \alpha} L_\beta) \cap A] = \pi[L_\alpha \cap A]$. □

12.5.4 The theorem

Theorem 12.5.8. *If V is a ZFL-universe, then $2^{\aleph_0} = \aleph_1$ holds in V.*

Proof. Under the hypothesis that V is a ZFL-universe, we prove that $\mathbb{P}(\omega) \subseteq L_{\aleph_1}$. As a result, $2^{\aleph_0} = |\mathbb{P}(\omega)| \leq |L_{\aleph_1}| = \aleph_1$, according to Proposition 12.4.5. Since the reverse inequality $\aleph_1 \leq 2^{\aleph_0}$ is clear from Cantor's theorem, this will prove the result.

Let us show that $X \subseteq \omega$ implies $X \in L_{\aleph_1}$. To achieve this, we let $Y = L_\omega \cup \{X\}$, which is countable by Proposition 12.4.5. Take a class sequence which includes the class $\mathcal{I}$; by Proposition 11.2.5, there exists an apt and extensional denumerable set A such that $Y \subseteq A$, the mentioned class sequence is A-absolute and the classes $C_0, \ldots, C_8, B_1, B_2, B_3, B_2^{-1}, B_3^{-1}$ are A-reliable. Thus A satisfies the hypotheses of Lemma 12.5.7. We now use the Mostowski's isomorphism $\pi : A \to M$ with M a transitive denumerable set. By the definition of Y, $\omega, X \subseteq A$ and $X \in A$, from which it follows that $\pi(X) = \pi[X \cap A] = \pi[X] = X \in M$ because every element of X is a finite ordinal, hence fixed under π.

Since $A \in L = V$, $A \subseteq L_\alpha$ for some ordinal α. By Lemma 12.5.7, $\pi[L_\alpha \cap A] = L_{\mu(\alpha)} = \pi[A] = M$. Thus $M = L_\lambda$ for a certain ordinal λ. Since M is denumerable, $|L_\lambda| = |\lambda| = \aleph_0$ by Proposition 12.4.5. Thus λ is countable and $\lambda < \aleph_1$. Hence $M = L_\lambda \subseteq L_{\aleph_1}$. This shows that $X \in L_{\aleph_1}$ as we had to see. □

An important observation is that the same result holds with any infinite cardinal κ substituted for $\aleph_0$. That is, the equation $2^\kappa = \kappa^+$ (i.e., the generalized continuum hypothesis) holds in any ZFL-universe. The proof is basically the same we have just seen for $\kappa = \aleph_0$. The only difference is that we must use Proposition 11.2.3 instead of Proposition 11.2.5. Then the same argument of Theorem 12.5.8 shows that any element of $\mathbb{P}(\kappa)$ belongs to L_{κ^+}.

12.6 Exercises

Unless stated otherwise, V will be assumed to be a ZF-universe.

1. Prove that for any set X, cns(X) is the smallest set that includes X and is closed under the Gödel operations.
2. Prove that if X is a transitive set, then cns(X) is also transitive (Hint: Prove inductively that if $x_1, \dots, x_n$ is an X-constructible sequence, then tc(x_i) consists of X-constructible elements for $i = 1, \dots, n$).
3. Give an example of a transitive set $X \in V$ and elements $a, b \in X$ with the property $a \times b \notin \mathcal{C}(X)$.
4. Let C be an X-constructible set. Define C^* and C_1, C_2 as in Proposition 3.1.14 and in Lemma 3.1.17 respectively. Prove, by similar methods to those used in the mentioned results, that C^* is X-constructible; and C_1 is X-constructible if and only if C_2 is.
5. Let A be an apt set and C a set consisting of ordered n-tuples ($n \geq 1$) which is $s(A)$-constructible. Show that the and transitive sets defined from C as in items (a), (b), (c), (d) of Lemma 3.1.18, but with the inserted element y running on A (for instance, in (a) we would have the set $\{\langle x_1, \dots, x_n, y\rangle | \langle x_1, \dots, x_n\rangle \in C \wedge y \in A\}$) are $s(A)$-constructible.
6. Consider the simple formulas of Propositions 3.2.1 and 3.2.2 with one or two free variables. Suppose that A is an apt and transitive set, and prove that the sets $\{\langle x_1, \dots, x_s\rangle \in A^s | P(x_1, \dots, x_s)\}$ (with P one of those simple formulas) are $s(A)$-constructible.
7. By Corollary 3.2.7, any formula with at least one free variable determines the class of the objects that satisfy the formula. Show by a similar argument that if A is an apt and transitive set and $P(x_1, \dots, x_n)$ ($n \geq 1$) is a formula with the free variables $x_1, \dots, x_n$ in which the only elements that may appear are elements of A (look at Proposition 3.2.1 to see how elements may appear in formulas) and the possible quantifiers are limited to elements of A (as $\forall x \in A$ or $\exists x \in A$), then $\{\langle x_1, \dots, x_n\rangle \in A^n | P(x_1, \dots, x_n)\}$ is $s(A)$-constructible.

8. Prove that $L_n = V_n$ for any $n \leq \omega$, but that $L_{\omega+1} \neq V_{\omega+1}$.
9. Prove that every cardinal of V is a cardinal of L.
10. Let λ be a limit ordinal. Prove that L_λ is closed under the Gödel operations; i.e., if $x, y \in L_\lambda$, then $G_i(x,y), G_j(x) \in L_\lambda$ for any of the Gödel functionals $G_2, \ldots, G_8$.
11. Complete the proof of Corollary 12.2.5 with the help of the preceding exercise.
12. Prove that if λ is a limit ordinal, then $L_\lambda = \cup_{\beta\in\lambda}\mathrm{cns}(L_\beta)$.
13. Justify that there is some subset of ω which does not belong to $\mathrm{cns}(\omega^\omega)$.
14. V will be now a ZFC-universe. By using Exercise 23 of Chapter 11, show that there exist ordinals $\alpha > \omega$ such that $|L_\alpha| = |V_\alpha|$. Prove that if $\alpha > \omega$ satisfies that equation, then α is a cardinal and $\alpha = \beth_\alpha$.
15. Let V be a ZF-universe and C a transitive V-class which is also a ZF-universe and has the same ordinals as V. Prove that the class L of the constructible sets of V is included in C, and that $L = L^C$, the class of the constructible sets of C.
16. Let V be a ZF-universe and C a transitive V-class which is also a ZF-universe. Prove that C has the same ordinals as V if and only if C is a big class.

A class A in the ZF-universe V is said to be Gödelian *when: (1) A is transitive; (ii) A is closed under Gödel operations; (iii) A is a big class (see Proposition 12.2.4).*

17. Prove that if A is a Gödelian class of V, then A is closed under intersections and unions of two elements; and under taking the inverse and taking the codomain of one element.
18. Prove that, if A is Gödelian, then $a \in A \Rightarrow (\cup a \in A \wedge \mathcal{I} \cap a \in A)$.
19. Prove that if A is Gödelian, then A satisfies axioms 1, 2, 3, 4, 7, 8 of ZF-universes.
20. Check that Gödelian classes satisfy also each instance of the axiom of replacement (Hint: Follow the proof of Proposition 12.3.2 and in item (v) use the V-rank instead of the constructible rank and apply the property that A is a big class).
21. Prove that any Gödelian class A contains all ordinals, by using the property (of the preceding exercise) that A satisfies the principle of separation. Deduce that A is a ZF-universe.
22. For $X \in L$ in a ZFC-universe V, show with detail that $\mathrm{cns}(X)^L = \mathrm{cns}(X)$ and conclude that $\mathcal{C}(X)^L = \mathcal{C}(X) \cap L$. In particular, show

that for $a, b \in L$, $(a \setminus b)^L = a \setminus b$, $(a \times b)^L = a \times b$, $\mathrm{D}(a)^L = \mathrm{D}(a)$, etc.

23. Complete the proof of Lemma 12.4.1 by describing the orderings of the sets $B_4 \setminus B_3, B_6 \setminus B_5, B_7 \setminus B_6, B_8 \setminus B_7$ and explaining why these are well-orderings.

24. Let $\{A_\alpha\}$ be a class of sets, indexed by ordinals, in a ZF-universe; and well-orderings W_α of A_α are given for each α. Assume that, for $\alpha \leq \beta$, the following conditions hold: (i) $A_\alpha \subseteq A_\beta$; (ii) $W_\alpha = W_\beta \cap (A_\alpha \times A_\alpha)$; and (iii) all pairs $\langle x, y \rangle$ with $x \in A_\alpha$ and $y \in A_\beta \setminus A_\alpha$ belong to W_β. Prove that $\cup W_\alpha$ is a well-ordering of $\cup A_\alpha$.

25. If $V = L$ is a ZFL-universe, the notation L_α could have two different meanings: the meaning used throughout the text is that L_α is the α-th step in the constructible hierarchy inside the universe; a second meaning would be the one given by the von Neumann hierarchy, with L_α denoting in this case the layer V_α of the universe. Do these meanings give the same values?

26. Let V be a ZFL-universe. By adapting the proof of Theorem 12.5.8, show that for any infinite ordinal α, $\mathbb{P}(\alpha) \subseteq L_\kappa$ where κ is the smallest cardinal such that $\alpha < \kappa$. Derive from this a proof of the generalized continuum hypothesis GCH.

27. Recall the definition of the sets H_κ of Exercise 14 of Chapter 11. Show that in any ZFL-universe, $L_\kappa = H_\kappa$ for every infinite cardinal κ (Hint: Suppose the contrary and let κ be minimum with $H_\kappa \neq L_\kappa$. Observe that $L_\kappa \subseteq H_\kappa$ and show that $\kappa = \mu^+$ for some cardinal μ, hence regular. Obtain $u \in H_\kappa \setminus L_\kappa$ with $u \subseteq L_\kappa$. Using the constructible rank of the elements of u, infer from the regularity of κ that $u \subseteq L_\rho$ for some $\rho < \kappa$. Repeat the steps of the preceding exercise to get that $u \in L_\kappa$).

13

Boolean Algebras

Boolean algebras will be an important tool in our next two chapters. This topic is more algebraic than set theoretic, but it has strong ties to set theory. Not only because of its role in forcing, which will be the main character in the rest of Part II of this book, but also because, like almost any other topic in algebra or analysis, it is properly set theory: the objects of interest are sets existing in some chosen ZF or ZFC-universe, and it is grounded on the axioms of set theory. We treat the subject with some more detail than is needed for later use, but this allows us to present central notions like duality and filters in a reasonably general setting.

13.1 Lattices

13.1.1 Basic properties

Throughout this chapter V will be a ZFC-universe satisfying property NNS. Recall Definition 4.2.17 for the dual notions of supremum and infimum of a subset in an ordered set. In particular, observe that supremum and infimum are unique when they exist.

Definition 13.1.1. Let $\langle \mathcal{L}, R \rangle$ be an ordered pair where R is an order relation in the set $\mathcal{L}$. $\langle \mathcal{L}, R \rangle$ is called a *lattice* when $\mathcal{L} \neq \emptyset$ and for every pair a, b of elements of $\mathcal{L}$, there exist $s, i \in \mathcal{L}$ which are, respectively, the supremum and infimum of $\{a, b\}$: $s = \sup(\{a, b\})$ and $i = \inf(\{a, b\})$.

Remarks and Examples 13.1.2. From now on, we shall write simply $\sup(a, b)$ or $\inf(a, b)$, instead of the correct notation. Following another usual simplification of language, we refer to the lattice $\langle \mathcal{L}, R \rangle$ as *the lattice* $\mathcal{L}$ (with the ordering R understood when possible). Also, it is customary to use $\leq$ for denoting the ordering, as this is the general habitude when dealing with order relations.

If R is a total ordering of the set $\mathcal{L}$, then $\mathcal{L}$ is a lattice. In this case, $\sup(a, b) = \max(a, b)$ which exists because either aRb or bRa; and $\inf(a, b) = \min(a, b)$.

DOI: 10.1201/9781003449911-13

Let $\mathcal{L} = \mathbb{Z}^+$ the set of positive integers; and let R be the divisibility relation; i.e., xRy means $x|y$. For any numbers x, y, $\sup(x, y) = \operatorname{lcm}(x, y)$, the least common multiple of x, y; because $x, y| \operatorname{lcm}(x, y)$; and $\operatorname{lcm}(x, y)$ is a divisor of any other common multiple of x, y. Likewise, $\inf(x, y) = \gcd(x, y)$: it is a "lower bound" of x, y (in the sense of the given relation, divisibility); and is the greatest such bound, again in the sense that it is "greater" in the considered relation than any other lower bound.

A basic example is the following. Let c be a set of the universe V, and consider $\mathcal{L} = \mathbb{P}(c)$. The inclusion is an ordering of $\mathbb{P}(c)$. Moreover, any two elements $x, y \in \mathbb{P}(c)$ (i.e., two subsets of c) have a supremum: the union $x \cup y$ is an upper bound for both x, y; and is included in any common upper bound of x, y; and they have an infimum: the intersection $x \cap y$.

Given a vector space, say $\mathbb{R}^n$, the inclusion is an order relation in the set of its subspaces, and gives a lattice. Here the infimum of two subspaces X, Y is their intersection $X \cap Y$; and the supremum is the sum subspace $X + Y$.

The concept of lattice could have been defined in a purely algebraic manner, in the sense that it can be seen as a set with two binary operations that satisfy certain properties. Let us show precisely what we mean.

Proposition 13.1.3. *Suppose $\langle \mathcal{L}, \leq \rangle$ is a lattice and define for arbitrary elements $a, b \in \mathcal{L}$:*

$$a + b = \sup(a, b), \quad a \cdot b = \inf(a, b)$$

Then the operations $+, \cdot$ satisfy the following properties for any elements $a, b, c \in \mathcal{L}$:

(1) Idempotency: $a + a = a \cdot a = a$.
(2) Commutativity: $a + b = b + a, \quad a \cdot b = b \cdot a$.
(3) Reduction: $a \cdot (b + a) = a + (b \cdot a) = a$.
(4) Associativity: $a + (b + c) = (a + b) + c, \quad a \cdot (b \cdot c) = (a \cdot b) \cdot c$.
(5) Absorption: $a \leq b \Leftrightarrow a + b = b \Leftrightarrow a \cdot b = a$.

The proof is left to the exercises. Let us however mention that item (4) of Proposition 13.1.3 can be obtained from the fact that $a + (b + c)$ is the supremum of $\{a, b, c\}$; and that it follows from Definition 13.1.1 that any finite subset of $\mathcal{L}$ has a supremum and an infimum, a property that can be shown inductively.

We now prove that a set with two binary operations $+, \cdot$ that satisfy the equations (1) to (4) of Proposition 13.1.3 can be seen as a lattice. In fact, such a set determines a lattice through the relation defined by condition (5) above.

Proposition 13.1.4. *Let $B \in V$ and $+, \cdot$ binary operations in B that satisfy conditions (1) to (4) of Proposition 13.1.3. Define a relation $\leq$ in B by: $x \leq y \Leftrightarrow x \cdot y = x$. Then $\langle B, \leq \rangle$ is a lattice.*

Proof. Observe first that for $x, y \in B$, $x \cdot y = x \Rightarrow x + y = (x \cdot y) + y = y$ by (2) and (3) of Proposition 13.1.3. The converse implication is entirely analogous;

and so we may define a relation $\leq$ in B by any of the conditions $x \cdot y = x$ or $x + y = y$. We will use freely any of these forms of the definition of $\leq$.

We see that $\leq$ is an ordering: reflexivity follows by (1) of Proposition 13.1.3; antisymmetry is a direct consequence of the definition of $\leq$, and item (4) implies transitivity. Finally, given $x, y \in B$, $x + y$ is an upper bound of x, y by (1) and (4) of Proposition 13.1.4, while if z is any upper bound of x, y, then $x + y \leq z$ by (4) and the definition of $\leq$. Thus $x + y$ is the supremum of x, y and similarly $x \cdot y$ is the infimum. □

Operations in lattices have also a monotony property.

Proposition 13.1.5. *If $+, \cdot$ are the operations of the lattice $\langle \mathcal{L}, \leq \rangle$, then for any $a, b, c \in \mathcal{L}$:*

$$a \leq b \Rightarrow a + c \leq b + c \text{ and } a \cdot c \leq b \cdot c$$

Proof. If $a \leq b$, then $b + c = (a + b) + (c + c) = (a + c) + (b + c)$ by Proposition 13.1.3. Hence $a + c \leq b + c$ by the definition of $\leq$. The second property may be obtained similarly. □

13.1.2 Filters, ideals and duality

Definition 13.1.6. Let C be an ordered set. A subset $F \subseteq C$ is a *filter* when it is not empty and the following conditions hold.

(i) $x, y \in F \Rightarrow \exists u \in F \ \ (u \leq x \wedge u \leq y)$.

(ii) $x \in F,\ y \in C,\ x \leq y \Rightarrow y \in F$.

Definition 13.1.7. Let C be an ordered set. A subset $I \subseteq C$ is an *ideal* when it is not empty and the following hold.

(i) $x, y \in I \Rightarrow \exists u \in I \ \ (x \leq u \wedge y \leq u)$.

(ii) $x \in I,\ y \in C,\ y \leq x \Rightarrow y \in I$.

Examples 13.1.8. If the ordered set C is a lattice, then C itself is both a filter and an ideal, called the *improper* filter or ideal. If an ordered set has a maximum m (respectively, a minimum u), then the set $\{m\}$ (resp., $\{u\}$) is a filter (resp., an ideal) which is the *trivial* filter (resp., *trivial*). More generally, if $a \in C$, then $\{x \in C | a \leq x\}$ (resp., $\{x \in C | x \leq a\}$) is a filter (resp., an ideal) of C: it is the *principal* filter (or ideal, respectively) generated by a.

For instance, in the lattice $\mathbb{Z}^+$ of positive integers with the divisibility relation $x|y$, the principal filter generated by 3 consists of all multiples of 3; while the principal ideal generated by n consists of all the divisors of n, hence it is a finite set. Observe that this terminology is at odds with the notion of the ideal (3) generated by 3 in the ring of integers, as (3) contains all multiples of 3, but (3) is indeed an ideal according to Definition 13.1.7, only it is an ideal with respect to the relation "x is a multiple of y".

If R is an order relation in a set C, then R^{-1} is also an ordering of C. When $\langle \mathcal{L}, R \rangle$ is a lattice, then $\langle \mathcal{L}, R^{-1} \rangle$ is a lattice, because the supremum and infimum of elements a, b are respectively the infimum and the supremum of a, b in the original lattice. This new lattice is called the *opposite or dual lattice* of $\mathcal{L}$ and written as $\mathcal{L}^{op}$. Of course, $(\mathcal{L}^{op})^{op} = \mathcal{L}$.

Remarks and Examples 13.1.9. Let us look at the dual lattices for some of the examples in Remarks 13.1.2. Consider for instance the divisibility relation $x|y$ in $\mathbb{Z}^+$. The inverse relation contains the ordered pairs $\langle x, y\rangle$ such that $y|x$. For this relation, aRb means that a is a multiple of b; so, in the sense of this relation, a multiple of b means that a is "below" b. Given positive integers n, m, the supremum of $\{n, m\}$ in this inverse relation, is their greatest common divisor $\gcd(n, m)$ because divisors are "above" their multiples. At the end of Examples 13.1.8 we have already hinted at this relation: relative to it, the multiples of 3 form an ideal while the divisors of 5 form a filter.

Analogous observations can be made for the other examples of lattices in Remarks 13.1.2; for the lattice $\mathbb{P}(c)$, $a \subseteq b \Leftrightarrow b \supseteq a$; thus in the dual lattice of $\mathbb{P}(c)$, the supremum of two subsets $A, B \subseteq c$ is $A \cap B$.

There is a connection between the properties of a lattice $\mathcal{L}$ and those of its dual lattice $\mathcal{L}^{op}$. Consider ω as a lattice with the membership ordering – since it is a total ordering, this gives a lattice as seen in Remarks 13.1.2. $\langle \omega, \leq\rangle$ has the property that there are no infinite strictly descending sequences (see, e.g., Exercise 5 of Chapter 8), but there are infinite strictly ascending sequences, like $2 < 4 < 7 < 15 < 83 < 101 < \dots$. The inverse relation to membership is $x \ni y$ which, as is usual in orderings, we may represent as $x < y$; and in this inverse lattice, there are no infinite strictly ascending sequences, but now $2 > 4 > 7 > 15 > 83 > 101 > \dots$ is an infinite strictly descending sequence.

Proposition 13.1.10. *Let $\mathcal{L}$ be a lattice, $+, \cdot$ the operations of the lattice. For $x, y, z \in \mathcal{L}$, $z = x+y$ in the lattice $\mathcal{L}$ if and only if $z = x \cdot y$ in the opposite lattice $\mathcal{L}^{op}$.*

Proof. Let $z = x+y$ in $\mathcal{L}$ and $\leq$ the corresponding order relation of the lattice. Thus $x, y \leq z$ so that $z \leq x, y$ in the lattice $\mathcal{L}^{op}$; that is, z is a lower bound of x, y. Also, if $z' \leq x, y$ in $\mathcal{L}^{op}$, then $x, y \leq z'$ in $\mathcal{L}$; since z is the smallest upper bound (in $\mathcal{L}$) of x, y, we deduce that $z \leq z'$ in $\mathcal{L}$ and $z' \leq z$ in $\mathcal{L}^{op}$. This means that z is (in $\mathcal{L}^{op}$) the greatest lower bound of x, y, hence $z = x \cdot y$. The converse property is proved analogously. □

Equations like $x = y + z$ or $a \cdot b = c$, along with inequalities as $x \leq y$ can be called *simple formulas for lattices* – though in fact they are abbreviations of more complex formulas of set theory that may include symbols for the set $\mathcal{L}$ and the relation $\leq$. Then we may call *formulas for lattices* to combinations of those simple formulas by means of the connective and quantifier symbols we have been using. Given such a formula P, its *dual formula* P^* is obtained by replacing the symbols $+, \cdot$ with the symbols $\cdot, +$ respectively; and replacing also $\leq$ with $\geq$ while leaving the elements, variables and logical symbols unchanged.

The same conventions apply to an order relation $\langle \leq, C\rangle$ with respect to the opposite or dual ordering $\geq$, but the only simple formulas in this context are inequalities $x \leq y$.

Proposition 13.1.11. *If a formula P for lattices holds in a lattice $\mathcal{L}$, then the dual formula P^* holds in the lattice $\mathcal{L}^{op}$. Similarly, if a formula for ordered sets holds in $\langle \leq, C \rangle$, then P^* holds in the opposite ordered set.*

Proof. This is easily seen for simple formulas from Proposition 13.1.10. Then it can be extended to formulas with connectives in a straightforward way by finite induction on the number of logical symbols. For instance, let $P = \exists x(Q)$ for some Q which satisfies the statement. Then $P^* = \exists x(Q^*)$; if P holds in the lattice $\mathcal{L}$, then some element $a \in \mathcal{L}$ satisfies Q. Thus a satisfies Q^* in $\mathcal{L}^{op}$ by the inductive assumption, hence P^* holds in $\mathcal{L}^{op}$. □

Examples 13.1.12. In the lattice $\langle \mathbb{Z}^+, | \rangle$ (with the divisibility relation), the following property (we speak normally of "property" in this context, in the sense of the property asserted by some formula) is easily seen to hold: $\forall x\ (\exists y\ (x|y \wedge x \neq y))$. The dual property is $\forall x\ (\exists y\ (y|x \wedge x \neq y))$. Note that this dual property does not hold in the given lattice. However, it holds in the opposite lattice where the relation is: "to be multiple of", as every element has some multiple. The reader will notice that both properties say in fact the same, with different terminology.

Definition 13.1.13. Let $\mathcal{C}$ be a class of lattices. We say that $\mathcal{C}$ is a *self-dual class* of lattices (also, that $\mathcal{C}$ is closed under duality) when $\mathcal{L} \in \mathcal{C} \Rightarrow \mathcal{L}^{op} \in \mathcal{C}$.

Proposition 13.1.14. *Let $\mathcal{C}$ be a class of lattices that is self-dual. If some property P holds in every lattice of $\mathcal{C}$, then its dual property P^* holds in every lattice of $\mathcal{C}$.*

Proof. Assume that P holds in every lattice of $\mathcal{C}$, and consider an arbitrary lattice $\mathcal{L} \in \mathcal{C}$; we show that P^* holds in $\mathcal{L}$. By hypothesis, P holds in $\mathcal{L}^{op}$ because $\mathcal{L}^{op} \in \mathcal{C}$. By Proposition 13.1.11, P^* holds in $(\mathcal{L}^{op})^{op} = \mathcal{L}$. □

This is the *principle of duality for lattices.* As a particular case of the preceding proposition, if the property P is valid in all lattices, then its dual P^* is also valid in all lattices, because the class of all lattices is trivially self-dual.

Filter and ideal are dual notions; that is, a non-empty subset $F \subseteq C$ is a filter if and only if F is an ideal of the opposite ordered set. Thus the properties of filters and ideals show this duality: if a property (given in terms of the ordering) is valid for all filters in all ordered sets, the dual property is valid for all ideals in all ordered sets. We accordingly apply this principle in the next result, where it is shown that the conditions for filter and ideal may be given in terms of the operations when the ordered set is a lattice.

Proposition 13.1.15. *Let $\mathcal{L}$ be a lattice, $\emptyset \neq F \subseteq \mathcal{L}$. Then F is a filter of $\mathcal{L}$ if and only if the following conditions hold.*

(1) $x, y \in F \Rightarrow x \cdot y \in F$.

(2) $x \in F, b \in \mathcal{L} \Rightarrow x + b \in F$.

Dually, for $\emptyset \neq I \subseteq \mathcal{L}$, I is an ideal if and only if the following conditions hold.

(1) $x, y \in I \Rightarrow x + y \in I$.

(2) $x \in I, b \in \mathcal{L} \Rightarrow x \cdot b \in I$.

Proof. Exercise. □

A filter or ideal of the lattice $\mathcal{L}$ which is $\neq \mathcal{L}$ is called *proper*. And a filter F is a *maximal filter* when it is proper and the following implication holds: (F' is a proper filter of $\mathcal{L}$ and $F \subseteq F'$) $\Rightarrow F' = F$. The same implication characterizes *maximal ideals.*

Examples 13.1.16. For an infinite set c, let us consider again the lattice $\mathbb{P}(c)$ of subsets of c with the relation $X \subseteq Y$. The subset of $\mathbb{P}(c)$ consisting of all the finite subsets of c is an ideal, according to Proposition 13.1.15: the union of two finite subsets is finite, and the intersection of a finite subset and any subset is finite again. By duality, the finite subsets of c form a filter with respect to the opposite relation $X \supseteq Y$. But another filter is more interesting, the *cofinite filter* of $\mathbb{P}(c)$: its elements are all the cofinite subsets $X \subseteq c$; i.e. the subsets X such that $c \setminus X$ is finite. The two properties of Proposition 13.1.15 are easily checked in this case. These are examples of ideals and filters which are not principal.

13.1.3 Distributive and complemented lattices

In an arbitrary lattice, the distributive property $a \cdot (b + c) = a \cdot b + a \cdot c$ need not hold. For example, the lattice of subspaces of a vector space is in general not distributive: take $\mathcal{L} = \mathbb{R}^3$ and the subspaces a, b, c being respectively, the plane $z = 0$ and the lines $x = y = 0$ and $x = 0, y = z$. The equation for distributivity for these subspaces would be $a \cap (b + c) = (a \cap b) + (a \cap c)$. Now, $a \cap b = a \cap c = (0)$, so the second term in the above equation gives the trivial subspace (0), but the sum of the two given lines is a plane, namely $x = 0$, and then $a \cap (b + c)$ is the intersection of two different planes, hence it is a line, namely, $x = z = 0$.

Despite this, some properties connected to distributivity hold in any lattice.

Proposition 13.1.17. *Let $\langle \mathcal{L}, \leq \rangle$ be a lattice. The following relations hold for any $a, b, c \in \mathcal{L}$.*

(i) $a \cdot b + a \cdot c \leq a \cdot (b + c)$

(ii) $(a + b) \cdot (a + c) \leq a + (b \cdot c)$.

Moreover, equality in (i) holds for every $a, b, c \in \mathcal{L}$ if and only if the equality in (ii) holds also for every a, b, c.

Proof. (i) Since $b \leq b + c$ by definition, $a \cdot b \leq a \cdot (b + c)$ by Proposition 13.1.5. Similarly, $a \cdot c \leq a \cdot (b + c)$ and this entails the result by the definition of the sum.

(ii) is obtained by duality.

For the final assertion, assume (i) holds generally. Then

$(a+b)\cdot(a+c) = (a+b)\cdot a + (a+b)\cdot c = a + (a+b)\cdot c = a + (a\cdot c + b\cdot c) = (a + (a\cdot c)) + b\cdot c = a + b\cdot c$

where we have used the hypothesis twice, along with the commutative, associative and reductive properties. This shows (ii), and the converse follows again by duality. □

Definition 13.1.18. A lattice $\mathcal{L}$ is *distributive* when the following holds for any $a, b, c \in \mathcal{L}$:

$$a\cdot(b+c) = a\cdot b + a\cdot c, \quad a + (b\cdot c) = (a+b)\cdot(a+c)$$

According to Proposition 13.1.17, the lattice $\mathcal{L}$ is distributive if and only if $a\cdot(b+c) \leq a\cdot b + a\cdot c$ for any $a, b, c \in \mathcal{L}$. On the other hand, a distributive lattice $\mathcal{L}$ always satisfies the *modular property*:

$$a \leq c \Rightarrow a + (b\cdot c) = (a+b)\cdot c$$

a property which is known to hold in lattices of submodules of a module, which are not necessarily distributive. Of course, if $\mathcal{L}$ is distributive, then

$$a \leq c \Rightarrow a + (b\cdot c) = (a+b)\cdot(a+c) = (a+b)\cdot c$$

because $a + c = c$ when $a \leq c$. Note that the dual of a distributive lattice is also distributive; therefore the class of distributive lattices is self-dual and the principle of duality (Proposition 13.1.14) is valid for distributive lattices.

If an element e of a lattice $\mathcal{L}$ is the minimum of $\mathcal{L}$, then obviously $e+x = x$ and $e\cdot x = e$ for each $x \in \mathcal{L}$. If it exists, this element will be denoted as 0, identified as the unique element satisfying any of the properties $0 + x = x \ (\forall x \in \mathcal{L})$ or $0\cdot x = 0 \ (\forall x \in \mathcal{L})$. Dually, an element $u \in \mathcal{L}$ that satisfies $u\cdot x = x$ for all $x \in \mathcal{L}$, or equivalently $u + x = u \ (\forall x \in \mathcal{L})$, will be written as 1 and is the maximum element of the lattice.

Definition 13.1.19. Let $\mathcal{L}$ be a lattice with minimum and maximum elements $0, 1$. Given $x \in \mathcal{L}$, an element $y \in \mathcal{L}$ is *a complement* of x if $x + y = 1$ and $x\cdot y = 0$. If every element of $\mathcal{L}$ has a complement, then we say that $\mathcal{L}$ is a *complemented* lattice.

By definition, a complemented lattice has the elements $0, 1$. When the lattice is distributive, a complement of x, if it exists, is unique so that we may speak of **the** complement of x: suppose $x+y = x+z = 1$ and $x\cdot y = x\cdot z = 0$. Then $y = y\cdot(x+z) = y\cdot x + y\cdot z = y\cdot z$, hence $y \leq z$. Dually one sees that $z \leq y$ whence $y = z$. The complement of $x \in \mathcal{L}$ (when it exists) will be written $x^{\perp}$.

Example 13.1.20. Let us consider again the divisibility relation of Remarks 13.1.2, but now in the set $\mathcal{L}$ of natural numbers instead of $\mathbb{Z}^+$. This lattice is distributive; and it has a maximum and a minimum, 0 and 1 respectively. Like this, the 0 of the lattice (in the terminology of Definition 13.1.19), is the number 1 and the 1 of the lattice is the number 0. But the lattice is not complemented: for each $n \in \mathbb{N}$ we may find several m with the property $n \cdot m = 0$ (that is, with numbers, $\gcd(n, m) = 1$). And if $n \neq 0, 1$, such elements m do not satisfy the property $n + m = 1$; this would mean $\text{lcm}(n, m) = 0$, which is only possible if $m = 0$, but then $\gcd(n, 0) = n \neq 1$.

Proposition 13.1.21. *Let $\mathcal{L}$ be a distributive and complemented lattice. Then the following hold for any $x, y \in \mathcal{L}$:*

(i) $(x^\perp)^\perp = x$. *Moreover,* $1^\perp = 0,\ 0^\perp = 1$.

(ii) $(x + y)^\perp = x^\perp \cdot y^\perp, \quad (x \cdot y)^\perp = x^\perp + y^\perp$.

(iii) $x \leq y \Leftrightarrow x \cdot y^\perp = 0 \Leftrightarrow x^\perp + y = 1 \Leftrightarrow y^\perp \leq x^\perp$.

Proof. (i) follows immediately from the definition of complement. The two properties in (ii) are dual to each other so we just prove the first one.

$$(x + y) \cdot (x^\perp \cdot y^\perp) = xx^\perp y^\perp + yx^\perp y^\perp = 0 + 0 = 0$$
$$(x + y) + (x^\perp \cdot y^\perp) = (x + y + x^\perp) \cdot (x + y + y^\perp) = 1 \cdot 1 = 1.$$

For (iii), it is enough to prove the implications to the right: the way to the left will follow from this, bearing in mind items (i) and (ii). Now, $x \leq y$ implies $x \cdot y^\perp \leq y \cdot y^\perp = 0$ by Proposition 13.1.5. If $x \cdot y^\perp = 0$, then $(x \cdot y^\perp)^\perp = 1 = x^\perp + y$ by item (ii). Finally, from $x^\perp + y = 1$ we get $y^\perp = y^\perp \cdot (x^\perp + y) = y^\perp \cdot x^\perp$ and the inequality follows by Proposition 13.1.4. □

13.2 Boolean Algebras

A Boolean algebra is a lattice which is distributive and complemented. But it is possible to give a definition solely in terms of the algebraic operations of the lattice. This is the definition we present now.

Definition 13.2.1. Let $\langle B, +, \cdot \rangle$ a triple where B is a set and $+, \cdot$ are binary operations in B. Then $\langle B, +, \cdot \rangle$ is called a *Boolean algebra* when the following properties hold:

(1) Conditions (1) to (4) of Proposition 13.1.3 are satisfied.

(2) The distributivity equations

$$a \cdot (b + c) = a \cdot b + a \cdot c, \quad a + (b \cdot c) = (a + b) \cdot (a + c)$$

also hold.

(3) B contains two elements $0 \neq 1$ such that, for every $x \in B$, $0 \cdot x = 0,\ 0 + x = x,\ 1 \cdot x = x,\ 1 + x = 1$.

(4) For each $x \in B$, there exists an element $x^\perp \in B$ such that $x \cdot x^\perp = 0$ and $x + x^\perp = 1$.

As usual, we will say *the Boolean algebra B* to refer to the first component of the triple giving the Boolean algebra, assuming that the operations may be understood. When B is a Boolean algebra, a (Boolean) *subalgebra* is any $A \subseteq B$ which contains 0 and is closed under sums, products and complements; that is, if $a, b \in A$ then $a + b, a \cdot b, a^\perp$ belong to A. It is easy to check that this gives indeed A as a Boolean algebra. A subalgebra of the Boolean algebra $\mathbb{P}(S)$ is called an *algebra of sets.*

Examples 13.2.2. The simplest example of a Boolean algebra is the set $\{0, 1\}$ where the lattice defined as in Proposition 13.1.4 has $0 < 1$ as the only strict inequality. But the more important general example is that of the power set $\mathbb{P}(S)$ of a set S: the operations, union and intersection, are mutually distributive by Proposition 2.4.5, $\emptyset$ and S are the 0 and 1, and the complement of $X \subseteq S$ is $S \setminus X$.

If $B = \mathbb{P}(S)$ is the Boolean algebra for an infinite set S, then the set of all the subsets of B that are either finite or cofinite (see Example 13.1.20) is a subalgebra: the intersection of a finite and a cofinite set is finite, their union is cofinite, the complement of a finite set is cofinite and vice versa.

Definition 13.2.3. A map $f : B_1 \to B_2$, with B_1, B_2 Boolean algebras, is a *homomorphism* (of Boolean algebras) when the following hold for every $x, y \in B$:

$$f(x + y) = f(x) + f(y), \quad f(x \cdot y) = f(x) \cdot f(y), \quad f(0) = 0, \quad f(1) = 1$$

If f is a homomorphism of Boolean algebras, then f preserves the orderings. For if $x \leq y$, then $x + y = y$ and $f(x + y) = f(y) = f(x) + f(y)$ whence $f(x) \leq f(y)$. It also preserves complements: $f(x) + f(x^\perp) = f(x + x^\perp) = f(1) = 1$, and similarly $f(x) \cdot f(x^\perp) = 0$; by the definition of complement, $f(x^\perp) = f(x)^\perp$. When the homomorphism $f : B_1 \to B_2$ is bijective, then it is an *isomorphism* (i.e., the inverse map $f^{-1} : B_2 \to B_1$ is a homomorphism too). The proof is easy and left to the reader.

We next turn to consider filters and ideals in Boolean algebras. When B is a Boolean algebra, then $1 \in F$, $0 \in I$ for any filter F or any ideal I. Also, the set $\{0\}$ is the *trivial* ideal and $\{1\}$ is the *trivial* filter. From a filter F we may form an ideal by taking $I(F) = \{x^\perp | x \in F\}$; and given an ideal I, there is a filter $F(I) = \{x^\perp | x \in I\}$. This association preserves inclusions and $I(F)$ is proper if and only if F is proper.

The formal identity between the definition of an ideal for a ring and the characterization of ideals in Proposition 13.1.15 is obvious. A further step in this connection is the concept of a prime ideal.

Definition 13.2.4. A proper ideal I of a Boolean algebra B is *prime* when for every $x, y \in B$, $x, y \notin I \Rightarrow x \cdot y \notin I$.

If I is a prime ideal, then $x \notin I \Rightarrow x^{\perp} \in I$ because $x \cdot x^{\perp} = 0 \in I$. Conversely, for an ideal I, the property $x \notin I \Rightarrow x^{\perp} \in I$ implies that I is prime or improper. Because then $x, y \notin I \Rightarrow x^{\perp}, y^{\perp} \in I$ and $x^{\perp} + y^{\perp} = (x \cdot y)^{\perp} \in I$; if I is proper, $x \cdot y \notin I$ and thus we get the implication $x, y \notin I \Rightarrow x \cdot y \notin I$, hence I is prime.

Prime ideals are maximal among proper ideals. Indeed, suppose that I is a prime ideal and $I \subseteq J$, with J an ideal. Take $x \in J \backslash I$ if possible, so $x^{\perp} \in I$. But then $x^{\perp} \in J$ and $x + x^{\perp} = 1 \in J$, hence $J = B$. Conversely, if I is a proper maximal ideal, then I is prime: if $x \notin I$, then $I + (x) = \{u | \exists v \in I, u \leq v + x\}$ is an ideal which includes I properly and hence $1 \leq v + x$ for some $v \in I$. But then $x^{\perp} \leq v$ by Proposition 13.1.21, and $x^{\perp} \in I$.

The dual concept and properties have an interest of their own.

Definition 13.2.5. A proper filter F of a Boolean algebra B is an *ultrafilter* when for every $x, y \in B$, $x, y \notin F \Rightarrow x + y \notin F$.

Proposition 13.2.6. *Let F be a proper filter of the Boolean algebra B. The following conditions are equivalent.*

(i) F is an ultrafilter.

(ii) $x \notin F \Rightarrow x^{\perp} \in F$.

(iii) F is maximal among all proper filters of B.

Proof. This is dual to the corresponding properties of ideals seen above. □

Theorem 13.2.7. *(The ultrafilter theorem) Every proper filter of a Boolean algebra B is included in an ultrafilter.*

Proof. Given the proper filter F, the proper filters of B which include F form a non-empty set $\mathcal{K}$, which is ordered by inclusion. If $\mathcal{C}$ is a chain of $\mathcal{K}$ with respect to this ordering, then $\cup\mathcal{C}$ is again a filter that includes F: if $x, y \in \cup\mathcal{C}$ then there are filters F_1, F_2 in $\mathcal{C}$ with $x \in F_1, y \in F_2$; and if, for example, $F_1 \subseteq F_2$, then $x \cdot y \in F_2$ and $x \cdot y \in \cup\mathcal{C}$. Moreover, $\cup\mathcal{C}$ is proper (hence it belongs to $\mathcal{K}$) because $0 \in \cup\mathcal{C}$ would give $0 \in F \in \mathcal{C}$, which is impossible; and $\cup\mathcal{C}$ is an upper bound of $\mathcal{C}$ in $\mathcal{K}$; thus Zorn's lemma can be applied, so that there exists a filter which is maximal among proper filters of B including F. It is immediate that this is a maximal filter, hence it is an ultrafilter by Proposition 13.2.6. □

Under the duality filters-ideals, ultrafilters correspond to prime ideals. As a consequence of Theorem 13.2.7, every proper ideal is included in a prime ideal. This is the *prime ideal theorem*, and both dual theorems are a noteworthy consequence of the axiom of choice. The next important result is *Stone's theorem* which shows that each Boolean algebra can be seen as an algebra of sets.

Theorem 13.2.8. *For any Boolean algebra B there exists a set S such that B is isomorphic to a subalgebra of the algebra $\mathbb{P}(S)$ of subsets of S.*

Proof. Consider the non-empty set S of all ultrafilters of B and define a map $p : B \to \mathbb{P}(S)$ given thus: $p(x) = \{F \in S | x \in F\}$. We see that p is an injective homomorphism of Boolean algebras.

Obviously, $p(0) = \emptyset$ and $p(1) = S$. By definition, $x+y \in F$ for an ultrafilter F if and only if $x \in F$ or $y \in F$. Therefore $p(x + y) = p(x) \cup p(y)$. By the definition of filters, $x \cdot y \in F \Leftrightarrow x, y \in F$ and thus $p(xy) = p(x) \cap p(y)$. Moreover, the homomorphism p is injective: given $x \neq y$, then $x \cdot y^{\perp} \neq 0$ or $y \cdot x^{\perp} \neq 0$, by Proposition 13.1.21. If, for example, $x \cdot y^{\perp} \neq 0$, then $\{u \in B | x \cdot y^{\perp} \leq u\}$ is a proper filter and is included in an ultrafilter $F \in S$ by Proposition 13.2.7. Thus $x \in F, y \notin F$, hence $p(x) \neq p(y)$.

It is easily seen that the codomain of a Boolean algebra homomorphism is a subalgebra of the second algebra. Since B is isomorphic to the codomain of the homomorphism p, the proof is complete. □

When B is a Boolean algebra, the subset $B^* = B \setminus \{0\}$ is an ordered set such that any two elements have a supremum. Moreover, each $x \in B^*$ other than 1 has a complement $x^{\perp} \in B^*$, which may be characterized by the property that $x^{\perp}$ is the minimum of all elements $y \in B^*$ such that $x + y = 1$, by Proposition 13.1.21. Also, if F is a proper filter of B then $F \subseteq B^*$ is also a filter of B^*, since the product of two elements of F is never 0. Conversely, if F is a filter of B^*, then it is a proper filter of B.

13.3 Complete Lattices and Algebras

13.3.1 Complete Boolean algebras

Definition 13.3.1. Let $\mathcal{L}$ be a lattice. $\mathcal{L}$ is a *complete lattice* if every non-empty subset of $\mathcal{L}$ has a supremum and an infimum. A Boolean algebra B is *complete* when the lattice $\langle B, \leq \rangle$ obtained from the algebra is a complete lattice.

Once again, our main example of a complete Boolean algebra is afforded by the power set $\mathbb{P}(S)$ of a non-empty set S. It is easily seen that the class of complete Boolean algebras is self-dual. The existence of supremum and infimum allows a generalization of the operations of sum and product in the lattice. For the complete lattice $\mathcal{L}$ and a subset $X \subseteq \mathcal{L}$, we may write

$$\sum_{u \in X} u = \sup(X), \quad \prod_{u \in X} u = \inf(X)$$

and of course these definitions generalize the known ones for X having only two elements. We then have an extension of the distributive property.

Proposition 13.3.2. *Let B be a complete Boolean algebra and $X, Y \subseteq B$. Then*

$$(\sum_{u\in X} u)\cdot(\sum_{v\in Y} v) = \sum_{\langle u,v\rangle\in X\times Y} u\cdot v, \quad (\prod_{u\in X} u)+(\prod_{v\in Y} v) = \prod_{\langle u,v\rangle\in X\times Y} (u+v)$$

Proof. We prove first the case when $Y = \{v\}$. Consider for instance the second equality and let $a = \prod_{u\in X} u$, that is, $a = \inf(X)$. We need to show that $a+v$ is the infimum b of all elements of the form $u+v$ for $u \in X$. By the choice of a and Proposition 13.1.5, $a+v \leq u+v$ for any $u \in X$; and suppose that $d \leq u+v$ for all $u \in X$; then $d \cdot v^{\perp} \leq u \cdot v^{\perp} \leq u$ for each $u \in X$. Thus $d \cdot v^{\perp} \leq a$. On the other hand, $d \cdot v \leq v$ and hence $d = (d\cdot v)+(d\cdot v^{\perp}) \leq a+v$, from which it follows that $a+v = b$ as we wanted to see.

For the general case of the statement, let $\prod_{v\in Y} v = c$, so that the first term is now $(\prod_{u\in X} u) + c = \prod_{u\in X}(u+c)$. The second term is $\prod_{u\in X}(\prod_{v\in Y}(v+u)) = \prod_{u\in X}(c+u)$, by the first part again. This shows the equality. Complete Boolean algebras are self-dual and thus the first equality follows by duality. □

Corollary 13.3.3. *Let B be a complete Boolean algebra and X a subset of B. Then $\prod_{u\in X} u^{\perp} = (\sum_{u\in X} u)^{\perp}$ and $\sum_{u\in X} u^{\perp} = (\prod_{u\in X} u)^{\perp}$.*

Proof. We use again the duality of complete Boolean algebras and prove only the first relation. By the uniqueness of complements seen after Definition 13.1.19, it will suffice to check that $(\prod_{u\in X} u^{\perp}) + (\sum_{u\in X} u) = 1$ and $(\prod_{u\in X} u^{\perp}) \cdot (\sum_{u\in X} u) = 0$.

By the distributivity properties (Proposition 13.3.2),

$$(\prod_{u\in X} u^{\perp})+(\sum_{u\in X} u) = \prod_{u\in X}(u^{\perp}+\sum_{u\in X} u) = 1$$

since each factor is $\geq u^{\perp}+u = 1$. Similarly,

$$(\prod_{u\in X} u^{\perp})\cdot(\sum_{u\in X} u) = \sum_{u\in X}(\prod_{u\in X} u^{\perp})\cdot u = 0$$

as every summand is $\leq u^{\perp}\cdot u = 0$. □

13.3.2 Separative orderings and complete algebras

We recall (see Definition 4.4.6) the notation $[x]$ for an element x of an ordered set P: $[x] = \{y \in P | y \leq x\}$. When necessary, we shall write $[x]_P$ to emphasize the field P of the relation.

Definition 13.3.4. Let P be an ordered set and $S_1, S_2 \subseteq P$. We say that S_1 is *dense* in S_2 when for every $x \in S_2$, there is $u \in S_1$ such that $u \leq x$. A subset $S \subseteq P$ is called a *dense subset* when S is dense in P.

This definition extends the usual one, which is the same but applied only in case $S_1 \subseteq S_2$. Density is a transitive relation: if S_1, S_2, S_3 are subsets of the ordered set P, and S_i is dense in S_{i+1} (for $i = 1, 2$), then S_1 is clearly dense in S_3.

Lemma 13.3.5. *Let $C \subseteq B$ and B a complete Boolean algebra. $C \setminus \{0\}$ is dense in $B \setminus \{0\} = B^*$ if and only if every $x \in B^*$ is the supremum (in B) of the set $C \cap [x]_B = \{u \in C | u \leq x\}$.*

Proof. Suppose $C \setminus \{0\}$ is dense in B^* and $x \in B^*$. Let $y \in B$ the supremum of the set $S = \{u \in C | u \leq x\}$. Obviously, $y \leq x$. If $x \neq y$, then $x \not\leq y$ so that $xy^\perp \neq 0$ by Proposition 13.1.21. By density, there is $0 \neq u \in C$ with $u \leq xy^\perp$. Thus $u \leq x$ hence $u \in S$ and $u \leq y$. But also $u \leq y^\perp$ and therefore $u = u \cdot y \cdot y^\perp = 0$, absurd. Therefore $y = x$.

Conversely, suppose the condition of the statement is satisfied and let $0 \neq b \in B^*$. Then b is the supremum of the set $S' = \{u \in C | u \leq b\}$ so that $S' \neq \emptyset, \{0\}$. Hence there exists $u \in C$ with $u \leq b$ and $u \neq 0$. □

Definition 13.3.6. Let P be a non-empty ordered set with the relation $\leq$. The ordering of P is *separative* if the following holds for each pair of elements $x, y \in P$: if $[x]$ is dense in $[y]$, then $y \leq x$.

Note that P is separative if and only if, for every $x, y \in P$, $x \leq y \Leftrightarrow [y]$ is dense in $[x]$. Observe also that if the ordered set P has a minimum element (e.g., if P is a Boolean algebra with a 0) then P is not separative unless it is trivially reduced to one element, i.e., it is the trivial non-empty ordered set. When an order relation $\leq$ in the set P is separative, we say also that P is separative, with the relation $\leq$ itself being understood. A trivial example of a separative ordered set is obtained by taking the identity relation $\mathcal{E}_C$ in any set C: then $[x] = \{x\}$, so $[x]$ is dense in $[y]$ if and only if $x = y$. A more interesting example follows.

Proposition 13.3.7. *Let D be a dense subset of B^* where B is a Boolean algebra. Then the induced ordering of D is separative.*

Proof. We use RAA and suppose $a, b \in D$ such that $[a]_D$ is dense in $[b]_D$ but $b \not\leq a$. Then $b \cdot a^\perp \neq 0$ by Proposition 13.1.21. By density, there is $u \in D$ such that $0 \neq u \leq b \cdot a^\perp$. Then $u \cdot a = 0$ so $[u]_D \cap [a]_D = \emptyset$ while $u \in [b]_D$, which contradicts the density of $[a]_D$ in $[b]_D$. □

Recall Definition 8.1.1: $S \subseteq C$ is a segment of the ordered set C when $(x \in S \wedge y \leq x) \Rightarrow y \in S$. In particular, $[x]_C$ is always a segment of C. We give now a refinement of the notion of segment.

Definition 13.3.8. Let C be a segment in an ordered set P. C is a *regular segment* when the following property holds for any $x \in P$: if C is dense in $[x]$ then $x \in C$.

Lemma 13.3.9. *Let P be a non-empty ordered set. Then P is separative if and only if $[x]$ is a regular segment for every $x \in P$.*

Proof. This is immediate from the definition of separative orderings. □

Lemma 13.3.10. *Let P be a separative ordered set with a dense subset C. If C is a regular segment, then $C = P$.*

Proof. Let $x \in P$. By hypothesis, C is dense in $[x]$ whence $x \in C$ by Definition 13.3.8. □

The main result of this section shows that any separative ordered set can be embedded in a complete Boolean algebra B in such a way that the set is dense in B^*. To simplify the exposition and also in view of future use, we introduce here some new terms for elements in ordered sets. If x, y are elements in an ordered set C, we say that x, y are (mutually) *compatible* when $[x] \cap [y] \neq \emptyset$. When x, y are not compatible we say x, y are *incompatible*; i.e., x, y incompatible means that $[x] \cap [y] = \emptyset$. It is immediate to characterize separative orderings in these terms, and this characterization is frequently useful: the ordered set C is separative if and only if the following implication holds for any $x, y \in C$: $x \nleq y \Rightarrow \exists u \in [x]$ $(u, y$ are incompatible$)$.

Definition 13.3.11. Let $f : P_1 \to P_2$ be a monotone map (recall Definition 8.2.1) of non-empty ordered sets. We say that f is a *preserving map* when for any $x, y \in P_1$, x, y are compatible if and only if $f(x), f(y)$ are compatible elements of P_2.

Note that the monotone map f is preserving if and only if f preserves compatible and incompatible elements; i.e., if x, y are compatible (respectively, incompatible) in P_1, then $f(x), f(y)$ are compatible (resp., incompatible) in P_2.

Lemma 13.3.12. *Let P_1, P_2 be separative ordered sets and $f : P_1 \to P_2$ a monotone map such that $f[P_1]$ is dense in P_2. Then f is a preserving map if and only if the following holds for any $x, y \in P_1$: $x \leq y \Leftrightarrow f(x) \leq f(y)$.*

Proof. It is straightforward to see that any monotone map takes compatible elements to compatible elements. Suppose now that f is preserving, but there are $x, y \in P_1$ with $f(x) \leq f(y)$ but $x \nleq y$. By Definition 13.3.6, $[y]$ is not dense in $[x]$, from which it follows that there exists $u \leq x$ with u, y incompatible. By preservation of incompatibility, $[f(u)] \cap [f(y)] = \emptyset$, but $f(u) \leq f(x) \leq f(y)$, a contradiction.

Conversely, suppose that $x \leq y \Leftrightarrow f(x) \leq f(y)$ holds, and take $x, y \in P_1$ such that $f(x), f(y)$ are compatible. Thus $[f(x)] \cap [f(y)] \neq \emptyset$ and there is $u \in P_2$ with $u \leq f(x), f(y)$. By the density hypothesis, there is $v \in P_1$ such that $f(v) \leq u$ and thus $f(v) \leq f(x), f(y)$. Hence $v \leq x, y$ by the hypothesis, and this shows that x, y are compatible. Therefore f is a preserving map. □

It follows from the lemma that a preserving map between separative sets is always injective, that is, a morphism of ordered sets.

Theorem 13.3.13. *Let P be a separative ordered set. There is a complete Boolean algebra B and a preserving map $f : P \to B^*$ such that $f[P]$ is a dense subset of B^*.*

Proof. We take B as the set of all regular segments of P and $f(x) = [x]$ which is a regular segment of P by Lemma 13.3.9. In B we consider the inclusion as the order relation. We will show that B is a complete lattice. First, $\emptyset$ and P are trivially regular segments and are the minimum and maximum of B, respectively. The intersection of any set of regular segments is a regular segment: if the set is $\{C_i\}_{i\in I}$, then $\cap_I C_i$ is clearly a segment; and if $\cap_I C_i$ is dense in $[x]$, then each C_i is dense in $[x]$ from which it follows that $x \in C_i$ for every index i. Thus $\cap_I C_i$ is the infimum of the set $\{C_i\}_{i\in I}$.

If C is any segment of P, then we take $\mathcal{R}$ as the (non-empty) set of all regular segments that include C; then $\cap\mathcal{R}$ is a regular segment, and therefore it is the smallest element of B which includes C. We denote it[1] as $\overline{C}$. Then $\overline{\cup_I C_i}$ is obviously the supremum of the set $\{C_i\}_{i\in I}$ of elements of B. Along with the easily checked properties (1) to (4) of Proposition 13.1.3, this proves that B is indeed a complete lattice.

As in every lattice, we may consider the operations $+$ and $\cdot$, and we see next that the lattice is distributive. By Proposition 13.1.17, we must only prove that $C_1 \cdot (C_2 + C_3) \subseteq C_1 \cdot C_2 + C_1 \cdot C_3$ for regular segments C_1, C_2, C_3. The reader will prove in the exercises that for whichever segments X, Y, we have $\overline{X} \cap \overline{Y} \subseteq \overline{X \cap Y}$. Thus in our case, $C_1 \cdot (C_2 + C_3) = C_1 \cap \overline{C_2 \cup C_3} \subseteq \overline{C_1 \cap (C_2 \cup C_3)} = \overline{(C_1 \cap C_2) \cup (C_1 \cap C_3)} = (C_1 \cdot C_2) + (C_1 \cdot C_3)$.

We show now that B is complemented. Given $C \in B$, let $C' = \{x \in P | [x] \cap C = \emptyset\}$. Clearly, C' is a segment. Suppose C' is dense in $[x]$ for $x \in P$ but $x \notin C'$. Then there is $u \leq x$ with $u \in C$ by the definition of C'. Since C' is dense in $[x]$, there is $v \in C'$ such that $v \leq u$, so $v \in C$. But $C \cap C' = \emptyset$ and the contradiction shows that C' is a regular segment.

We see that $C' = C^{\perp}$ by showing that $C + C'$ is dense in P from which $C + C' = P$ by Lemma 13.3.10. It is enough to show that $C \cup C'$ is dense in P. If $x \in P \setminus C'$ then there exists $y \leq x$ such that $y \in C \subseteq C \cup C'$ and we are done.

So B is a complete Boolean algebra with $0 = \emptyset$. Now, if $\emptyset \neq C \in B$, then for $x \in C$, $[x] = f(x) \subseteq C$ and $[x] \neq 0$. This shows that $f[P]$ is dense in B^*. It follows obviously from the definition of B that for any $x, y \in P$, $x \leq y \Leftrightarrow [x] \subseteq [y] \Leftrightarrow f(x) \leq f(y)$, so that f is preserving by Lemma 13.3.12. $\square$

---

[1] This notation collides with its having been used with other meanings; for instance, to note elements in a quotient set. But it is generally used to identify closures in different contexts and here it is also a kind of closure what is meant. Anyway, this will not create any problem, since the notation will only be used in this proof and related arguments.

If P is the trivial separative ordered set with just one element, then the corresponding complete Boolean algebra is the trivial algebra $\{0, 1\}$. The complete Boolean algebra B of Theorem 13.3.13 with the preserving map f is called the *canonical completion algebra* of the ordered set P. Other complete algebras with the same properties may exist, but there is a high degree of uniqueness.

Proposition 13.3.14. *Let P be a separative ordered set, A_1, A_2 complete Boolean algebras and $f : P \to A_1^*$, $g : P \to A_2^*$ preserving maps, so that $f[P], g[P]$ are dense subsets of A_1^*, A_2^* respectively. Then there is an isomorphism $h : A_1 \to A_2$ of Boolean algebras such that $h \circ f = g$.*

Proof. It will be enough to show that if $f : P \to B^*$ is the map of the canonical completion, A is a complete Boolean algebra and $g : P \to A^*$ is a preserving map with $g[P]$ dense in A^*, then there is an isomorphism $h : B \to A$ satisfying $h \circ f = g$. The statement of the proposition will then be an easily deduced consequence.

Let us start by proving the following claim, under the hypotheses of the proposition. Let S be a segment of P, $z \in A^*$ is the supremum of $g[S]$, and $u \in P$. Then S is dense in $[u] \Leftrightarrow g(u) \leq z$.

Suppose first that S is dense in $[u]$ but $g(u) \not\leq z$; by separativity of A^* and density of $g[P]$ in A^*, there is some $u' \in P$ with $g(u') \leq g(u)$ and $g(u'), z$ incompatible. Since g is preserving, $u' \leq u$; and $g(u')$ is incompatible with every element of $g[S]$, from which it follows that u' is incompatible with every element of S. But this contradicts the hypothesis that S is dense in $[u]$, hence in $[u']$.

For the converse, take $u \in P, g(u) \leq z$, and suppose that S is not dense in $[u]$. Then, there exists $u' \leq u$ such that u', x are incompatible for every $x \in S$. Therefore $g(u'), g(x)$ are incompatible in A^* and, working in the algebra A, $g(u')g(x) = 0$ (for every $x \in S$). Then $g(u') \cdot (\sum_{x \in S} g(x)) = \sum_{x \in S}(g(u') \cdot g(x)) = 0 = g(u') \cdot z$. But $g(u') \leq z$, hence $g(u') = 0$ which is absurd. This proves our claim.

We now define $h : B^* \to A$, setting $h(S) = \sup(\{g(u) | u \in S\})$. h is clearly a monotone map. Let us show that it is surjective. Let $z \in A^*$ (we have $g(\emptyset) = 0$); by Lemma 13.3.5, $z = \sup(T)$ with $T = \{g(x) | g(x) \leq z \wedge x \in P\}$. By the claim, T is a regular segment of P, hence $z = h(T)$. The equation $h \circ f = g$ is immediate. Finally, h is preserving: if $h(S_1) = z_1 \leq z_2 = h(S_2)$ and $u \in S_1$, then $g(u) \leq z_1 \leq z_2$ and S_2 is dense in $[u]$, whence $u \in S_2$ because S_2 is regular; that is, $h(S_1) \leq h(S_2)$ implies $S_1 \subseteq S_2$.

This proves that h is an isomorphism of ordered sets, and this plainly entails that it is an algebra isomorphism, because the operations are determined by the order. □

13.3.3 The completion of an ordered set

As just shown, separative ordered sets can be seen as dense subsets of A^* for a complete Boolean algebra A. This same idea of *completing to* *algebras* can be applied to arbitrary ordered sets. We start with the necessary definition.

Definition 13.3.15. Let P be a non-empty ordered set without minimum element, A a complete Boolean algebra and $f : P \to A^*$ a monotone map. The pair $\langle A, f\rangle$ is a *completion algebra* (or a completion) of P when f is a preserving map and $f[P]$ is a dense subset of A^*.

Usually, we speak of the algebra A as the *completion* of P. When P is separative, the existence of the completion algebra of P is established in Theorem 13.3.13. In particular, if B is a Boolean algebra, then B^* has a completion algebra A according to that theorem and Proposition 13.3.7. Note that Definition 13.3.15 could have been given for arbitrary non-empty ordered sets. But in that case the algebra $\{0, 1\}$ would be a completion algebra of any Boolean algebra B, because all elements of B are compatible, as $0 \in B$; and thus the constant map $B \to \{0, 1\}$ with value 1 is preserving. By limiting the definition to ordered sets without a minimum, we avoid this possibility. In exchange, with a slight abuse of language, we call the *completion algebra* of the Boolean algebra B to the completion algebra, in the sense of Definition 13.3.15 of the ordered set B^*.

Thus, our objective is to extend Theorem 13.3.13 to ordered sets with no minimum. We will reach our goal by canonically associating to any such ordered set a separative ordered set. And to this end, we need some simple additional considerations about orderings. Let us recall Exercise 17 of Chapter 5: a *preorder* R in a set S is a relation in S that is reflexive and transitive. Elements x, y of S are *associated* when xRy and yRx, and association is an equivalence relation giving the quotient set $\overline{S}$. Then, if we define the relation $\hat{R}$ in $\overline{S}$ as $\overline{x}\ \hat{R}\ \overline{y} \Leftrightarrow xRy$, $\hat{R}$ is well defined and an ordering of $\overline{S}$.

Proposition 13.3.16. *Let P be a non-empty ordered set with no minimum element. The relation $x \preceq y :\Leftrightarrow [y]$ is dense in $[x]$ is a preorder relation of P. The corresponding order relation in $\overline{P}$ will be also denoted by $\preceq$, and it is a separative ordering of $\overline{P}$.*

Proof. Let us denote the original order relation in P as $\leq$. Thus $x \preceq y \Leftrightarrow [y]_\leq$ is dense in $[x]_\leq$. This is clearly a preorder, as density is transitive. As mentioned before, $\overline{x} \preceq \overline{y}$ if and only if $x \preceq y$, and this is an ordering of $\overline{P}$. We prove that it is separative.

Suppose $x, y \in P$ and $[\overline{y}]_\preceq$ is dense in $[\overline{x}]_\preceq$. We show first that this implies that $[y]_\leq$ is dense in $[x]_\leq$ in P. Take $u \leq x$; then $u \preceq x$ and thus $\overline{u} \preceq \overline{x}$. Since $[\overline{y}]_\preceq$ is dense in $[\overline{x}]_\preceq$ by hypothesis, there exists v such that $\overline{v} \preceq \overline{y}$ and $\overline{v} \preceq \overline{u}$. This entails that $v \preceq y$ and $v \preceq u$. By the definition of $\preceq$, there exists $u_0 \leq u$ such that $u_0 \leq v$; and there is $y_0 \leq y$ such that $y_0 \leq u_0$. Since $u_0 \leq u$, this shows that $[y]_\leq$ is dense in $[x]_\leq$ as promised. But this means $x \preceq y$ and therefore $\overline{x} \preceq \overline{y}$ as needed. □

$\overline{P}$ with the ordering given in Proposition 13.3.16 is called *the separative quotient* of the ordered set P. We study now the projection map $P \to \overline{P}$.

Proposition 13.3.17. *Let P, $\overline{P}$ as in Proposition 13.3.16, and $p : P \to \overline{P}$ the canonical projection with $p(x) = \overline{x}$. Then p is a monotone map that is preserving.*

Proof. It is clear that p is a monotone map: if $x \leq y$ in P, then $[y]$ is dense in $[x]$ and hence $x \preceq y$ and $\overline{x} \preceq \overline{y}$.

It remains to see that p preserves incompatible elements; or, equivalently, that $p(x), p(y)$ compatible, implies x, y compatible. So, suppose that $p(x), p(y)$ are compatible, hence there is $u \in P$ such that $p(u) \preceq p(x)$ and $p(u) \preceq p(y)$. This entails that $u \preceq x$ and $u \preceq y$, which means that both $[x]$ and $[y]$ are dense in $[u]$. Therefore, there exists $x_0 \leq x$ such that $x_0 \leq u$; and then $y_0 \leq y$ such that $y_0 \leq x_0 \leq x$, so x, y are compatible. □

We now obtain the extension of Theorem 13.3.13 to non-separative ordered sets.

Corollary 13.3.18. *Every non-empty ordered set without minimum element has a completion algebra.*

Proof. Let $p : P \to \overline{P}$ be the projection of P onto its separative quotient. By Proposition 13.3.13, there is a preserving map $f : \overline{P} \to A^*$ with $f[\overline{P}]$ dense in A^*, A being a complete Boolean algebra. It is immediate that the composition of preserving maps is preserving and thus $i = f \circ p : P \to A^*$ is preserving. Then $i[P]$ is a dense subset of A^* because p is surjective, hence $i[P] = f[\overline{P}]$. Therefore $i : P \to A^*$ gives a completion of P. □

13.4 Generic Filters

Let V be a ZFC-universe and $P \in V$ an ordered set. Throughout this section, we shall assume that G is a collection of elements of P, but not necessarily a set. It is clear that the conditions (i) and (ii) for G being a filter in Definition 13.1.6 make sense also when G is not a set, and so we will use freely the term "filter" for G when the collection G satisfies those conditions.

Definition 13.4.1. Let V be a ZFC-universe, $P \in V$ an ordered set, and G a subcollection of P that is a filter. Then G is a *generic filter* of P if for every dense subset $D \subseteq P$, we have $D \cap G \neq \emptyset$.

We may obtain a characterization of generic filters by slightly relaxing condition (i) of Definition 13.1.6.

Lemma 13.4.2. *Let P be an ordered set without minimum element, and H a subcollection of P that satisfies: (a) $x \in H$, $y \in P$, $x \leq y \Rightarrow y \in H$; (b) for every dense subset $D \subseteq P$, $D \cap H \neq \emptyset$; (c) if $x, y \in H$ then $[x] \cap [y] \neq \emptyset$. Then H is a generic filter of P.*

Proof. We only have to prove condition (i) of Definition 13.1.6. Let $x, y \in H$ and take $D = \{u \in P | ([u] \cap [x] = \emptyset) \vee ([u] \cap [y] = \emptyset) \vee (u \in [x] \cap [y])\}$, which is easily seen to be a set, as its definition makes no mention of H. D is a dense subset of P: if $p \in P \setminus D$, then there is $u_1 \leq p, x$; if $u_1 \notin D$, then there is $u_2 \leq u_1, y$, hence $u_2 \in D$ with $u_2 \leq p$.

By (b), there exists $h \in H \cap D$. By (c) this entails that $[h] \cap [x] \neq \emptyset$ and $[h] \cap [y] \neq \emptyset$. Therefore $h \leq x, y$, which shows the condition we needed. □

Generic filters may appear, for instance, when V is a universe included in another ZFC-universe W, and $G \in W$, $G \subseteq P \subseteq V$ but $G \notin V$. We are now going to see some important properties of generic filters. The first one says that a generic filter of A^* for A a Boolean algebra is basically an ultrafilter – except because G is possibly not a set, hence not a subset of A^*.

Proposition 13.4.3. *Let A be a Boolean algebra, G a generic filter of A^* and $b \in A^*$. Then if $b \notin G$ then $b^\perp \in G$.*

Proof. Let $b \in A^*$ and define $S = \{x \in A^* | x \leq b \vee x \leq b^\perp\}$, which is a set. Take $u \in A^*$ arbitrary and suppose $u \not\leq b$. Since A^* is separative, $[b]$ is not dense in $[u]$ and there is $0 \neq v \leq u$ such that v is incompatible with b. This says that $v \cdot b = 0$ and hence $v \leq b^\perp$ (by Proposition 13.1.21), and $v \in S$. Consequently, S is dense in A^*. Since G is generic, there is $g \in G$ which belongs to S and hence either $b \in G$ or $b^\perp \in G$. □

Proposition 13.4.4. *Let A be a complete Boolean algebra and $G \subseteq A^*$ a generic filter of A^*. Then for any $X \in V$ such that $X \subseteq G$, its infimum $\prod_{u \in X} u$ belongs to G.*

Proof. Let $X \subseteq G$ be a set of V. We define the subset Y of A^* as follows: $u \in Y \Leftrightarrow u$ is a lower bound of X or u is incompatible with some element of X. We check that Y is dense in A^*.

Suppose that $x \in A^* \setminus Y$. Then there is some $z \in X$ with $x \not\leq z$; by separativity of A^*, there is $u \leq x$ such that u, z are incompatible. By construction $u \in Y$, and this justifies the claim that Y is dense.

Since G is generic, there is $a \in G \cap Y$. Since $X \subseteq G$, it follows from condition (c) of Lemma 13.4.2 that a is compatible with all elements of X. By the definition of Y, a is a lower bound of X and thus $a \leq \prod_{u \in X} u$. Hence $\prod_{u \in X} u$ belongs to G, as we had to show. □

Corollary 13.4.5. *Let A be a complete Boolean algebra and G a generic filter of A^*. If $X \in V$ is a subset of A, and $\sum_{u \in X} u \in G$, then some of the elements $u \in X$ belongs to G.*

Proof. Suppose, to the contrary, that $\sum_{u \in X} u \in G$ and pretend that for each $u \in X$, $u \notin G$. By Proposition 13.4.3, $u^\perp \in G$ and $\prod_{u \in X} u^\perp \in G$ by Proposition 13.4.4. By Corollary 13.3.3, $(\sum_{u \in X} u)^\perp \in G$, but this would imply $0 \in G \subseteq A^*$, absurd. □

Proposition 13.4.6. *Let P be a non-empty ordered set without minimum element, and G a generic filter of P. Let $i : P \to A$ denote the Boolean completion of P as in Corollary 13.3.18. Consider $G' = \{x \in A | \exists g \in G \ (i(g) \leq x)\}$. Then G' is a generic filter of A^*.*

Proof. (Since G' is defined through G, G' need not be a set, unless G is a set). That G' satisfies conditions (a) and (c) of Lemma 13.4.2 is clear. So, it remains to show condition (b) of the lemma. Let D be a dense subset of A^* and take $D' = \{x \in P | i(x) \leq u \text{ for some } u \in D\}$, so D' is a set. We see now that D' is dense in P: for any $x \in P$ there is $u \in D$ such that $0 \neq u \leq i(x)$, by density of D. But $i[P]$ is dense in A^*, therefore there is $0 \neq y \in P$ with $i(y) \leq u$ and hence $i(y) \leq i(x)$. It follows that x, y are compatible because i is a preserving map; so that there is $z \in P$ with $z \leq x, y$. Then $i(z) \leq i(y) \leq u \in D$, hence $z \in D'$. But also $z \leq x$, which shows that D' is a dense subset of P.

Consequently, there is $g \in G$ with $g \in D'$, so that $i(g) \leq u$ for some $u \in D$. Since $g \in G$, $u \in G' \cap D \neq \emptyset$, and this proves (b) of Lemma 13.4.2 for G'. ☐

13.5 Exercises

1. Prove Proposition 13.1.3.
2. Justify that any finite set of elements in a lattice has a supremum and an infimum.
3. Consider the lattice $\mathbb{Z}^+$ with the divisibility relation of Remarks 13.1.2. Prove that the following holds for all elements: $x \leq z \Rightarrow x + (y \cdot z) = (x + y) \cdot z$.
4. Prove that the lattice of the preceding exercise is distributive.
5. Consider the subset $\{1, 2, 3, 4, 12\} \subseteq \mathbb{Z}^+$ with the divisibility relation as a lattice. Prove that the modular property ($x \leq z \Rightarrow x + (y \cdot z) = (x + y) \cdot z$ for all x, y, z) does not hold in this lattice.
6. Prove Proposition 13.1.15.
7. If a lattice is complemented, is its dual lattice also a complemented lattice?
8. If $\mathcal{L}$ is a complemented lattice (but possibly not distributive), is there uniqueness for the complement of an element? Prove that this is so, or give a counterexample.
9. Let $f : B_1 \to B_2$ be a homomorphism of Boolean algebras, and denote in the same form B_1, B_2 the associated lattices. Prove that f is a morphism of ordered sets if it is injective. Give also an example where $f : B_1 \to B_2$ is a morphism of ordered sets but not a homomorphism of Boolean algebras.

10. Prove that the codomain of a homomorphism $f : B_1 \to B_2$ of Boolean algebras is a subalgebra of B_2.
11. Let I be an ideal in a Boolean algebra and define the filter $F(I)$ as in the observations before Definition 13.2.4. Give the proof that $F(I)$ is a filter of F; and show that it is proper if and only if I is a proper ideal.
12. Let I be an ideal in a Boolean algebra. Prove that $I \cup F(I)$ is a subalgebra and state and prove the dual assertion.
13. Let B be a Boolean algebra, I an ideal of B, $x \in B$. Prove that $I + (x) = \{u \in B | \exists v \in I \ (u \leq v + x)\}$ (as defined just before Definition 13.2.5) is an ideal of B.
14. In the proof of Stone's theorem, the following result is needed. For any Boolean algebra B, there is a set which consists of all the ultrafilters of B. Show precisely how this can be deduced from one instance of the principle of separation (Proposition 5.1.11), say the C-instance; and describe the class C.
15. Let B be a Boolean algebra and X a subset of B. Show that there is a smallest subalgebra of B including X; and give a description of the elements of that subalgebra when $X \neq \emptyset$.
16. Find a Boolean algebra B with a subset C which is a Boolean algebra relative to the same ordering (more precisely, its restriction) but is not a subalgebra.
17. Prove that the inverse of an isomorphism of Boolean algebras is a homomorphism.
18. Prove that if $\mathcal{L}$ is a lattice and $s \in \mathcal{L}$, then the set $\{x \in \mathcal{L} | s \leq x\}$ is a filter of $\mathcal{L}$. These are the *principal filters* presented in Examples 13.1.8.
19. Let S be an infinite set and consider the Boolean algebra $\mathbb{P}(S)$. Take $F \subseteq \mathbb{P}(S)$ consisting of all subsets $X \subseteq S$ such that $S \setminus X$ is finite. Show that F is a filter of $\mathbb{P}(S)$. This is the *cofinite filter* of Example 13.1.16. Prove that every ultrafilter that extends the cofinite filter is not principal.
20. Let B be a Boolean algebra, (s) a principal proper filter of B. Prove that (s) is an ultrafilter if and only if s is a minimal element of the ordered set B^*.
21. Let B be a Boolean algebra and C a dense subset of B^* which is closed under sums. Prove that if $x < y$ in B with $x \in C$, then there exists $x' \in C$ such that $x < x' \leq y$.
22. For B a Boolean algebra, prove that B is complete if every subset of B has a supremum.

23. Prove directly, without using the principle of duality, the first equality of Proposition 13.3.2.

24. Let C be a segment of a separative ordered set P. Prove that its corresponding regular segment $\overline{C}$ (see the proof of Theorem 13.3.13) is the set $\{x \in P | C \text{ is dense in } [x]\}$, by showing that this set is a regular segment that includes C and is the smallest such set.

25. Let P be a separative ordered set. Prove that if X, Y are segments of P, then $\overline{X} \cap \overline{Y} = \overline{X \cap Y}$ (the preceding exercise may be helpful).

26. With reference to the proof of Theorem 13.3.13, complete the proof that B is a complete lattice by proving that it satisfies conditions (1) to (4) of Proposition 13.1.3.

27. Let $g: P \to S$ be a preserving map with $g[P]$ dense in S and S separative. Prove that there exists a preserving map $h : \overline{P} \to S$ such that $h \circ p = g$, p being the projection of P onto its separative quotient $\overline{P}$. Derive from this the uniqueness of the Boolean completion of an ordered set without minimum element.

28. Explain that D', as defined in the proof of Proposition 13.4.6, is a set.

29. Let P be an ordered set in the ZFC-universe V, and $F \subseteq P$ a collection of objects of P (but not necessarily a set). Suppose that F satisfies the conditions for being a filter of P. Prove that F is a generic filter if and only if $F \cap D \neq \emptyset$ for every subset $D \subseteq P$ with the following property: every $p \in P$ is compatible with some element of D.

30. Let G be a generic filter of the ordered set P. Prove that if C is a subset of P with $C \cap G = \emptyset$, then there exists some element $x \in G$ which is incompatible with every element of C.

31. Let P be an ordered set, D a subset of P and $p \in P$. We say that D is *dense below* p (in P) when for every $x \leq p$ there exists $y \in D$ such that $y \leq x$. Prove that if G is a generic filter of P, $p \in G$ and D is dense below p, then $G \cap D \neq \emptyset$.

32. Let P be an ordered set in the ZFC-universe V with the following property: if $p \in P$, then there exist $x, y \in P$ such that $x, y \leq p$ and x, y are incompatible. Prove that no generic filter G of P is a set (Hint: Suppose that G is a set, and consider the set $P \setminus G$).

14

Generic Extensions of a Universe

In previous chapters, we have constructed certain sub-universes inside a given universe (the sub-universe of pure sets, the sub-universe of well-founded sets, the sub-universe of constructible sets) to serve different purposes. In the last case, the purpose is to obtain a universe having some property that is not assumed to hold in the whole universe. With the same idea of constructing universes with different properties, it is also possible to consider constructions in the opposite direction: to extend a given ZFC-universe V and obtain a new universe which includes V. It must be noted that in order to introduce this new universe, we need to assume that V is already included in another bigger universe W so that we may extend V by adding to V some of the elements of W. It turns out that in this way (using a technique which is known as *forcing*) we get not only a new universe from a given ZFC-universe V, as was the case with the other examples, but many universes that extend V, and this makes the technique of forcing very flexible and powerful.

14.1 The Basics

The extensions of a ZFC-universe that are produced by the forcing method are called *generic extensions*, and we present their construction in two steps. Starting with a ZFC-universe M and a complete Boolean algebra A in M, the first step yields a (non-transitive) class of M, the *Boolean-valued universe* M^A. It is not its condition of being a sub-universe of M what is relevant about the Boolean-valued universe; what counts is the fact that M^A contains the seeds that will produce new universes which are extensions of M; the construction of these new universes will be our second step.

14.1.1 Boolean universes

Proposition 14.1.1. *Let M be a ZFC-universe and let $A \in M$ a complete Boolean algebra. Consider the following functional F: for any set $X \in M$, $F(X) = \{f | f$ is a function, $D(f) \subseteq X$ and $cD(f) \subseteq A\}$. Then there is a*

DOI: 10.1201/9781003449911-14

functional H with domain ***Ord*** *such that $H(\emptyset) = \emptyset$ and for each ordinal α and each limit ordinal λ,*

$$H(\alpha + 1) = F(H(\alpha)), \quad H(\lambda) = \cup_{\beta\in\lambda} H(\beta).$$

Proof. This is obtained by recursion (Theorem 8.4.5), in the same form of the proof of Proposition 9.2.1. □

There is a parallelism between this construction and that of the von Neumann universe (Proposition 9.2.1): we use the functional F above exactly as we used the functional $\mathbb{P}$ (for power set) for defining the von Neumann universe. As it happened with the von Neumann universe, we may show by ordinal induction that $H(\beta) \subseteq H(\alpha)$ whenever $\beta \leq \alpha$. And also in this case we have a more usual writing of these sets: M_α^A for $H(\alpha)$. Like this,

$$M_0^A = \emptyset, \quad M_\lambda^A = \cup_{\beta\in\lambda} M_\beta^A \text{ for } \lambda \text{ limit},$$
$$M_{\alpha+1}^A = \{f | f \text{ is a function, } \mathrm{D}(f) \subseteq M_\alpha^A \text{ and } \mathrm{cD}(f) \subseteq A\}$$

M^A will be the union of all the M_α^A and thus $M^A \subseteq M$. M^A is known as a *Boolean-valued universe* of M; we sometimes abbreviate this expression to *Boolean universe.*

Examples 14.1.2. Let us compute the first steps in constructing Boolean-valued universes. For an arbitrary complete Boolean algebra A, we have:

$M_0^A = \emptyset$.

M_1^A contains functions whose domain is a subset of M_0^A; thus the only such function is the empty function $\emptyset$, hence $M_1^A = \{\emptyset\}$.

The elements of M_2^A must be functions whose domain is a subset of M_1^A, so just two subsets are possible: if the domain is $\emptyset$, we get again the empty function. If the domain is $\{\emptyset\}$, we have as many functions as elements of A. Each such function is $\{\langle\emptyset, a\rangle\}$ for $a \in A$. If we call $s_a = \{\langle\emptyset, a\rangle\}$, then M_2^A contains $\emptyset$ and all the elements s_a.

M_3^A consists of all functions having domain $\subseteq M_2^A$ (and codomain $\subseteq A$). This includes all the elements of M_2^A (these are the functions with domain $\emptyset$ or $\{\emptyset\}$), but also any other choice of a subset of M_2^A, with a choice of the A-values for the elements of the subsets. Too long to write down, but let us consider a simpler case.

Suppose that A is the trivial algebra $\{0, 1\}$. M_0^A, M_1^A are exactly as above. Now, $M_2^A = \{\emptyset, \{\langle\emptyset, 0\rangle\}, \{\langle\emptyset, 1\rangle\}\}$, so it has three elements. In M_3^A we have the empty function, as above; then functions whose domain is a subset of M_2^A with one element: there are three choices for the domain, two choices for the image in each function, which makes a total of six functions. Besides the two non-empty functions already in M_2^A, we have four more functions, like $\{\langle\{\langle\emptyset, 0\rangle\}, 0\rangle\}$ or $\{\langle\{\langle\emptyset, 0\rangle\}, 1\rangle\}$. Then we have functions with a domain consisting of two elements of M_2^A; there are three choices for the domain and four choices for the images of the two elements, so 12 functions in total. One of these functions is $\{\langle\emptyset, 1\rangle, \langle\{\langle\emptyset, 1\rangle\}, 0\rangle\}$ and three other functions have the same domain. Finally, there are the functions whose domain is all of M_2^A; here we have to choose the image for each element of the domain, so there are 8 choices and thus 8 functions with that domain. One of these is $\{\langle\emptyset, 1\rangle, \langle\{\langle\emptyset, 0\rangle\}, 1\rangle, \langle\{\langle\emptyset, 1\rangle\}, 0\rangle\}$. In total, M_3^A has $1+6+12+8 = 27$ elements.

Pursuing this line of thought does not seem to be a good idea for getting useful information about the sets M_α^A: they consist of functions whose domains are sets of functions whose domains are sets of functions, etc.; and, of course, this goes beyond the finite. But remember that also the von Neumann universe consists of sets whose elements are sets whose elements are sets whose elements are sets, and so on; and on top of this, all those sets are constructed from the empty set. We may lack an intuitive idea of how the elements of, say, V_{ω^2} are constructed from the empty set, but still we may reason about those elements abstractly. The same occurs with Boolean-valued universes: we know precisely how each $M_{\alpha+1}^A$ is constructed from M_α^A and we can reason from that.

Paradoxically, while M^A is an M-class so that $M^A \subseteq M$, it is also a kind of expansion of M, as shown in the next result.

Proposition 14.1.3. *There is an injective functional $\Phi : M \to M^A$ such that for each $x \in M$, $\Phi(x)$ is the function with domain $D(\Phi(x)) = \Phi[x]$ and constant value $1 \in A$ on its domain.*

Proof. The informal notation above ($\Phi : M \to M^A$ meaning that the functional Φ has domain M and codomain included in M^A) repeats the notation for maps; it will be used without further mention and is self-explanatory.

We define maps $\Phi_\alpha : M_\alpha \to M_\alpha^A$ for all ordinals $\alpha \in M$ by recursion, but for a technical necessity, we define these as relations $\Phi_\alpha \subseteq M_\alpha \times M_\alpha^A$, and prove later that they are functions. Φ_0 is necessarily the empty function. Then $\Phi_{\alpha+1}(x)$ (for $x \in M_{\alpha+1}$) is the constant function with value $1 \in A$ and domain $\Phi_\alpha[x] \subseteq M_\alpha^A$. If λ is limit, $\Phi_\lambda = \cup_{\beta\in\lambda}\Phi_\beta$.

We prove inductively the following properties: (1) Each Φ_α is a function with domain M_α and $\mathrm{cD}(\Phi_\alpha) \subseteq M_\alpha^A$; (2) If $\beta < \alpha$, then $\Phi_\beta \subseteq \Phi_\alpha$; (3) For each $x \in M_\alpha$, $\Phi_\alpha(x)$ has domain $\Phi_\alpha[x]$ and values $1 \in A$. To this end, assume these properties for α and let us prove them for $\alpha + 1$.

(1) is immediate by the definition of $\Phi_{\alpha+1}$ given above. For (2), it is enough to show that $\Phi_\alpha \subseteq \Phi_{\alpha+1}$, so let $u \in M_\alpha$; by hypothesis (3) $\mathrm{D}(\Phi_\alpha(u)) = \Phi_\alpha[u] = \mathrm{D}(\Phi_{\alpha+1}(u))$, and thus $\Phi_\alpha(u) = \Phi_{\alpha+1}(u)$ by the definition. This proves (2) and in turn shows that, for $x \in M_{\alpha+1}$, $\mathrm{D}(\Phi_{\alpha+1}(x)) = \Phi_\alpha[x] = \Phi_{\alpha+1}[x]$ from which condition (3) for $\alpha + 1$ follows.

Consider now the case when λ is a non-zero limit and every $\alpha < \lambda$ satisfies (1), (2), (3). By (2) and Lemma 6.4.1, Φ_λ is a function whose domain is the union of the domains M_β (for $\beta \in \lambda$) and therefore $\mathrm{D}(\Phi_\lambda) = M_\lambda$: also the codomain is the union of the codomains and it follows that $\mathrm{cD}(\Phi_\lambda) \subseteq M_\lambda^A$. This shows condition (1), and (2) is obvious. Finally, let $x \in M_\lambda$; there exists some $\beta \in \lambda$ such that $x \in M_\beta$ and $\Phi_\lambda(x) = \Phi_\beta(x), \Phi_\lambda[x] = \Phi_\beta[x]$, and condition (3) follows immediately.

By setting $\Phi = \cup_{\alpha \in \mathbf{Ord}} \Phi_\alpha$, we see that Φ is a functional from M to M^A and condition (3) shows that the description of $\Phi(x)$ given in the proposition is correct. $\square$

There is a more usual notation for $\Phi(x)$: we write $\check{x}$ for $\Phi(x)$. Thus

$$D(\check{x}) = \{\check{t} | t \in x\}, \text{ and } \check{x}(\check{t}) = 1 \tag{14.1}$$

Furthermore, Φ is injective. This can be proved by showing inductively that each Φ_α is injective. Assuming this for α, let $x, y \in M_{\alpha+1}$ and $\Phi_{\alpha+1}(x) = \Phi_{\alpha+1}(y)$. In particular, $\Phi_\alpha[x] = \Phi_\alpha[y]$. So, if $u \in x$, then $\Phi_\alpha(u) = \Phi_\alpha(v)$ for some $v \in y$. But Φ_α is injective by the inductive hypothesis, therefore $u = v$. This shows that $x \subseteq y$, and we may see in the same manner that $y \subseteq x$, thus $x = y$. Injectivity for limit ordinals is now obvious.

The fact that Φ is injective means that M^A is a kind of extension of M; or, more exactly, of the copy of M given as $\mathrm{cD}(\Phi)$. But Φ does not preserve membership: if $x \in y$ we may deduce that $\langle \check{x}, 1 \rangle \in \check{y}$, but in general $\check{x} \notin \check{y}$. M is somehow deconstructed when we identify it as the class $\mathrm{cD}(\Phi)$ inside $M^A \subseteq M$ – under the natural identification $x \mapsto \check{x}$. One could say that M "lives" inside M^A, but that M^A is written in some encrypted language; and that makes it difficult to recognize the relations between the elements of M under such disguise.

14.1.2 The construction of a generic extension

Remarks 14.1.4. The injective functional $\Phi : M \to M^A$ has a kind of inverse functional when A is the trivial algebra $A = \{0, 1\}$. For defining this inverse functional, we basically consider each function $f \in M^A_\alpha$ as a characteristic function, which selects the elements x of $\mathrm{D}(f)$ such that $f(x) = 1$. To be more precise we would have to use a recursive definition, starting with $\emptyset^* = \emptyset$ for the empty function; and generally, for $f \in M^A_{\alpha+1}$, define $f^* = \{x^* | x \in \mathrm{D}(f) \wedge f(x) = 1\}$. Let us check how this works for the first steps M^A_α that we described in Examples 14.1.2.

M^A_0 has no elements, and M^A_1 has the only element $\emptyset$, and we agree that $\emptyset^* = \emptyset$. Now, M^A_2 has three elements, $\emptyset, \{\langle \emptyset, 0 \rangle\}, \{\langle \emptyset, 1 \rangle\}$; calling f_1, f_2 to these two non-empty functions, we get: $f_1^* = \emptyset, f_2^* = \{\emptyset^*\} = \{\emptyset\}$. Thus this level M^A_2 produces two elements, $\emptyset$ and $\{\emptyset\}$.

M^A_3 has 27 functions, so let us examine only some of them. We know that $\emptyset, f_1, f_2 \in M^A_2$, and these give the images $\emptyset, \{\emptyset\}$, as we have seen. Moreover, let $g_1 = \{\langle f_2, 1 \rangle\}$: we get $g_1^* = \{f_2^*\} = \{\{\emptyset\}\}$. Let now $g_2 = \{\langle \emptyset, 1 \rangle, \langle f_1, 0 \rangle, \langle f_2, 1 \rangle\}$; then $g_2^* = \{\emptyset^*, f_2^*\} = \{\emptyset, \{\emptyset\}\}$. So, we have four elements which are images of elements of M^A_3: $\emptyset, \{\emptyset\}, \{\{\emptyset\}\}, \{\emptyset, \{\emptyset\}\}$. No other M-set can be obtained from a function of M^A_3; because the domain of each such function is a subset of M^A_2 and the elements with image 1 must form a subset of M^A_2 as well; so the sets we obtain then by applying $()^*$ must be subsets of $\{\emptyset^*, f_1^*, f_2^*\} = \{\emptyset, \{\emptyset\}\}$.

It is not difficult to understand that the process must go on this same way. In fact, the four-elements set we have obtained with the images of M_3^A give nothing else than M_3, a level of the von Neumann universe. The next level must consist of subsets of M_3, hence elements of M_4; and all the elements of M_4 can be obtained from an adequate characteristic function like those we are examining. In all, this gives an inverse functional for Φ, as $(\check{x})^* = \{(\check{t})^* | t \in x\}$ and it follows inductively that $(\check{t})^* = t$. Following the analogy at the end of the preceding section, passing from f to f^* is a kind of deciphering of the elements of M^A: from being functions whose domains are functions, they turn to recognizable ordinary sets. And the elements of M are recovered as themselves.

But of course this works just for the trivial algebra A. The general case is much more fruitful. From this point on, we will work in this chapter under the following hypothetical assumptions and the associated notations.

(1) M is a ZFC-universe, $A \in M$ is a complete Boolean algebra in the universe M and $A^* = A \backslash \{0\}$. We keep the notations M^A and Φ of Propositions 14.1.1 and 14.1.3. Additionally, M is a transitive class of another ZFC-universe V.

(2) G is a generic filter of A^* in the universe M, as defined in Section 13.4. That is, G is a collection of elements of $A^* \subseteq M$, but G is not necessarily a set in M. We also assume that G is an element of the universe V. Since the concept of generic filter is central to the theory, let us repeat and update its definition in the current context. $G \in V$ is a *M-generic filter* of the ordered M-set P when: (i) $G \subseteq P$; (ii) $x, y \in G \Rightarrow (\exists u \in P\ (u \leq x, u \leq y, u \in G)$; (iii) $(x, y \in P, x \leq y, x \in G) \Rightarrow y \in G$; and (iv) if $D \in M$ is a dense subset of P, then $G \cap D \neq \emptyset$.

We will globally refer to these assumptions as the *(GE) hypotheses* – GE for "generic extensions".

In the situation of the hypotheses (GE), we would like to repeat the construction above that for each f in some M_α^A defines $f^* = \{x^* | x \in \mathrm{D}(f) \wedge f(x) = 1\}$, but with the condition $f(x) = 1$ replaced with $f(x) \in G$. In this way, even though $f \in M$, the set $f^{-1}[G]$ need not belong to M if $G \notin M$; and this makes it possible to have also $f^* \notin M$, giving elements of $V \backslash M$ which form part of a new universe that properly extends M. As suggested above, the "ciphered" elements of M^A are transformed through this construction into ordinary elements of the universe V. And this transformation is a recursive process: if we accept that the elements $x \in M_\alpha^A$ have been transformed into new elements $x^* \in V$, then for $f \in M_{\alpha+1}^A$, we select those $x \in \mathrm{D}(f)$ such that $f(x) \in G$, and f^* will be the set consisting of the transformed elements x^* of those $x \in \mathrm{D}(f)$. With a slightly different notation, this is what is developed in the next result.

Proposition 14.1.5. *Under the hypotheses (GE), there is a functional class of V, $\Psi : M^A \to V$ such that for each $f \in M^A$, $\Psi(f) = \{\Psi(t) | t \in D(f) \wedge f(t) \in G\}$.*

Proof. This construction is carried out in the universe V and is similar to that of the M-functional Φ of Proposition 14.1.3. In the recursive definition of the $\Psi_\alpha : M_\alpha^A \to V_\alpha$ we will use the ordinals of M so that, in case these ordinals are just the elements of some limit ordinal $\rho \in V$ (see Corollary 9.2.18), the recursion is limited to those ordinals.

With $\Psi_0 = \emptyset$, we set for $f \in M_{\alpha+1}^A$, $\Psi_{\alpha+1}(f) = \Psi_\alpha[f^{-1}[G]] = \{\Psi_\alpha(t) | f(t) \in G\}$; and for limit ordinals, $\Psi_\lambda = \cup_{\beta\in\lambda}\Psi_\beta$. Again, these are defined as relations, $\Psi_\alpha \subseteq M_\alpha^A \times V_\alpha$; and we prove they are functions.

Our induction hypothesis now is: (1) Ψ_α is a function, $\mathrm{D}(\Psi_\alpha) = M_\alpha^A$ and $\mathrm{cD}(\Psi_\alpha) \subseteq V_\alpha$; (2) if $\beta < \alpha$, then $\Psi_\beta \subseteq \Psi_\alpha$; (3) if $f \in M_\alpha^A$, then $\Psi_\alpha(f) = \Psi_\alpha[f^{-1}[G]]$. We prove these conditions for $\alpha + 1$.

(1) is clear by the definition of $\Psi_{\alpha+1}$. The inductive hypothesis (3) entails immediately that $\Psi_\alpha \subseteq \Psi_{\alpha+1}$, showing (2) for $\alpha + 1$. Then condition (3) follows now directly from (2).

For limit ordinals, the proof is again the same as that of Proposition 14.1.3. Then Ψ is the union of the Ψ_α and the properties of the statement follow. □

If $\Phi : M \to M^A$ is the functional of Proposition 14.1.3, then $\Psi(\Phi(x)) = x$ for all $x \in M$. In fact $\Psi_\alpha \circ \Phi_\alpha$ is the identity on the elements of M_α. This can easily be shown inductively: if $x \in M_{\alpha+1}$, $\Phi_{\alpha+1}(x)$ gives the constant function 1 on $\Phi_\alpha[x]$. Consequently, $(\Psi_{\alpha+1} \circ \Phi_{\alpha+1})(x) = \Psi_\alpha[\Phi_\alpha[x]] = x$ by the inductive hypothesis, since $1 \in G$.

As Φ, Ψ has a more compact and usual notation: $\Psi(x)$ is written as x^G for $x \in M^A$. Using this, Proposition 14.1.5 may be stated thus:

$$x^G = \{t^G | t \in D(x) \wedge x(t) \in G\}, \quad (\check{x})^G = x \tag{14.2}$$

The codomain of the functional Ψ will be denoted as $M[G]$ and called a *generic extension* of M. Since $\Psi \circ \Phi$ is the identity on M, M is included in $M[G]$ so that this is a faithful extension of M, as announced. We prove next that $G \in M[G]$ and thus $M[G]$ is a proper extension of M, provided $G \notin M$.

Proposition 14.1.6. *Under the hypotheses (GE), $G \in M[G]$.*

Proof. Suppose that $A \in M_{\alpha+1}$ hence $A \subseteq M_\alpha$. Like this, for each $a \in A$ we have $\check{a} \in M_\alpha^A$. We consider $h \in M_{\alpha+1}^A$ with $\mathrm{D}(h) = M_\alpha^A$ given by the conditions $h(\check{a}) = a$ for all $a \in A$; and $h(x) = 0$ for any other $x \in M_\alpha^A$. Then $h^G = G$ by the definition of Ψ in Proposition 14.1.5. □

14.2 The Generic Universe is a ZFC-Universe

14.2.1 First axioms

Hypothesis (GE) is assumed for the rest of the chapter. Our objective is to establish that $M[G]$ is a ZFC-universe. With this aim, we prove now that the first axioms are satisfied in $M[G]$. We will make systematic use of the results of Chapter 13 on generic filters, frequently without an explicit mention; namely Propositions 13.4.3 and 13.4.4 and Corollary 13.4.5. Note that the application of these results requires that the concerned sum or product is taken over a set of elements of the algebra A **that belongs to the universe** M.

Proposition 14.2.1. *$M[G]$ is a 2-universe.*

Proof. First, $\emptyset = \emptyset^G$ so $M[G]$ contains an empty set. Second, any element of $M[G]$ is, by definition, of the form x^G for some $x \in M^A$. So, let $x, y \in M^A$, giving $x^G, y^G \in M[G]$. If $x, y \in M^A_\alpha$, then there is $h \in M^A_{\alpha+1}$ with $\mathrm{D}(h) = \{x, y\}$ satisfying $h(x) = h(y) = 1$. Clearly $h^G = \{x^G, y^G\}$ and this shows that the axiom of pairs holds in $M[G]$. □

Proposition 14.2.2. *$M[G]$ is a transitive class of V. Therefore $M[G]$ satisfies the axioms of extension and regularity.*

Proof. If $x^G \in M[G]$ and $y \in x^G$, then $y = u^G$ for some $u \in \mathrm{D}(x)$ with $x(u) \in G$, by equation 14.2. Thus $y \in M[G]$ and $M[G]$ is transitive. This immediately implies that $M[G]$ satisfies extension and regularity. □

Proposition 14.2.3. *The axiom of unions holds in $M[G]$.*

Proof. Let $x^G \in M[G]$ with some $x \in M^A_{\alpha+1}$. We have $\mathrm{D}(x) \subseteq M^A_\alpha$ and thus $E_x = \{u | \exists z\, (z \in \mathrm{D}(x) \wedge u \in \mathrm{D}(z))\} \subseteq M^A_\alpha$ is a M-subset of M^A_α. We may define h with codomain included in A and domain E_x by setting $h(u) = \sum_{\{z \in D(x) | u \in D(z)\}} z(u)x(z)$. This gives that $h \in M$ is a function and $h \in M^A_{\alpha+1}$.

By Proposition 7.1.9, we need to show that the union in V, $\cup(x^G)$ belongs to $M[G]$; and we prove this by showing that $\cup(x^G) = h^G$. By Corollary 13.4.5, the elements of h^G are those u^G with the property that there exists $z \in \mathrm{D}(x)$ with $x(z), z(u) \in G$. These conditions imply $z^G \in x^G$ and $u^G \in z^G$, thus $h^G \subseteq \cup(x^G)$. Conversely, any element of $\cup(x^G)$ is an element of some z^G with $x(z) \in G$; and therefore it has to be some t^G with $t \in \mathrm{D}(z)$ and $z(t) \in G$. Hence $t \in E_x$, $h(t) \in G$ and $t^G \in h^G$. Thus $\cup(x^G) = h^G \in M[G]$, completing the proof. □

Remark 14.2.4. Some readers might feel uneasy about this proof by the fact that we are using simultaneously different universes, M, $M[G]$ or V. They could possibly ask where must we consider the domain of an element: for instance, is $\mathrm{D}(x)$ to be taken in M or in V? In this particular case, this makes no difference. But we know that sometimes constructions in one or another universe can give different results, for instance taking the power set. However, leaving aside those constructions, the fact that M is transitive in V entails that any set $x \in M$ has the same elements in M or in V and hence the simultaneous consideration of M and V as the universe where x belongs is not by itself a source of ambiguity. When doubts may arise, we will try to specify the universe where some construction is being carried on.

The preceding proof may be seen as a typical example of these situations. Since we want to prove the axiom of unions in $M[G]$, we must start with an arbitrary element of $M[G]$ which, by definition, has the form x^G with $x \in M^A \subseteq M$. The argument that follows is made entirely in M, until we find the element $h \in M^A$. Then we check that h^G is the union (in the universe V) of x^G; and this part requires the use of the functional Ψ and consequently needs $M[G]$ as the reference frame, but this does not forbid us the use of elements of M, like h, z, t together with elements of $M[G]$; after all, they all live in a common universe.

14.2.2 The axiom of the power set

The relationship between M^A (a class of M) and $M[G]$ (a class of V) is at the center of the arguments in the above proofs. But this relationship is not a smooth one in what concerns the basic relations of membership and inclusion: for $x, y \in M^A$, $x^G \in y^G$ (or $x^G \subseteq y^G$, or $x^G = y^G$) does not imply $x \in y$ (or, respectively, $x \subseteq y$, or $x = y$). It turns out that in order to prove the axiom of the power set for $M[G]$ (and for subsequent developments) we are forced to look for a complete characterization of membership and inclusion in $M[G]$ in terms of properties of M^A. With this aim, we will introduce now three functionals defined on pairs of elements of M^A, and with values in A. This will be done entirely within the universe M. Namely, we look for functionals of M

$$B, E, I : M^A \times M^A \to A$$

with the properties, for any $x, y \in M^A$:

$$x^G \in y^G \Leftrightarrow B(x,y) \in G, \quad x^G = y^G \Leftrightarrow E(x,y) \in G, \quad x^G \subseteq y^G \Leftrightarrow I(x,y) \in G$$

We define the functional I first, and then B and E will be defined from I. I will be the union of functions $I_\alpha \subseteq M^A_\alpha \times M^A_\alpha \times A$, and these I_α are defined by recursion (Theorem 8.4.5). The ordinals are, of course, the ordinals of M.

From I_α we want to define $I_{\alpha+1} \subseteq M^A_{\alpha+1} \times M^A_{\alpha+1} \times A$. For $x, y \in M^A_{\alpha+1}$,

$$\langle x, y, a\rangle \in I_{\alpha+1} \Leftrightarrow a = \prod_{t \in D(x)} [x(t)^\perp + (\sum_{u \in D(y), \langle t,u,b\rangle, \langle u,t,b'\rangle \in I_\alpha} y(u) \cdot b \cdot b')]$$

The I_α are in fact functions, and this gives a much more readable expression for the above relation:

$$I_{\alpha+1}(x,y) = \prod_{t \in D(x)} [x(t)^\perp + \sum_{u \in D(y)} y(u) I_\alpha(t,u) I_\alpha(u,t)]$$

For λ limit, $I_\lambda = \cup_{\beta \in \lambda} I_\beta$. This definition can be justified similarly to that of Φ in Proposition 14.1.3. In analogy to that proof, the inductive hypothesis is now: (1) Each I_α is a function; (2) $\beta < \alpha \Rightarrow I_\beta \subseteq I_\alpha$; and (3) when $x, y \in M^A_\alpha$,

$I_\alpha(x,y) = \Pi_{t\in D(x)}[x(t)^\perp + (\sum_{u\in D(y)} y(u)I_\alpha(t,u)I_\alpha(u,t))]$; this new condition has the same relation with (2) than in the proof of Proposition 14.1.3. The proof is similar to that of the proposition and the details are left as an exercise.

$I = \cup_{\alpha\in\mathbf{Ord}} I_\alpha$ is then a functional, and we may write generally:

$$I(x,y) = \prod_{t\in D(x)} [x(t)^\perp + (\sum_{u\in D(y)} y(u)I(t,u)I(u,t))] \tag{14.3}$$

When $x = 0$, this equation shows that $I(\emptyset, y) = 1$ since it would give an empty product, which equals 1 in a complete Boolean algebra. On the other hand, $I(x,\emptyset) = \Pi_{t\in D(x)} x(t)^\perp$, since an empty sum in a complete Boolean algebra is 0. In particular, $I_0 = \emptyset$ by the definition for limit ordinals; and then $I_1(\emptyset,\emptyset) = 1$, as it is an empty product.

We now define the other two functionals, both having the same domain $M^A \times M^A$.

$$E(x,y) = I(x,y)I(y,x), \quad B(x,y) = \sum_{u\in D(y)} y(u)I(x,u)I(u,x)$$

It follows from these definitions that the following equations hold in all cases.

$$B(x,y) = \sum_{t\in D(y)} y(t)E(x,t), \quad I(x,y) = \prod_{t\in D(x)} (x(t)^\perp + B(t,y)) \tag{14.4}$$

These functionals are constructed inside M, but the next argument occurs in V, as we have to use the M-generic filter G.

Proposition 14.2.5. *The following hold for any $x, y \in M^A$:*

$B(x,y) \in G \Leftrightarrow x^G \in y^G, \quad E(x,y) \in G \Leftrightarrow x^G = y^G,$
$I(x,y) \in G \Leftrightarrow x^G \subseteq y^G$

Proof. We use induction on the M^A_α to prove the relation for I, the other two follow then easily. It is true that we are now working in V, so we must use induction in V; while the ordinals for the induction are those of M. As we did in the proof of Proposition 14.1.5, we use induction in V up to the ordinal ρ which contains all the ordinals of M, if this is the case. Also M^A or I are V-classes (Proposition 7.3.6), and the M^A_α or the values $I(x,y)$ are also objects of V, so the arguments that follow have V as their frame, even if M^A or the functionals I, B, E are defined within M.

Suppose the relation true up to all elements of $M^A_\alpha \times M^A_\alpha$, and take any $\langle x,y\rangle \in M^A_{\alpha+1} \times M^A_{\alpha+1}$. Assume first that $I(x,y) \in G$. An arbitrary element of x^G will be t^G for some $t \in \mathrm{D}(x)$ with $x(t) \in G$. By Proposition 13.4.4, $x(t)^\perp + \sum_{u\in D(y)} y(u)I(t,u)I(u,t) \in G$, hence there is some $u \in \mathrm{D}(y)$ with $y(u), I(t,u), I(u,t) \in G$, according to Corollary 13.4.5. Thus $y(u) \in G$ so that

$u^G \in y^G$; and moreover $t^G = u^G$ by the inductive hypothesis. So $t^G \in y^G$, proving that $x^G \subseteq y^G$.

For the converse, suppose $I(x, y) \notin G$. Again by Proposition 13.4.4, there is some $t \in \mathrm{D}(x)$ such that $x(t)^{\perp} + \sum_{u \in D(y)} y(u)I(t, u)I(u, t) \notin G$. Therefore $x(t) \in G$ (hence $t^G \in x^G$) and, as before, $y(u)I(t, u)I(u, t) \notin G$ (for every $u \in \mathrm{D}(y)$). Thus $y(u) \in G \Rightarrow u^G \neq t^G$ again by the inductive hypothesis and Proposition 13.4.4. Consequently, $t^G \notin y^G$ and $x^G \not\subseteq y^G$. The case of a limit ordinal is now trivial.

The property for E is then immediate by the principle of extension. And then $B(x, y) = \sum_{t \in D(y)} y(t)E(x, t) \in G$ implies that there is $t \in \mathrm{D}(y)$ with $y(t), E(x, t) \in G$, so that $x^G \in y^G$; the converse is also immediate. □

Proposition 14.2.6. *The axiom of the power set holds in the universe $M[G]$.*

Proof. First, to any given inclusion $x^G \subseteq y^G$ in $M[G]$, there is some $z \in M^A$ such that $\mathrm{D}(z) = \mathrm{D}(y)$ and $z^G = x^G$. To see this, let $z : \mathrm{D}(y) \to A$ (thus an element of $M^A \subseteq M$) given as $z(u) = B(u, x)$. The inclusion $z^G \subseteq x^G$ is clear by Proposition 14.2.5; for the converse inclusion, let $t^G \in x^G$ with some $t \in \mathrm{D}(x)$. We see from the hypothesis that $t^G \in y^G$ and hence there is $u \in \mathrm{D}(y)$ such that $u^G = t^G$ and $y(u) \in G$. Thus $u^G \in x^G$ and $B(u, x) \in G$ again by Proposition 14.2.5. It follows that $t^G = u^G \in z^G$.

Take $y \in M^A_{\alpha+1}$ so that $\mathrm{D}(y) \subseteq M^A_\alpha$. According to Proposition 7.1.9 we need to prove that $\mathbb{P}(y^G) \cap M[G] \in M[G]$, with the power set taken in V. For this, we define $p : M^A_{\alpha+1} \to A$ by $p(x) = I(x, y)$; and show that $p^G = \mathbb{P}(y^G) \cap M[G]$.

Any element of p^G is a x^G with $x \in \mathrm{D}(p)$ and $p(x) \in G$; so that $I(x, y) \in G$ and $x^G \subseteq y^G$ by Proposition 14.2.5; this shows that $p^G \subseteq \mathbb{P}(y^G)$. Conversely, if $x^G \subseteq y^G$, there is $z \in M^A_{\alpha+1}$ with $z^G = x^G$ by the observation at the beginning of this proof. Thus $z \in \mathrm{D}(p)$ and $I(z, y) \in G$, whence $z^G \in p^G$ and $x^G \in p^G$. This shows that $M[G] \cap \mathbb{P}(y^G) \subseteq p^G$. We thus get the equality $p^G = \mathbb{P}(y^G) \cap M[G]$ and we are done. □

14.2.3 Elements and classes of a generic extension

We are still assuming hypothesis (GE), and we want to consider different ways of representing elements of the generic extension $M[G]$. If $x \in M^A_\alpha$, then $x^G \in M[G]$ by definition. In this case, x is a function and $x \subseteq M^A_\alpha \times A$. But also for subsets $S \subseteq M^A_\alpha \times A$, we get elements of $M[G]$. In such situation, let us call call $S^G = \{x^G | \exists a \in G \ (\langle x, a \rangle \in S)\}$ – this notation does not conflict with the already known f^G for elements $f \in M^A$; it is simply a generalization.

Lemma 14.2.7. *Let $S \subseteq M^A \times A$ be a set of M. Then $S^G \in M[G]$.*

Proof. $\mathrm{D}(S) \subseteq M^A$ is a set. By using the functional that assigns to each element of M^A its rank in the sequence of the M^A_α, it is easy to see that there is an ordinal β such that $\mathrm{D}(S) \subseteq M^A_\beta$. Thus we may define a map f :

$\mathrm{D}(S) \to A$ by setting $f(x) = \sum\{a \in A | \langle x, a\rangle \in S\}$ so that $f \in M^A$. It then follows from Corollary 13.4.5 that $S^G = f^G \in M[G]$. □

From the M-functionals B, E, I of equation 14.4, we define now other M-functionals which are also of interest. They are defined for arbitrary elements of M^A.

$S_0(x, u) = B(u, x)(\Pi_{y \in D(x)} x(y)^{\perp} + E(u, y))$.
$P_0(x, u, v) = B(u, x)B(v, x)(\Pi_{y \in D(x)}(x(y)^{\perp} + E(u, y) + E(v, y)))$.
$Q_0(x, b, c) = \sum_{u,v \in M^A} P_0(x, u, v)S_0(u, b)P_0(v, b, c)$.
$T_0(x, a, b, c) = \sum_{u \in M^A} Q_0(u, a, b)Q_0(x, u, c)$.

It could seem that some of these functionals are not well defined because they use sums of the elements of a proper class, namely M^A. For instance, $Q_0(x, u, v)$ is defined in terms of a sum whose domain is $M^A \times M^A$. But this presentation is only a way of speaking. More accurately we should have given that sum as the sum of one set: the M-set of those $a \in A$ for which there exists $u, v \in M^A$ with the property that $a = P_0(x, u, v)S_0(u, b)P_0(v, b, c)$. This is an M-set because it has been defined within M; and it is important to note this in view of the applications of Propositions 13.4.3 and 13.4.4, and Corollary 13.4.5. The same observations can be applied in similar situations to several of the definitions of functionals that will follow. The next property refers to the universe $M[G]$.

Lemma 14.2.8. *(i)* $S_0(x, u) \in G \Leftrightarrow x^G = \{u^G\}$.
(ii) $P_0(x, u, v) \in G \Leftrightarrow x^G = \{u^G, v^G\}$.
(iii) $Q_0(x, b, c) \in G \Leftrightarrow x^G = \langle b^G, c^G\rangle$.
(iv) $T_0(x, a, b, c) \in G \Leftrightarrow x^G = \langle a^G, b^G, c^G\rangle$.

Proof. We prove just (i) and (iii) and leave the rest as an exercise.

(i) Let $x, u \in M^A$ and $S_0(x, u) \in G$. By Proposition 13.4.4, this entails $B(u, x) \in G$ so that $u^G \in x^G$ by Proposition 14.2.5. The same reasons show that if $y^G \in x^G$, then $y^G = u^G$. Hence u^G is the only element of x^G and $x^G = \{u^G\}$. The converse is based on the same results.

(iii) Let $x^G = \langle b^G, c^G\rangle = \{\{b^G\}, \{b^G, c^G\}\}$ for some elements $x, b, c \in M^A$. Since $M[G]$ is a 2-universe, there exist elements $u, v \in M^A$ such that $u^G = \{b^G\}$ and $v^G = \{b^G, c^G\}$. Consider the summand corresponding to u, v in the equation giving $Q_0(x, b, c)$. We have $S_0(u, b) \in G$ by (i); $P_0(v, b, c) \in G$ by (ii), and then $P_0(x, u, v) \in G$ again by (ii). Thus this summand belongs to G and $Q_0(x, b, c) \in G$. The converse is similar. □

Corollary 14.2.9. *Let* $f : S \to A$ *a function that belongs to* M *with* $S \subseteq M^A_\alpha \times M^A_\alpha$. *Denote* $f^G = \{\langle x^G, y^G\rangle | f(x, y) \in G\}$. *Then* $f^G \in M[G]$.

Proof. From f, we define $h : M^A_{\alpha+2} \to A$ as $h(u) = \sum_{x,y \in M^A_\alpha} Q_0(u, x, y)f(x, y)$. h belongs to M, as it is defined in M and $f \in M$. An element of f^G has, by definition, the form $\langle x^G, y^G\rangle$ with $x, y \in M^A_\alpha$ and $f(x, y) \in G$. With the same proof of item (iii) of Lemma 14.2.8, we see that there exists some $u \in M^A_{\alpha+2}$

such that $u^G = \langle x^G, y^G \rangle$. Then $h(u) \in G$ and $u^G \in h^G$, hence $f^G \subseteq h^G$. Conversely, let $u^G \in h^G$, so that we may assume that $u \in M^A_{\alpha+2}$ and $h(u) \in G$. Then there are $x, y \in M^A_\alpha$ such that $Q_0(u, x, y) \in G$ and $f(x, y) \in G$. So, $u^G = \langle x^G, y^G \rangle \in f^G$ and $h^G \subseteq f^G$. □

Note that the same trick of Lemma 14.2.7 would serve to prove the following: if $S \subseteq M^A \times M^A \times A$ is an M-set, and we denote $S^G = \{\langle x^G, y^G \rangle | \exists a \in G\ (\langle x, y, a \rangle \in S)\}$, then $S^G \in M[G]$. It is also clear that, by using the functional T_0 instead of Q_0, one could prove the property analogous to Corollary 14.2.9, but for ordered triples instead of pairs.

For any $x \in M^A$ and $y \in \mathrm{D}(x)$, we know that $x(y) \in G \Rightarrow y^G \in x^G$, but the converse does not hold in general – it could be that $x(y) \notin G$ but $x(u) \in G$ for some $u \in \mathrm{D}(x)$ satisfying $u^G = y^G$. However, it is possible to find a substitute $w \in M^A$ for x with the properties that $w^G = x^G$ and $y^G \in w^G \Leftrightarrow w(y) \in G$ for any $y \in \mathrm{D}(w)$. To this, take simply $w(t) = B(t, x)$ with $\mathrm{D}(w) = \mathrm{D}(x)$. We call this function w the *completion function* of x. We now see an analogous property for functional M-classes.

Definition 14.2.10. Let F be a functional M-class with $\mathrm{D}(F) = M^A$ and $\mathrm{cD}(F) \subseteq A$. The functional M-class $H : M^A \to A$ with $H(x) = \sum_{u \in M^A} F(u)E(x, u)$ is called the *completion functional* of F.

We shall use the following notation. If $F : M^A \to A$ is a functional M-class, then:

$$F^G = \{x^G \in M[G] | F(x) \in G\} \tag{14.5}$$

F^G is a V-class because F is also a V-class by Proposition 7.1.12.

Lemma 14.2.11. *If H is the completion functional of a functional M-class $F : M^A \to A$, then: (1) for any $x \in M^A$, $x^G \in H^G \Leftrightarrow H(x) \in G$; and (2) $H^G = F^G$.*

Proof. If $H(x) \in G$ then $x^G \in H^G$ by definition; conversely, if $x^G \in H^G$, then there is some t such that $H(t) \in G$ and $t^G = x^G$; and hence there is $u \in M^A$ with $u^G = t^G$ and $F(u) \in G$. Therefore $H(x) \geq F(u)E(x, u)$ belongs to G by Proposition 14.2.5. The same reasons prove now condition (2). □

14.2.4 The axiom of replacement for generic extensions

Before proving the replacement axioms, we have to determine which are the classes of the universe $M[G]$.

Proposition 14.2.12. *Let C be an $M[G]$-class. Then one may find a functional class F of M with domain M^A and values in A such that $C = F^G$.*

Proof. C will be a term in a class sequence (Definition 3.1.2) $C_1, \ldots, C_n = C$ of $M[G]$. We will construct, by finite induction, functional classes $F_1, \ldots, F_n$

of M such that $F_i^G = C_i$. So, we will have one proof of the statement for each particular class C. The constructions below will be made inside M, though the property we want to prove is a consequence obtained in $M[G]$. Lemma 14.2.11 will be applied several times when replacing some of the functionals with their completions.

1) If $C_i = x$ is any set of $M[G]$, then $x = c^G$ for some $c \in M^A$. We may easily extend c to a functional defined on M^A by adding values 0.

2) Let C_i be the membership relation of $M[G]$. We know from Proposition 14.2.5 that $B^G = \mathcal{M}^{M[G]} = C_i$. We define $F : M^A \to M$ by $F(x) = \sum_{u,v \in M^A} Q_0(x,u,v)B(u,v)$. It is straightforward to verify (in $M[G]$) that $F^G = C_i$. The case of the equality relation is analogous.

3) Suppose $C_i = C_j \setminus C_k$ so that $C_j = F^G$ and $C_k = H^G$ by the inductive assumption. First, we replace H with its completion functional, which we still write as H. Then the functional $U(x) = F(x) \cdot H(x)^{\perp}$ gives $C_i = U^G$ by Proposition 13.4.3.

4) Let $C_i = C_j \times C_k$ with $C_j = F^G$ and $C_k = H^G$. We define a functional U on $M^A \times M^A$ by setting $U(x,y) = F(x)H(y)$. We then transform U into the functional $K : M^A \to A$ by setting $K(x) = \sum_{u,v \in M^A} Q_0(x,u,v)U(u,v)$; and we easily see that $K^G = C_j \times C_k$ in $M[G]$.

5) Suppose now that $C_i = \mathrm{D}(C')$ for some $M[G]$-class C', where $C' = F^G$. Let us define (in M) the functional $H : M^A \to A$ with $H(x) = \sum_{y,z \in M^A} F(z)Q_0(z,x,y)$. We check that $H^G = C_i$.

Any element of H^G is x^G such that there are $y, z \in M^A$ with $F(z) \in G$ and $z^G = \langle x^G, y^G \rangle$. But then $\langle x^G, y^G \rangle \in C'$ and $x^G \in C_i$. Conversely, suppose x^G is an element of C_i, so that $\langle x^G, y^G \rangle \in C'$ for some $y \in M^A$. Since $C' = F^G$, there is $u \in M^A$ such that $F(u) \in G$ and $u^G = \langle x^G, y^G \rangle$. By Lemma 14.2.8, $Q_0(u,x,y) \in G$ and it follows that $H(x) \in G$. Hence $x^G \in H^G$.

6) We consider just one of the cases for classes of triples, the other one being similar. Let $C_i = F^G$ and we write $\sigma[C_i]$ to denote the class of all triples $\langle a,b,c \rangle \in M[G]$ such that $\langle b,a,c \rangle \in C_i$. We define $U(x) = \sum_{t,u,v,w \in M^A} F(t)T_0(t,u,v,w)T_0(x,v,u,w)$. Again, it is easy to check that the elements of U^G are those of $\sigma[C_i]$. □

Proposition 14.2.13. *$M[G]$ satisfies each instance of the replacement axiom.*

Proof. Let C be a class of triples of $M[G]$, and take $u, s \in M[G]$. We have to show that $\mathrm{cD}(C|_{\{u\} \times s})$ is a set of $M[G]$, under the assumption that for each $x \in s$, there is at most one triple $\langle u,x,y \rangle \in C$. According to Proposition 14.2.12, there is a functional class F of M with $F^G = C$. We also set $s = v^G$ and $u = w^G$ for certain elements $v, w \in M^A$; and by taking the completion of v, we may assume that, for $x \in \mathrm{D}(v)$, $v(x) \in G \Leftrightarrow x^G \in s$. Our strategy for the proof is simple: working in M we define a certain element $t \in M$ with $t \subseteq M^A \times A$. By Lemma 14.2.7, $t^G \in M[G]$. Then we check in the universe $M[G]$ that $t^G = \mathrm{cD}(C|_{\{u\} \times s})$ and this proves the C-instance of the axiom.

Our first steps allow us to transform the M-functional F satisfying $F^G = C$ into another M-functional L with the properties: (i) $L^G \subseteq F^G$; and (ii) if $\langle x^G, y^G, z^G\rangle \in F^G$ and z^G is unique with that property for the given x^G, y^G, then $\langle x^G, y^G, z^G\rangle \in L^G$.

From F we construct a functional $F_0 : M^A \times M^A \times M^A \to A$ by defining $F_0(x, y, z) = \sum_{u\in M^A} F(u)T_0(u, x, y, z)$. Similarly to Lemma 14.2.11, we may assume that F_0 is a completion functional and that $F_0(x, y, z) \in G \Leftrightarrow \langle x^G, y^G, z^G\rangle \in C$.

The constructions that follow are made in the ZFC-universe M. From F_0 we define the M-class L given thus: $\langle x, y, z, a\rangle \in L$ if and only if (i) $F_0(x, y, z) = a$ and (ii) $(z' \in M^A \wedge F_0(x, y, z') = a) \Rightarrow \mathrm{rk}(z) \leq \mathrm{rk}(z')$. Since $L \subseteq F_0$, L is a functional and $L(x, y, z) \in G$ implies that $\langle x^G, y^G, z^G\rangle \in C$. Note also that if $F_0(x, y, z) = a$, then there is some $z_0 \in M^A$ with the property $\langle x, y, z_0, a\rangle \in L$.

Consider the M-class K of all pairs of the form $\langle\langle x, y, a\rangle, z\rangle$ with $\langle x, y, z, a\rangle \in L$: by the principle of separation (Proposition 5.1.11), if the values x_0, y_0, a_0 are fixed, the corresponding values z form a set, because all those z belong to $M_{\alpha+1}$ with α the smallest possible rank of z in elements $\langle x_0, y_0, z, a_0\rangle \in F_0$.

We next construct from L another M-class H of triples.

$$\langle x, \langle a, \langle y, a'\rangle\rangle, \langle z, aa'\rangle\rangle \in H \Leftrightarrow (a' \in A) \wedge \langle x, y, z, a\rangle \in L$$

By the preceding observation, if we fix the first two components of H, then the corresponding values of the third component form a set. Thus we may apply the strong form of the axiom of replacement (Proposition 5.1.7) in M for the class H. Fixing $x = w$ and with the second component ranging over the set $A \times v$, the axiom entails that $\{\langle z, a_0\rangle | \exists y, a, a'(\langle w, y, z, a\rangle \in L \wedge \langle y, a'\rangle \in v \wedge a_0 = aa')\}$ is a set $t \in M$, with $t \subseteq M^A \times A$.

Henceforth we work in the universe V to check that $t^G = \mathrm{cD}(C|_{\{u\}\times s})$. By Lemma 14.2.7, this will give $t^G \in M[G]$, completing the proof.

Let us take any $z^G \in t^G$ with $\langle z, a_0\rangle \in t$ and $a_0 \in G$; there are $\langle w, y, z, a\rangle \in L$ and $\langle y, a'\rangle \in v$ such that $aa' = a_0 \in G$ whence $a, a' \in G$; and thus $y^G \in v^G = s$ and $\langle w^G, y^G, z^G\rangle = \langle u, y^G, z^G\rangle \in C$ because $L(w, y, z) \in G$. Therefore $z^G \in \mathrm{cD}(C|_{\{u\}\times s})$.

Conversely, any element of $\mathrm{cD}(C|_{\{u\}\times s})$ will be z^G for $\langle w^G, y^G, z^G\rangle \in C$ with $y^G \in s = v^G$ and $y \in \mathrm{D}(v)$; by the completion hypothesis about v, $v(y) = a' \in G$. We must see that $z^G \in t^G$. We have $\langle w, y, z, a\rangle \in F_0$ with $a \in G$. By the definition of L, there is $z_0 \in M^A$ such that $\langle w, y, z_0, a\rangle \in L$, hence $\langle w^G, y^G, z_0^G\rangle \in C$, from which it follows that $z_0^G = z^G$ by the functionality of C restricted to $\{u\} \times s$. On the other hand, $\langle w, \langle a, \langle y, a'\rangle\rangle, \langle z_0, aa'\rangle\rangle \in H$ and hence $\langle z_0, aa'\rangle \in t$. Since $aa' \in G$, $z_0^G = z^G \in t^G$, as we had to show. □

Remark 14.2.14. The above proof for the particular class C uses only finitely many instances of the axiom of replacement for the universe M. Specifically, the proof that $M[G]$ satisfies the C-instance of the axiom requires that in M

the principle of separation is satisfied for the class K defined in the proof; and that the strong form of the axiom of replacement is satisfied for the class H constructed from F in that same proof. In turn, the principle of separation for K is satisfied in M if the K'-instance of replacement (for K' obtained as in the proof of Proposition 5.1.11) holds in M; and the strong form of the axiom of replacement for the class H is satisfied in M if the H_1-instance of replacement (for H_1 obtained as in the proof of Proposition 5.1.7) holds in M.

14.2.5 Infinity and choice

Proposition 14.2.15. *$M[G]$ is a ZF-universe.*

Proof. In view of the previous results, it only remains to prove that $M[G]$ satisfies the axiom of infinity. (Generally, we are assuming property NNS which implies the axiom of infinity; and it is trivial that this property in M implies the property in $M[G]$. But we give a proof which is independent of assertion NNS). Since $M \subseteq M[G]$, $\omega \in M[G]$ and $\omega \subseteq M[G]$. Now, for each $x \in \omega$, $s(x) = x \cup \{x\}$ is the successor of x also in $M[G]$, hence ω is inductive in $M[G]$. □

Before we may focus on the remaining axiom of ZFC-universes, some facts must be taken into account.

Lemma 14.2.16. *Every ordinal of M is also an ordinal of $M[G]$.*

Proof. Let α be an ordinal of M. By Proposition 9.2.16, α is transitive, membership in α is transitive and every two distinct elements of α are comparable for membership. Since M is transitive in V, every element of α is M-visible and α is a transitive element of $M[G]$. Also, if $x \in y$ as elements of M, then $x \in y$ in $M[G] \subseteq V$and thus membership is transitive in $\alpha \in M[G]$, and distinct elements are comparable. Proposition 9.2.16 implies that α is an ordinal of the ZF-universe $M[G]$. □

Proposition 14.2.17. *M and $M[G]$ have the same ordinals. Moreover, let $\Psi : M^A \to M[G]$ be the functional of Proposition 14.1.5. Then for each ordinal $\alpha \in M$, $\Psi[M^A_\alpha] = M[G]_\alpha$, the α-level in the von Neumann hierarchy of the ZF-universe $M[G]$.*

Proof. The first property is obvious if M contains all the ordinals of V. If not, it follows from Proposition 9.2.17 that there is a limit ordinal $\lambda \in V$ such that the ordinals of M are all the elements of λ. By Proposition 9.2.17 and Lemma 14.2.16, either these are the ordinals of $M[G]$ or else $\lambda \in M[G]$. We show now, by ordinal induction in V up to λ, that the equation of the statement holds.

We know that $\Psi(x) = x^G = \{u^G | u \in \mathrm{D}(x) \wedge x(u) \in G\}$. When $\alpha = 0$, the equation holds trivially. Assume that $\Psi[M^A_\alpha] = M[G]_\alpha$. To prove that $\Psi[M^A_{\alpha+1}] \subseteq M[G]_{\alpha+1}$, let $x \in M^A_{\alpha+1}$ so that $\mathrm{D}(x) \subseteq M^A_\alpha$. By the inductive hypothesis $x^G \subseteq M[G]_\alpha$, thus $x^G \in M[G]_{\alpha+1}$. For the converse, let $c^G \in M[G]_{\alpha+1}$ so that $c^G \subseteq M[G]_\alpha$ with $c \in M^A$. Working in M, we define $g : M^A_\alpha \to A$ as

$$g(x) = \sum_{z \in D(c)} E(x, z)\, c(z)$$

and we show (in V now) that $g^G = c^G$ from which it follows that $c^G \in \Psi[M^A_{\alpha+1}]$. Any element of c^G has to be z^G for some $z \in \mathrm{D}(c)$ with $c(z) \in G$. By the inductive hypothesis, there is $u \in M^A_\alpha$ such that $u^G = z^G$. Then $g(u) \in G$ by Proposition 14.2.5 and $z^G = u^G \in g^G$. Conversely, let $u \in M^A_\alpha$ such that $u^G \in g^G$ and $g(u) \in G$; there is $z \in \mathrm{D}(c)$ with $E(z, u) \in G$ and $c(z) \in G$. But then $z^G \in c^G$ and $z^G = u^G$, completing the proof of the second inclusion. The property for a nonzero limit ordinal α is now easy, bearing in mind that $\cup_{\beta \in \alpha} M^A_\beta$ is the same in M or V by Proposition 7.1.9. By the same reason, we get

$$\Psi[M^A] = M[G] = \Psi[\cup_{\alpha\in\lambda} M^A_\alpha] = \cup_{\alpha\in\lambda} M[G]_\alpha \tag{14.6}$$

If we had $\lambda \in M[G]$, then $\lambda \in M[G]_\alpha$ for some $\alpha < \lambda$ by equation 14.6. But then the $M[G]$-rank of the ordinal λ is $< \lambda$, which contradicts the property $\mathrm{rk}(\lambda) = \lambda$ in ZF-universes (see, e.g., Exercise 10 of Chapter 9). Thus the ordinals of $M[G]$ and of M are the same.

The same inductive proof just given for the equation of the statement works also in case the ordinals of M are all the ordinals of V. □

Proposition 14.2.18. *$M[G]$ is a ZFC-universe.*

Proof:. First of all, we claim the following property of ZF-universes: if α is an ordinal and $g : \alpha \to S$ is a surjective map, then S can be well-ordered. To prove this, define $h : S \to \alpha$ with $h(x) = \min(g^{-1}[\{x\}]) \subseteq \alpha$. By the hypothesis, the sets $g^{-1}[\{x\}]$ are non-empty and pairwise disjoint, and this entails that h is an injective map. Then $h[S]$ is well-ordered because it is a subset of a well-ordered set, and then S inherits the well ordering of $h[S]$ by simply setting $x \leq y \Leftrightarrow h(x) \leq h(y)$.

To prove that the axiom of choice holds in the ZF-universe $M[G]$, it will be enough, according to Propositions 14.2.15 and 10.3.5, to prove that any $x^G \in M[G]$ can be well-ordered. Since $x \in M$, $D = \mathrm{D}(x) \subseteq M^A$ is a set of M and it can be well-ordered because M is a ZFC-universe. By Proposition 10.3.5, there is a bijective map in M, $f : \alpha \to D$ for some ordinal $\alpha \in M$.

$D \in M$ gives $\check{D} \in M^A$; and if we call D_1 to the domain of $\check{D}$, $D_1 = \{\check{t} | t \in D\} \subseteq M^A$. The following is then a set X of M: $\{\langle\langle \check{t}, t\rangle, 1\rangle\}$, 1 being the supremum of the algebra A. We have $X \subseteq M^A \times M^A \times A$, and we may apply Corollary 14.2.9 in its version for subsets included in $M^A \times M^A \times A$ that is mentioned after the proof of that corollary; the consequence is that $X^G \in M[G]$. But X^G consists of all ordered pairs $\langle(\check{t})^G, t^G\rangle$ for $t \in D$. That is, $Y = \{\langle t, t^G\rangle | t \in D\}$ belongs to $M[G]$. Thus $Y : D \to \Psi[D]$ is a surjective map, and by composition we get a surjective map $\alpha \to \Psi[D]$ in $M[G]$, with α being an ordinal of $M[G]$ by Lemma 14.2.16. Therefore, $\Psi[D]$ can be well-ordered in $M[G]$ by the claim above.

Now, $x^G = \{t^G | t \in D(x) \wedge x(t) \in G\}$ is a subset of $\Psi[D]$, and hence it can be well-ordered as we wanted to see. □

14.3 Cardinals of Generic Extensions

We keep in this section the assumptions (GE) and thus $M[G]$ is a ZFC-universe by Proposition 14.2.18. We observe that cardinals of $M[G]$ are also cardinals of M: since cardinals are ordinals, any cardinal α of $M[G]$ is an ordinal in M by Proposition 14.2.17. If α is not a cardinal in M, then there is $\beta < \alpha$ and a bijection (in M) between β and α. But the bijection also belongs to $M[G]$ because $M \subseteq M[G]$, and therefore α is not a cardinal of $M[G]$.

The converse is harder: we prove it under special conditions and only for regular cardinals, to start. First we give another lemma.

Lemma 14.3.1. *Let $z \in M$ be a function with values in A, such that $z \subseteq M^A \times M^A \times A$; and let $p_z = \prod_{\langle s_1,t_1\rangle,\langle s_2,t_2\rangle \in D(z)} (z(s_1,t_1)^\perp + z(s_2,t_2)^\perp + E(s_1,s_2)^\perp + E(t_1,t_2)) \in A$. For any M-generic filter H of A^*, $p_z \in H \Leftrightarrow z^H$ is a function of $M[H]$*[1].

Proof. By Corollary 14.2.9, $z^H = \{\langle x^H, y^H\rangle | z(x,y) \in H\}$ is an element of $M[H]$, if H is a M-generic filter of A^*. Suppose that $p_z \in H$ and that $\langle u^H, x^H\rangle, \langle u^H, y^H\rangle \in z^H$. By the definition of z^H and choosing $u, x, y \in M^A$ adequately, this entails that there are elements $b_1, b_2 \in H \subseteq A^*$ and some $u' \in M^A$ so that $\langle u, x, b_1\rangle, \langle u', y, b_2\rangle \in z$; and moreover $(u')^H = u^H$. Now, each factor of p_z belongs to H and thus $z(u,x)^\perp + z(u',y)^\perp + E(u,u')^\perp + E(x,y) \in H$. It follows then that $E(x,y) \in H$, and thus $x^H = y^H$ by Proposition 14.2.5, proving that Z^H is functional. The converse is similar. □

Definition 14.3.2. Let S be an ordered set. An *antichain* in S is a subset $L \subseteq S$ such that for any $x, y \in L$, $x \neq y \Rightarrow x, y$ are incompatible.

The ordered set S is said to satisfy the *countable chain condition* (or ccc, for short) when every antichain of S is countable.

Proposition 14.3.3. *In the hypotheses (GE), suppose that the following two conditions hold: (1) The countable chain condition holds for A^* in M; (2) For any $x \in A^*$, there exists a M-generic filter $H \in V$ of A^* such that $x \in H$.*

If μ is a regular cardinal of M, then μ is also a regular cardinal of $M[G]$.

Proof. If $\mu \leq \omega$, the property is obvious and so we assume that $\mu \geq \aleph_1$ (in M). To show that μ is a regular cardinal of $M[G]$ it will be enough to prove

[1] It is in this sense that the element $p_z \in A$ *forces* z to be a function.

that if $\beta \in \mu$ and $f : \beta \to \mu$ is a map in $M[G]$, then f is not cofinal: this will prove that μ has cofinality μ (in $M[G]$) and hence it is a cardinal by the observation (2) after Definition 10.6.1. What we do with this purpose is to construct a subset $T \subseteq \mu$ with $T \in M$ which is not cofinal in μ (and this property is equivalent in M or $M[G]$ because the elements and the ordering of μ are the same in both universes), and then to check that $\mathrm{cD}(f) \subseteq T$, hence f is not cofinal in μ.

So, let $f : \beta \to \mu$ be a map in $M[G]$ with $\beta < \mu$. Then $f = x^G$ for some $x \in M^A$ so that x^G consists of ordered pairs. From now, and with just one exception for a brief observation, the proof runs completely inside the universe M until we reach the conclusion that T is not cofinal in μ.

Say that $\mathrm{D}(x) \subseteq M^A_\alpha$; we define a function z from $\mathrm{D}(\check{\beta})\times$ $\mathrm{D}(\check{\mu})$ to A, with $z(\check{r}, \check{s}) = \sum_{t\in D(x)} x(t)Q_0(t, \check{r}, \check{s})$, so that $z \in M^A$. The brief observation in $M[G]$ is now this: $z^G = x^G$ by the definition of z (and with z^G as in Corollary 14.2.9). Taking p_z (for this particular z) as defined in Lemma 14.3.1, we obtain from that lemma that $p_z \in G$ because z^G is a function.

In particular, $p_z \neq 0$. We next define a map of M, $h : \beta \times \mu \to A$, with $h(r, s) = z(\check{r}, \check{s}) \cdot p_z$. We consider, for each $r \in \beta$, the set T_r of elements $s \in \mu$ such that $h(r, s) \neq 0$ and note that if $s_1 \neq s_2 \in T_r$, then $h(r, s_1), h(r, s_2)$ are incompatible (in A^*). Because if they are compatible, then $h(r, s_1) \cdot h(r, s_2) \neq 0$. By the hypothesis (2) above, there is a M-generic filter H of A^* such that $h(r, s_1) \cdot h(r, s_2) \in H$. Then $p_z \in H$ so that z^H is a function by Lemma 14.2.7; and $z(\check{r}, \check{s_1}), z(\check{r}, \check{s_2}) \in H$ which entails that $z^H(\check{r}^H) = \check{s_1}^H = \check{s_2}^H = s_1 = s_2$. Thus different elements of T_r give incompatible elements of A^*, and hence T_r has only countably many elements by the ccc hypothesis.

Then $T = \cup_{r\in\beta} T_r$ is a union of countable M-sets indexed by β. Since $|\beta| < \mu$, $|T| < \mu$ by Proposition 10.6.9 because μ is regular in M; and therefore T cannot be cofinal in μ.

It remains to show that $\mathrm{cD}(f) \subseteq T$, so we must now take the elements of $M[G]$ into account. If $s \in \mathrm{cD}(f)$ and $f(r) = s$, then $\langle r, s\rangle \in x^G = z^G$. Thus $z(\check{r}, \check{s}) \in G$ because this is the only pair in $\mathrm{D}(\check{\beta}) \times \mathrm{D}(\check{\mu})$ giving the elements r, s. Therefore $h(r, s) \in G$ so that $h(r, s) \neq 0$ and $s \in T_r \subseteq T$. Thus $\mathrm{cD}(f) \subseteq T$ and hence $\mathrm{cD}(f)$ is not cofinal in μ (in $M[G]$). It follows that μ is a regular cardinal of $M[G]$. □

Corollary 14.3.4. *Under the same hypotheses of Proposition 14.3.3, the cardinals of M and of $M[G]$ are the same.*

Proof. We have already seen that if κ is a cardinal of $M[G]$, then it is a cardinal of M. For the converse, let ρ be a cardinal of M. To prove that it is a cardinal of $M[G]$, use RAA and assume that ρ is the smallest cardinal of M which is not a cardinal of $M[G]$. By Proposition 14.3.3, ρ is singular. Let $|\rho|_{M[G]} = \kappa$, $|\rho|_{M[G]}$ being the cardinal of ρ in $M[G]$. Then $\kappa < \rho$ (in M and in $M[G]$: both have the same ordinals by Proposition 14.2.17).

In M, κ is a cardinal and it has a successor cardinal κ^+ which is regular by Proposition 10.6.7, and hence it cannot be ρ. Therefore $\kappa < \kappa^+ < \rho$. But

κ^+ is a cardinal of $M[G]$ by Proposition 14.3.3 and thus $\kappa^+ \leq |\rho|_{M[G]} = \kappa$, which is absurd. □

14.4 Exercises

1. Referring to the construction of Proposition 14.1.1, prove that $M^A_\beta \subseteq M^A_\alpha$ if $\beta \leq \alpha$.
2. The recursive definition in that same Proposition 14.1.1 has been given for a particular complete Boolean algebra A. Change the universal class used in that result for the definition of the functional F, so that it gives a recursive definition valid for any choice of the Boolean algebra A.
3. Let A be a complete Boolean algebra of the ZFC-universe M. Show that for each ordinal α of M, there exists a smallest ordinal β such that $M^A_\alpha \subseteq M_\beta$. If α is a limit ordinal, is the corresponding β necessarily a limit ordinal?
4. Construct the proof of Proposition 14.1.3 making explicit how it is based on the recursion theorem; i.e., identify the functionals F, G in the statement of Theorem 8.4.5 that are applied for the recursive definition in the proposition.
5. Let A be a complete Boolean algebra of the ZFC-universe M. Show that every ordinal $\alpha \in M$ satisfies that $\check{\alpha} \in M^A_{\alpha+1}$ but $\check{\alpha} \notin M^A_\alpha$.
6. Give an explicit recursive definition of the functional I of Section 14.2.2, identifying the classes to which the recursion theorem (Theorem 8.4.5) is applied.

In the exercises that follow, hypothesis (GE) is assumed.

7. Working in the universe V, show that a functional $\Delta : M[G] \to M^A$ can be defined (recursively) with the property that $\Psi \circ \Delta$ is the identity on $M[G]$.
8. Prove that the functionals P_0, T_0 of Lemma 14.2.8 satisfy the properties stated in the lemma.
9. In item (2) of the proof of Proposition 14.2.12, it is shown how to find a functional F on one variable such that $F^G = B^G$, where B is the membership functional. Generalize this idea by extending the notation of equation 14.5 to functionals on two variables; and then show how to construct a functional F on one variable for an arbitrary functional H on two variables so as $F^G = H^G$; and also construct a functional on two variables which corresponds in a similar way to a given arbitrary functional on one variable.

10. Define the completion functional of a functional on two or three variables, as in Definition 14.2.10.

11. Let F be a filter in an ordered set $P \in M$ – but possibly $F \notin M$. Prove that F is a M-generic filter if and only if $K \cap F \neq \emptyset$ for every $K \in M$ which is a maximal antichain of P.

12. Let $A \in M$ be a complete Boolean algebra. Prove that A, viewed as an object of V, is also a Boolean algebra. Explain why A might fail to be a complete Boolean algebra as an object of V.

13. A is a complete Boolean algebra of M. Exhibit an algebra homomorphism $A \to \{0, 1\}$ induced by G (and hence existing in V but possibly not in M), which preserves supremum and infimum of M-subsets of A.

14. Let $L \subseteq A$ be a filter of A^* which satisfies the properties that Propositions 13.4.3 and 13.4.4 and Corollary 13.4.5 attribute to G. Let E be the functional defined in Section 14.2.2. Consider the binary relation between elements of M^A given thus: $xRy \Leftrightarrow E(x, y) \in L$. Prove that R is an equivalence relation in the universe M. Prove also that each equivalence class of M^A modulo R is a proper class (Hint: For $x \in M^A$, choose some element in $M^A \setminus \mathrm{D}(x)$ and add it to the domain of x).

15. In the preceding exercise, use the idea of Exercise 4 of Chapter 9 for reducing the class M^A so that the equivalence R gives now a quotient class, that we improperly denote M^A/R. Assume now the hypotheses (GE) with $G = L \in V$, and consider the quotient M^A/R as a V-class. Describe the relationship between the quotient M^A/R and the universe $M[G]$.

16. In item (6) of the proof of Proposition 14.2.12, it is considered the case of the transformation of rule 6 of Definition 3.1.2 to obtain the class F with $F^G = C_i$. Give the proof for the case of rule 7 of that same definition.

17. Fix an element $u \in M^A$, such that $u^G \neq \emptyset$. Prove that the collection of all elements $x \in M^A$ with $\mathrm{D}(x) = \mathrm{D}(u)$ and such that $x^G = u^G$, is a set of the ZFC-universe M if and only if $G \in M$ (Hint: Define for each $a \in A$, the set $G(a)$ of all $z \in M^A$ with $\mathrm{D}(z) = \mathrm{D}(u)$ and such that $a, 0$ are the only values of the functions z. Use $G(a)$ to distinguish, in M, the elements $a \in G$).

18. For elements $x, y \in M^A$, prove that $x^G \cup y^G = (x \cup y)^G$ – here $x \cup y$ is not in general an element of M^A, it is just a subset of $M^A \times A$; thus the meaning of $(x \cup y)^G$ is the same given in Lemma 14.2.7 for S^G.

19. Let p, q be incompatible elements of A^*. Check that for any $x \in M^A$, $y = \{\langle \{\langle x, p\rangle\}, q\rangle, \langle \check{0}, 1\rangle\}$ gives $y^G = \{\emptyset\}$.

20. Let $x \in M^A$ be such that $\mathrm{D}(x) \subseteq \mathrm{D}(\check{\omega})$. Let us define the subset y of $\mathrm{D}(x) \times A$ consisting of all pairs $\langle \check{n}, p\rangle$ such that $p, x(\check{n})$ are compatible. Prove that $z = (\mathrm{D}(\check{\omega}) \times A) \setminus y$ satisfies $z^G = \omega \setminus x^G$ (z^G is defined as in Lemma 14.2.7).

21. Let C be a transitive class of V such that $M \subseteq C$ and $G \in C$. Prove that, if C is a ZFC-universe, then $M[G] \subseteq C$.

22. Let P be an ordered set without minimum element, A a complete Boolean algebra, and $i : P \to A^*$ the completion algebra (Definition 13.3.15). Let G be a M-generic filter of P, and let $G' \subseteq A^*$ be the M-generic filter of A^* obtained from G as in Proposition 13.4.6. Prove that if C is a transitive class of V which is a ZFC-universe and $G \in C$, then $M[G'] \subseteq C$ if $M \subseteq C$.

23. Suppose that instead of defining the sets M^A_α as sets of functions, we define another construction where each $M^{(A)}_{\alpha+1}$ is the set of all subsets (in M) of $M^{(A)}_\alpha \times A$. The union of these sets will give the class $M^{(A)}$; and, given the M-generic filter G of A^*, define for each ordinal α and $x \in M^{(A)}_\alpha$ the set $x^{(G)} = \{u^{(G)} | u \in \mathrm{D}(x) \wedge (\exists a \in G(\langle u, a\rangle \in x))\}$. This would give $M(G)$ in a way analogous to that giving $M[G]$. Prove that $M^A_\alpha \subseteq M^{(A)}_\alpha$ and define functions $f_\alpha : M^{(A)}_\alpha \to M^A_\alpha$ so that f_α is the identity when restricted to M^A_α. Is $M[G] = M(G)$?

24. As in in the preceding exercise, we change the definition of the sets M^A_α by setting $M^{\langle A\rangle}_{\alpha+1}$ as the set of all functions with domain $M^{\langle A\rangle}_\alpha$ and codomain included in A; and $x^{\langle G\rangle}$ is defined exactly as x^G, giving in this case $M\langle G\rangle$. Try to make a study similar to that of the previous exercise, and compare the universes $M[G]$ and $M\langle G\rangle$.

25. Let G be a M-generic filter of A^* and π an algebra automorphism of A in the universe M. If $H = \pi[G]$, prove that H is M-generic. Then, define the π-induced maps $M^A_\alpha \to M^A_\alpha$ and deduce that $M[G] = M[H]$.

26. In Proposition 14.2.18 a set $D \subseteq M^A_\alpha$ has produced an $M[G]$-set $Y = \{\langle t, t^G | t \in D\}$. Show that this can be done generally, so that any Ψ_α is in fact an element of $M[G]$.

27. Formulas were introduced in Section 3.2; and a *closed formula* is a formula without free variables. There is a way of assigning to any closed formula P an element of A, called the value $||P||$ of the formula. We do this for formulas where elements of $M^A \subseteq M$ may appear, in a recursive way; that is, we define $||P||$ by assuming that the values of closed formulas shorter than P are known. Thus we set $||P \wedge Q|| = ||P|| \cdot ||Q||$; $||P \vee Q|| = ||P|| + ||Q||$; $||\neg(P)|| = (||P||)^\perp$; and $||\exists x\ (P(x))|| = \sum_{u \in M^A} ||P(u)||$; $||\forall x\ (P(x))|| = \prod_{u \in M^A} ||P(u)||$. Finally, $||u \in v|| = B(u, v)$, and $||u = v|| = E(u, v)$, where B, E are the functionals defined in Section 14.2.2.

For a M-generic filter G of A^*, show that the closed formula P is true for $M[G]$ (replacing each element $u \in M^A$ with u^G) if and only if $||P|| \in G$.

28. Assume condition (2) of the statement of Proposition 14.3.3. With reference to the previous exercise, prove that if P is a closed formula which translates an axiom of ZFC-universes, then $||P|| = 1$ (Hint: Use RAA, suppose $||P|| = x \neq 1$, and apply the assumption).

29. In the hypotheses of Lemma 14.3.1, define an element $q_z \in A$ having the property that for any M-generic filter H of A^*, $q_z \in H \Leftrightarrow z^H$ is an injective function.

30. Prove that if A satisfies the conditions of Proposition 14.3.3, then every ordinal has the same cofinality in M and in $M[G]$.

15

Independence Proofs

15.1 Pointed Universes

In the beginning of the book we postulated a universe, i.e., a collection of objects; by imposing axioms about the objects of that universe, we arrived to the notion of a ZF-universe or a ZFC-universe. Let us imagine now that we had started differently, by postulating not just a universe, but a *pointed universe*; that is, a collection of objects $\mathcal{U}$ with an special, distinguished object $M \in \mathcal{U}$. Thus, $\mathcal{U}$ and M would constitute our initial data, and we might proceed as before, stating axioms that the objects of the universe, including M, would have to fulfill. We do precisely this now, and arrive at a new theory which, in correspondence with the first case, will give *pointed ZFC⁺-universes* (or, simply, ZFC$^+$-universes). The idea of pointed universes and the theorem that follows are adapted from Weaver's book [24]. Let us state the axioms for this new theory where $\mathcal{U}$ is the universe and M the distinguished object.

(1) $\mathcal{U}$ satisfies all the axioms of ZFC-universes.

(2) The distinguished object M satisfies three additional axioms as an object of $\mathcal{U}$: M is a transitive set, M is denumerable, and $\omega \in M$.

(3) Considered as a universe by itself, M satisfies all the axioms of ZFC-universes.

When the pair $\mathcal{U}, M$ satisfy these axioms, we say that $\mathcal{U}$, with the distinguished object M, is a *pointed ZFC$^+$-universe*. In that case $\mathcal{U}$ is a ZFC-universe and also M is a ZFC-universe. The most useful relationship between ZFC and ZFC$^+$-universes is the following metatheoretic result.

Theorem 15.1.1. *If the theory of ZFC-universes is consistent, then the theory of ZFC$^+$-universes is consistent.*

Proof. Suppose that from the axioms of ZFC$^+$-universes, a contradiction $P \wedge \neg(P)$ is deduced. Since a deduction is a finite argument, only finitely many axioms will have been used in that proof giving $P \wedge \neg(P)$. In particular, only finitely many instances of the replacement axioms for the universe M will have been necessary for constructing the supposed proof. According to Proposition 5.1.10, we may assume that the instances used in the proof correspond to the neat classes $C_1, \ldots, C_n$; that is, only the instances of the axiom involving the M-classes $C_1^M, \ldots, C_n^M$ will be needed for the proof (recall from the discussion

DOI: 10.1201/9781003449911-15

just before Theorem 11.4.4 that neat classes may be seen abstractly, as classes that can be constructed in the same way in any possible universe. In this sense, each neat class is "the same" in every ZFC-universe). And this entails that $P \wedge \neg(P)$ can be deduced from the hypotheses that $\mathcal{U}$ is a ZFC-universe, $M \in \mathcal{U}$ is a transitive denumerable set containing ω, and M is a ZFC*-universe where the C_i^M-instances of replacement are satisfied.

Start now with a ZFC-universe V. By Theorem 11.3.4, there exists $M \in V$ with $\omega \in M$, which is a denumerable transitive set and a ZFC*-universe such that the C_i-instances of the axiom of replacement hold in M. We therefore may see the pair V, M as a pointed universe where all the axioms for ZFC$^+$-universes hold, except replacement for M, but M satisfies the C_i^M-instances of replacement. Then the supposed proof of $P \wedge \neg(P)$ can be carried out under those hypotheses. Like this, from the hypothesis that V is a ZFC-universe we arrive to a contradiction. That is, the ZFC-theory would be inconsistent if the supposed proof of a contradiction from the ZFC$^+$-axioms exists. □

We now make our first steps in the ZFC$^+$-theory, giving a simple but absolutely fundamental result.

Proposition 15.1.2. *Let V a ZFC$^+$-universe with $M \in V$ the distinguished set. Let $P \in M$ be an ordered set, and $p \in P$. Then there exists $G_0 \in V$ which is a M-generic filter of P such that $p \in G_0$.*

Proof. In the universe M, consider all the dense subsets $D \subseteq P$. Viewed in V, they form a countable set because M is countable. We thus may consider them ordered (in V) as $D_1, \ldots, D_k, \ldots$; and we may assume that this is an infinite sequence by repeating the sets D_k if necessary. By using the axiom of choice of the universe V, we define recursively an infinite sequence $p_0, p_1, \ldots$ of elements of P in the following form:

$$p_0 = p,\ p_1 \in D_1 \cap \{q \in P | q \leq p_0\}, \ldots, p_{k+1} \in D_{k+1} \cap \{q \in P | q \leq p_k\}, \ldots.$$

Then we define $G_0 = \{q \in P | \exists n\, (p_n \leq q)\}$. G_0 is clearly a filter with $p \in G_0$; and for each dense subset $D_k \in M$ there is $p_k \in D_k \cap G_0$. Thus G_0 is a M-generic filter of P – but its construction has been made in V, thus G_0 is possibly not in M. □

This shows that if V, M are the data for a pointed ZFC$^+$-universe, then item (2) of the hypotheses of Proposition 14.3.3 holds, because Proposition 15.1.2 can be applied to the ordered set B^* of a complete Boolean algebra B. We consider next the relationship between these generic filters and corresponding filters in the completion algebra.

Proposition 15.1.3. *Let V a ZFC$^+$-universe with $M \in V$ the distinguished set. Let $P \in M$ be an ordered set without minimum element and let $i : P \to A$ be the completion algebra of P in M whose existence is shown in Corollary 13.3.18. If G_0 is a M-generic filter of P and G is the M-generic filter of A^* shown in Proposition 13.4.6 (i.e., $G = \{x \in A^* | \exists p \in G_0\, (i(p) \leq x)\}$), then $G_0 = i^{-1}[G]$ and $G_0 \in M[G]$.*

Proof. Note first that, according to Proposition 15.1.2, the data V, M, A, G satisfy the hypotheses (GE), so that $M[G]$ is a ZFC-universe. Now, the property $G_0 \subseteq i^{-1}[G]$ is obvious. Conversely, let $p \in P$ and $i(p) \in G$. By definition of G, there is $q \in G_0$ with $i(q) \leq i(p)$. It follows from Lemma 13.3.12 that $\overline{q} \preceq \overline{p}$ as elements of the separative quotient $\overline{P}$; and hence $[p]$ is dense in $[q]$ as subsets of P, by Proposition 13.3.16. If we define the set $D \in M$ which is the union of $[p]$ and the set of all elements of P which are incompatible with q, it can be easily seen that D is dense in P. Then $D \cap G_0 \neq \emptyset$. From the facts that elements of G_0 cannot be incompatible and $q \in G_0$, we infer that some $g \in G_0$ belongs to $[p]$, hence $p \in G_0$ (This can also be seen as a consequence of Exercise 31 of Chapter 13, because the condition "$[p]$ is dense in $[q]$" means the $[p]$ is dense below q and $q \in G_0$).

Both i and G belong to the ZFC-universe $M[G]$, and it follows that $i^{-1}[G] = G_0 \in M[G]$. □

We may summarize what is going to be most interesting for its use in this chapter, in the following statement:

If V and $M \in V$ form a pointed ZFC$^+$-universe and $A \in M$ is a complete Boolean algebra in the universe M, then there exists some M-generic filter G of A^*, and V, M, A, G satisfy the hypotheses (GE).

15.2 Cohen's Theorem on the CH

15.2.1 Overview

The continuum hypothesis (CH) is the assertion that the equality $\aleph_1 = 2^{\aleph_0}$ holds in all ZFC-universes. As explained in the introduction to Part II, an statement like CH about the objects of ZFC-universes is called *independent* relative to the axioms of ZFC-universes when both CH and $\neg$(CH) are consistent with those axioms. Like this, saying that the CH is independent in the theory of ZFC-universes means that, provided that the theory is consistent, there is no proof of CH, and there is also no proof of the contrary assertion, i.e., of $\aleph_1 \neq 2^{\aleph_0}$.

Our objective in the first part of this chapter will be to show that CH is indeed independent in that sense. In fact, one half of this task is a direct consequence of the results of Chapter 12. Namely, we may easily show that there is no proof of $\aleph_1 \neq 2^{\aleph_0}$ for all ZFC-universes (if the theory is consistent): this is *Gödel's theorem on the CH*. But we need a piece of metatheory.

Proposition 15.2.1. *Let (S) be an assertion about sets, and suppose that there is a proof that (S) holds in every ZF(C)-universe. From that proof, it is possible to obtain a neat class sequence $W_1, \ldots, W_n$, such that (S) holds in* *every ZF(C)***-universe which satisfies each of the* W_i*-instances of the axiom of replacement.*

Proof. This is obtained by the same argument used in the proof of Theorem 15.1.1: the proof of (S) uses only finitely many instances of the axiom of replacement and thus is valid in ZF(C)*-universes which satisfy replacement for the corresponding neat classes. □

We proceed now with the promised proof that CH is consistent with the ZFC-axioms: suppose we had a proof of $\aleph_1 \neq 2^{\aleph_0}$. By Proposition 15.2.1, there exists a neat class sequence $W_1, \ldots, W_n$ such that $\aleph_1 \neq 2^{\aleph_0}$ holds in every ZFC*-universe for which the W_i-instances of replacement hold. But if L is the constructible universe of a ZFC-universe V, we know from Corollary 12.4.4 that L is a ZFC-universe. This entails that there is a proof that L satisfies the W_i-instance of replacement for $i = 1, \ldots, n$, and L is a ZFC*-universe too. And therefore $\aleph_1 \neq 2^{\aleph_0}$ must hold in L. But Theorem 12.5.8 shows that $\aleph_1 = 2^{\aleph_0}$ in L, and we have reached a contradiction.

The reader may justifiably be surprised by the twist in the argument. Would it not be simpler to point out that since L is a ZFC-universe, the inequality of the cardinals in L would follow directly from the supposed proof and give a contradiction with Theorem 12.5.8? The problem, however, is that Corollary 12.4.4 has no (finite) proof: being a statement which involves all classes, there is a proof that L satisfies the replacement axioms for any finite list of classes, but no proof for all instances of the axiom. And the final goal of the argument is to produce a contradiction: that is, a **proof** that the inequality holds in L along with a proof that the equality holds in L. This is why we need a (finite) proof of those facts.

Back to the above argument, the contradiction arises by assuming that there is a proof of the negation $\neg$(CH) of CH. Consequently, the conclusion is: either there is no proof of $\neg$(CH) or else the axioms of the ZFC-universes produce a contradiction. If the theory of ZFC-universes is consistent, then there is no proof of the inequality $\aleph_1 \neq 2^{\aleph_0}$.

To show that CH is an independent statement in the ZFC-theory we must now show the other half, i.e., that CH is not provable from the axioms of ZFC-universes. Essentially, this will be done through the same steps of the above argument; we postulate that there is a proof of CH and arrive to a contradiction by finding a certain universe which gives a counterexample to CH. But there is a significant difference: we do not show a ZFC or ZFC*-universe where CH fails; instead, we work in a ZFC$^+$-universe where the hypotheses (GE) hold, and obtain that CH fails in the ZFC-universe $M[G]$. Then, if there existed a proof of CH for ZFC-universes we would reach a contradiction from the ZFC$^+$-theory. Now, with the help of Theorem 15.1.1, a contradiction would be obtained in the ZFC-theory.

15.2.2 A counterexample to CH

The strategy for finding some ZFC-universe $M[G]$ where CH fails is to look for an injective map $g : \aleph_2 \to {}^{\aleph_0}2$. This will entail that $\aleph_2 \leq 2^{\aleph_0}$ from which it follows that $2^{\aleph_0} > \aleph_1$. The process needs three components: (1) The function g will be constructed naturally from a function $f : \aleph_2 \times \aleph_0 \to \{0,1\}$: if $\mu \in \aleph_2$ is left fixed, $g(\mu) = f(\mu, -)$ gives a function $\aleph_0 \to \{0,1\}$. (2) The function f is obtained as the union of a filter of finite functions whose domains are included in $\aleph_2 \times \aleph_0$, and the codomains are included in $\{0,1\}$; and the set of all such finite functions will be the ordered set P that allows the construction of the generic extension $M[G]$. (3) If A is the completion algebra of the set P, then A^* will satisfy the ccc – see Definition 14.3.2. This guarantees that M and $M[G]$ have the same cardinals: thus if $\aleph_\alpha$ is the α-th cardinal of M, it is also the α-th cardinal of $M[G]$.

We start our quest. V and M form a ZFC$^+$-universe. We define the ordered M-set P. For our convenience, we admit condition NNS about ω. This is not strictly necessary, but like this we are sure that finite sets have the usual meaning.

Let $P \in M$ be the set of all functions f with finite domain, $D(f) \subseteq \aleph_2 \times \aleph_0$ and $cD(f) \subseteq \{0,1\}$. An ordering of P is defined by reverse inclusion; i.e., $f \leq g \Leftrightarrow g \subseteq f$.

Thus $f \leq g$ precisely when the function f is an extension of g. Consequently, f, g are incompatible if and only if f, g have no common extension, i.e., there is some $u \in \mathrm{D}(f) \cap \mathrm{D}(g)$ such that $f(u) \neq g(u)$.

The hypotheses (GE) are now met with the ZFC$^+$-universe V and the denumerable transitive ZFC-universe $M \in V$. A is the completion algebra of P; and by Proposition 15.1.2, there is a M-generic filter G_0 of P, which by Proposition 15.1.3 determines the M-generic filter G of A^*; and $G_0 \in M[G]$. Thus we know that $M[G]$ is a ZFC-universe by Proposition 14.2.18. Now, we check the ccc for P, then for A^*.

Proposition 15.2.2. *The ordered set P satisfies the ccc in M.*

Proof. Suppose $C \subseteq P$ is an antichain. If each of the sets $\{f | f \in C \wedge |\mathrm{D}(f)| = n\}$ (for each $n \in \omega$) were countable, then C would be a countable union of countable sets, hence countable. Therefore we may prove that P satisfies the ccc by showing that, for every natural number n, there is no uncountable antichain C of P with the property $f \in C \Rightarrow |\mathrm{D}(f)| = n$. We show this by natural induction, starting with $n = 1$. If all functions of C have a one-element domain and are pairwise incompatible, all of them have the same domain, so there exist at most two functions in C.

Let us admit as inductive hypothesis that there is no uncountable antichain of P all of whose elements have a domain with n elements ($n \geq 1$). Take now, if possible, an uncountable antichain C such that every element of C has a domain with $n+1$ elements. Choose and fix some $f \in C$ with domain $\{u_0, u_1, \ldots, u_n\}$ and for $i = 0, \ldots, n$, let $C_i = \{g \in C | u_i \in \mathrm{D}(g) \wedge g(u_i) \neq$

$f(u_i)\}$. Since every $g \in C\backslash\{f\}$ must be incompatible with f, $C\backslash\{f\} = \cup_{i=0}^n C_i$ and it follows that there is an index k such that C_k is uncountable. Let C'_k be the set of restrictions of the functions $g \in C_k$ to the domain $\mathrm{D}(g) \setminus \{u_k\}$. Since all elements of C_k agree on u_k, any two elements of C'_k are incompatible and C'_k is an uncountable antichain of elements of P whose domain has n elements, which is impossible by the inductive hypothesis. □

Note that the ordered set P is separative. Because if $f \not\leq g$ then there is some $\langle u, b\rangle \in g \setminus f$. Therefore, for $b' \neq b$, either $\langle u, b'\rangle$ belongs to f or to some extension of f which will be incompatible with g.

Corollary 15.2.3. *Let P be the set of Proposition 15.2.2. If $i : P \to A$ is the completion algebra of P, then A^* satisfies the ccc.*

Proof. Let C be an antichain of A^*. By the separativity of P, the preserving map $i : P \to A$ is injective (see Lemma 13.3.12) and $i[P]$ is a dense subset of A^*. This density allows us to define, by using the axiom of choice, a map $\varphi : C \to P$ with $i(\varphi(x)) \leq x$; moreover, different x's are incompatible by hypothesis, and hence the $\varphi(x)$ are incompatible because i is preserving; in particular, φ is an injective map. Therefore $\varphi[C]$ is an antichain of P so that it is countable by Proposition 15.2.2. By the injectivity of φ, C is countable. □

We now know that the cardinal $\aleph_2 \in M$ used for the definition of the ordered set P, is also the cardinal $\aleph_2$ of $M[G]$ by Corollary 14.3.4. The next step is to find the function f promised in item (2) at the beginning of this section.

Proposition 15.2.4. *Let P be the ordered set of Proposition 15.2.2 and let G_0 be a M-generic filter of P. Then $\cup G_0 \in M[G]$ is functional.*

Proof. By Proposition 15.1.3, $G_0 = i^{-1}[G]$ belongs to $M[G]$. Suppose $\langle u, b\rangle \in f \in G_0$ but $\langle u, b'\rangle \in g \in G_0$ with $b \neq b'$. Since G_0 is a filter, there is $h \in G_0$ which extends both f and g, but this is impossible. This implies that $\cup G_0$ is functional. If we call $f = \cup G_0$, $\mathrm{D}(f) \subseteq \aleph_2 \times \aleph_0$ in the ZFC-universe $M[G]$. □

We finally use the property that G_0 is an M-generic filter to get the desired consequence.

Proposition 15.2.5. *With the same notation above, let $f = \cup G_0 \in M[G]$. Then $D(f) = \aleph_2 \times \aleph_0$. Moreover, the map $g : \aleph_2 \to {}^{\aleph_0}2$ induced by f in the natural way (that is, $g(\alpha)$ is the function $\aleph_0 \to \{0, 1\}$ with $g(\alpha)(n) = f(\alpha, n)$) is injective.*

Proof. Take any $u \in \aleph_2 \times \aleph_0$. The set $X_u = \{f \in P | u \in \mathrm{D}(f)\} \in M$ is a dense subset of P; because for any $g \in P$, there is clearly an extension f of g belonging to X_u: we use $\langle u, b\rangle \in g$ if it exists, or else we add to g the pair $\langle u, 0\rangle$. Since G_0 is generic, there is $f \in X_u \cap G_0$, so $\langle u, b\rangle \in \cup G_0$ for some b, hence $u \in \mathrm{D}(\cup G_0)$.

Given $\alpha, \beta \in \aleph_2$ with $\alpha \neq \beta$, let $D = D_{\alpha,\beta} = \{p \in P | \exists t\ (p(\alpha, t) \neq p(\beta, t))\}$. D is an M-set dense in P: any $h \in P$ can be extended by taking some $k \in \omega$ with $\langle\alpha, k\rangle, \langle\beta, k\rangle \notin \mathrm{D}(h)$, and adding $\langle\langle\alpha, k\rangle, b\rangle$ and $\langle\langle\beta, k\rangle, b'\rangle$ with $b \neq b'$; and this extension is an element of D. Now, since G_0 is generic, there is $p \in D \cap G_0$ and hence, for some k, $g(\alpha)(k) = f(\alpha, k) = p(\alpha, k) \neq p(\beta, k) = f(\beta, k) = g(\beta)(k)$, which shows that $g(\alpha) \neq g(\beta)$. Thus g is an injective map of $M[G]$. □

Remark 15.2.6. From these proofs, it is clear that there is nothing special about the cardinal $\aleph_2$; any other cardinal $\aleph_\alpha$ would have worked in the same manner. So, for each ordinal α there is some generic extension $M[G]$ with an injective map $\aleph_\alpha \rightarrow {}^{\aleph_0}2$ – but, of course, different α's need different universes.

Theorem 15.2.7. *(Cohen's theorem on the CH) If the theory of ZFC-universes is consistent, then there is no proof of the continuum hypothesis CH in that theory.*

Proof. We will show that if a proof of CH from the axioms of the ZFC-theory exists, then the theory of ZFC$^+$ universes is inconsistent. By Theorem 15.1.1, it will follow that also the ZFC-theory is inconsistent, and this justifies the theorem.

So, suppose that there is a proof of CH for ZFC-universes. By Proposition 15.2.1, there exists a neat class sequence $W_1, \ldots, W_k$ such that the CH holds in every ZFC*-universe in which the W_i-instances of replacement are satisfied. Working in the theory of ZFC$^+$-universes, we have just constructed V, M, P, A, G with the property that $g : \aleph_2 \rightarrow {}^{\aleph_0}2$ is an injective map in the ZFC-universe $M[G]$. By Proposition 10.1.6 this entails that $\aleph_2 \leq 2^{\aleph_0}$ and from this we infer that $\aleph_1 < 2^{\aleph_0}$. But $M[G]$ is a ZFC*-universe in which the W_i-instances of replacement are satisfied and hence CH holds in $M[G]$.

This contradiction shows that, under that supposition, the theory of ZFC$^+$ pointed universes is inconsistent, as it was to be proved. □

15.3 Intermediate Generic Extensions

15.3.1 Motivation

Our next task will be proving that AC is independent of the axioms of ZF-universes. As was the case with CH for ZFC-universes, one half of this result is an easy consequence of the theory of the constructible universe, namely that AC is consistent with the ZF-axioms. For suppose that there is a proof of $\neg$(AC) from the ZF-axioms. By Proposition 15.2.1, there are neat classes $W_1, \ldots, W_n$ such that it can be deduced that $\neg$(AC) holds in every ZF*-universe where the W_i-instances of replacement are satisfied. Since L satisfies those conditions, $\neg$(AC) must hold in L. But, on the other hand, AC holds in L by Corollary 12.4.4. Thus we have inferred a contradiction from the ZF-axioms, hence the ZF-theory is inconsistent and hence AC is consistent with the ZF-axioms. This is *Gödel's theorem on the AC*.

So we must now see that also $\neg$(AC) is consistent with ZF. But, before embarking on that proof of independence, we need to extend our theory on generic extensions. Besides the extensions $M[G]$ we will introduce now a second kind of extensions, $M(G, \mathcal{F})$ which will stand between M and $M[G]$; that is, $M \subseteq M(G, \mathcal{F}) \subseteq M[G]$ with possibly several of these intermediate universes for each $M[G]$. What we are going to develop in the second part of this chapter is like a miniature copy of what was developed in Chapters 13, 14, and the first part of this chapter. The first section to come is an algebraic study which continues that of Boolean algebras of Chapter 13, now within the frame of the Boolean universe M^A; the next section parallels Chapter 14 in that we construct the extension $M(G, \mathcal{F})$ and find its properties, like we did with $M[G]$; and the last section corresponds to the previous proof of the independence of CH from the ZFC-axioms: now, it shows that AC cannot be proved from the ZF-axioms, if these are consistent.

15.3.2 Automorphisms of the algebra A

For all of this section, M will be a ZFC-universe, A a complete Boolean algebra in this universe, and M^A the Boolean universe of Proposition 14.1.1. We show how to transform an automorphism of the complete Boolean algebra A (i.e., an algebra isomorphism $A \to A$) into a permutation of the class M^A.

Proposition 15.3.1. *Let $\varphi : A \to A$ be an automorphism of the complete Boolean algebra A. There is a bijective functional $\tilde{\varphi} : M^A \to M^A$ (that is, $\tilde{\varphi}$ is an injective functional with domain and codomain equal to M^A) such that for each $x \in M^A$, $D(\tilde{\varphi}(x)) = \tilde{\varphi}[D(x)]$ and for each $t \in D(x)$, $\tilde{\varphi}(x)(\tilde{\varphi}(t)) = \varphi(x(t))$.*

Proof. We follow a path similar to that of Proposition 14.1.3. We define recursively the relations $\varphi_\alpha \subseteq M^A_\alpha \times M^A_\alpha$ for ordinals $\alpha \in M$. For limit ordinals $\varphi_\lambda = \cup_{\alpha\in\lambda}\varphi_\alpha$, and in particular $\varphi_0 = \emptyset$; $\varphi_{\alpha+1}$ is defined as a function $M^A_\alpha \to M^A_\alpha$, with $\varphi_{\alpha+1}(x)$ having domain $\varphi_\alpha[\mathrm{D}(x)]$ and $\varphi_{\alpha+1}(x)(\varphi_\alpha(t)) = \varphi(x(t))$, for $x \in M^A_{\alpha+1}$ and $t \in \mathrm{D}(x)$.

We claim that the following holds: (1) Each φ_α is a function with domain M^A_α and codomain included in M^A_α; (2) $\beta < \alpha \Rightarrow \varphi_\beta \subseteq \varphi_\alpha$; (3) For $x \in M^A_\alpha$, $\mathrm{D}(\varphi_\alpha(x)) = \varphi_\alpha[\mathrm{D}(x)]$ and $\varphi_\alpha(x)(\varphi_\alpha(t)) = \varphi(x(t))$ for $x \in M^A_\alpha, t \in \mathrm{D}(x)$.

The inductive proof of the claim is entirely analogous to those of Propositions 14.1.3 and 14.1.5, and it is left as an exercise. Then $\tilde{\varphi}$ will be the union of the φ_α, hence a functional.

The bijectivity can be proved by showing inductively that each $\varphi_\alpha : M^A_\alpha \to M^A_\alpha$ is a bijective map. Let ψ be the inverse homomorphism to φ; we define the functions ψ_α in the same way, step by step. It will then be enough to show that $\psi_\alpha \circ \varphi_\alpha$ is the identity (as well as $\varphi_\alpha \circ \psi_\alpha$) for obtaining that the φ_α are indeed bijections. As a consequence, $\tilde{\varphi}$ will be bijective.

Let us assume the property up to α, so that $\psi_\alpha, \varphi_\alpha$ are inverse bijections $M^A_\alpha \to M^A_\alpha$. Then $\mathrm{D}(\psi_{\alpha+1}(\varphi_{\alpha+1}(x)) = \psi_\alpha[\mathrm{D}(\varphi_{\alpha+1}(x)] = \psi_\alpha[\varphi_\alpha[\mathrm{D}(x)]] = \mathrm{D}(x)$ for $x \in M^A_{\alpha+1}$. Also, if $t \in \mathrm{D}(x)$,

$$\psi_{\alpha+1}(\varphi_{\alpha+1}(x))(\psi_\alpha(\varphi_\alpha(t))) = \psi(\varphi_{\alpha+1}(x)(\varphi_\alpha(t)) = \psi(\varphi(x(t)) \tag{15.1}$$

and $\psi(\varphi(x(t)) = x(t)$, from which it follows that $\psi_{\alpha+1}(\varphi_{\alpha+1}(x)) = x$ and $\psi_{\alpha+1} \circ \varphi_{\alpha+1}$ is the identity, as is $\varphi_{\alpha+1} \circ \psi_{\alpha+1}$. The property at limit steps is obvious. □

Recall the notation $\check{x} \in M^A$ for $x \in M$. We see that any automorphism leaves $\check{x}$ unchanged.

Proposition 15.3.2. *If φ is an automorphism of A and $x \in M$, then $\tilde{\varphi}(\check{x}) = \check{x}$.*

Proof. $\mathrm{D}(\check{x}) = \{\check{t} | t \in x\}$. By making induction on the rank of x, we may assume that $\tilde{\varphi}(\check{t}) = \check{t}$ for every $t \in x$; then $\mathrm{D}(\tilde{\varphi}(\check{x})) = \tilde{\varphi}[\mathrm{D}(\check{x})] = \mathrm{D}(\check{x})$. Now, $\tilde{\varphi}(\check{x})(\tilde{\varphi}(\check{t})) = \varphi(\check{x}(\check{t})) = 1 = \check{x}(\check{t})$ and hence $\varphi(\check{x}) = \check{x}$. □

Lemma 15.3.3. *Any isomorphism $f : A_1 \to A_2$ of complete Boolean algebras preserves the supremum and the infimum of each set of elements.*

Proof. Let X a subset of A_1, $u_1 = \sum_{x \in X} x \in A_1$ and $u_2 = \sum_{x \in X} f(x) \in A_2$. For $x \in X$, $x \leq u_1$ hence $f(x) \leq f(u_1)$ and $u_2 = \sum_{x \in X} f(x) \leq f(u_1)$. By applying this inequality to the isomorphism f^{-1} and the set $f[X]$, we obtain $u_1 \leq f^{-1}(u_2)$ and $f(u_1) \leq u_2$, so that the equality follows. The case of the infimum is similar. □

In the preceding chapter (Proposition 14.2.12), we have learned how to construct, for a given generic filter G of A^*, a sequence of functional M-classes $F_1, \ldots, F_n$ from a class sequence $C_1, \ldots, C_n$ of $M[G]$, so that $F_i^G = C_i$. To achieve this, one follows the rules given in the proposition (which only depend on the universe M); and it turns out that these functionals $F_i : M^A \to A$ commute with many of the automorphisms of the algebra A. We start showing this for the functionals I, B, E and later we will consider the general situation.

Proposition 15.3.4. *If F is any of the functionals I, B, E of M, φ is an automorphism of A and $x, y \in M^A$, then $\varphi(F(x, y)) = F(\tilde{\varphi}(x), \tilde{\varphi}(y))$.*

Proof. We use induction to prove the property for the functional I. So, assume that it holds for the elements of M^A_α and let $x, y \in M^A_{\alpha+1}$. Then, by using Lemma 15.3.3 and the properties of homomorphisms,

$$\varphi(I(x,y)) = \varphi(\textstyle\prod_{t \in D(x)}(x(t)^\perp + \sum_{u \in D(y)} y(u) I(u,t) I(t,u))) =$$
$$\textstyle\prod_{t \in D(x)}(\varphi(x(t))^\perp + \sum_{u \in D(y)} \varphi(y(u)) \varphi(I(u,t)) \varphi(I(t,u)))$$

By the inductive hypothesis, $\varphi(I(u,t)) = I(\tilde{\varphi}(u),\tilde{\varphi}(t))$. Moreover $\varphi(x(t)) = \tilde{\varphi}(x)(\tilde{\varphi}(t))$. Thus

$$\varphi(I(x,y)) = \prod_{t\in D(x)}(\tilde{\varphi}(x)(\tilde{\varphi}(t))^{\perp} + \sum_{u\in D(y)} \tilde{\varphi}(y)(\tilde{\varphi}(u))I(\tilde{\varphi}(u),\tilde{\varphi}(t))$$
$$I(\tilde{\varphi}(t),\tilde{\varphi}(u)))$$

And since $\mathrm{D}(\tilde{\varphi}(x)) = \tilde{\varphi}[\mathrm{D}(x)]$, this gives $\varphi(I(x,y)) = I(\tilde{\varphi}(x),\tilde{\varphi}(y))$.

The property for the functional E is then immediate. As for B,

$$\varphi(B(x,y)) = \varphi(\sum_{u\in D(y)} y(u)E(x,u)) = \sum_{u\in D(y)} \varphi(y(u))\varphi(E(x,u)) =$$
$$= \sum_{u\in D(y)} \tilde{\varphi}(y)(\tilde{\varphi}(u))E(\tilde{\varphi}(x),\tilde{\varphi}(u)) = B(\tilde{\varphi}(x),\tilde{\varphi}(y)).$$ □

Equation 15.1, extended to all the elements of M^A, can be obtained by the same argument for two arbitrary automorphisms ψ,φ of the algebra A. This shows that $\widetilde{\psi\circ\varphi} = \tilde{\psi}\circ\tilde{\varphi}$. That is, the assignment $\varphi \mapsto \tilde{\varphi}$ has the properties of a group homomorphism from the group of automorphisms of A to the permutations of M^A. We take adavantage of this property and, by abusing notation, we identify the automorphism φ and the permutation $\tilde{\varphi}$, so that we write also φ to denote this permutation. Hopefully, this will not cause confusion: when applied to elements of A, it has the first meaning; when to elements of M^A, it has the second.

15.4 The Construction of the Extension

15.4.1 Another Boolean universe

Again, M is a ZFC-universe, and $A \in M$ is a complete Boolean algebra. We will assume at this point that the reader knows some basic concepts of group theory, namely subgroup and conjugation. Now, $\mathcal{G}$ will be a group of automorphisms of A under composition (and thus it may also be seen as a group of permutations of M^A), and $\mathcal{F} \in M$ will be a filter in the ordered set of subgroups of $\mathcal{G}$, the ordering being inclusion. We recall from Definition 13.1.6 that this means that $\mathcal{F} \neq \emptyset$; and (i) if L is a subgroup of $\mathcal{G}$ and $H \in \mathcal{F}$, then $H \subseteq L \Rightarrow L \in \mathcal{F}$; and (ii) $H_1, H_2 \in \mathcal{F} \Rightarrow H_1 \cap H_2 \in \mathcal{F}$. We add to these conditions an additional property: $\mathcal{F}$ is closed under conjugation; that is, if $H \in \mathcal{F}$ and $\varphi \in \mathcal{G}$, then the conjugate subgroup $\varphi^{-1}\circ H\circ\varphi$ belongs to $\mathcal{F}$. This will be the setting for all of the section.

Definition 15.4.1. Let $M, A, \mathcal{G}, \mathcal{F}$ as above. An element $f \in M^A$ is called *$\mathcal{F}$-stable* if its stabilizer subgroup $\mathrm{Stb}(f)$ belongs to $\mathcal{F}$.

Recall that $\mathrm{Stb}(f) = \{\varphi \in \mathcal{G} | \varphi(f) = f\}$. The following is quite obvious.

Lemma 15.4.2. *With the preceding assumptions and notations, if $f \in M^A$ and $\varphi \in Stb(f)$, then φ induces a permutation of the elements of $D(f)$.*

Proof. Since $\varphi(f) = f$, $\varphi[\mathrm{D}(f)] = \mathrm{D}(\varphi(f)) = \mathrm{D}(f)$. By Proposition 15.3.1, φ gives a permutation of $\mathrm{D}(f)$. □

The next definition generalizes the one included in Proposition 14.1.1. In fact, the Boolean universe M^A is obtained in the same form we are to show, when $\mathcal{F}$ is the improper filter.

Definition 15.4.3. Let $M, A, \mathcal{G}, \mathcal{F}$ be as in Definition 15.4.1. We define recursively a functional as follows.

$M_0^{(A,\mathcal{F})} = \emptyset, \quad M_\lambda^{(A,\mathcal{F})} = \cup_{\beta \in \lambda} M_\beta^{(A,\mathcal{F})}$, if λ is limit.

$M_{\alpha+1}^{(A,\mathcal{F})} = \{f | f \in M^A,\ \mathrm{D}(f) \subseteq M_\alpha^{(A,\mathcal{F})}$ and f is $\mathcal{F}$-stable$\}$.

Then $M^{(A,\mathcal{F})}$ is the union of the sets $M_\alpha^{(A,\mathcal{F})}$. $M^{(A,\mathcal{F})}$ is the class of *hereditarily stable* elements of M^A with respect to $\mathcal{F}$. This means that $f \in M^A$ belongs to $M^{(A,\mathcal{F})}$ when f is $\mathcal{F}$-stable and the elements of $\mathrm{D}(f)$ are hereditarily stable.

Lemma 15.4.4. *Given $f \in M^A$, we have $f \in M^{(A,\mathcal{F})}$ if and only if $D(f) \subseteq M^{(A,\mathcal{F})}$ and f is $\mathcal{F}$-stable. Moreover, for each ordinal $\alpha \in M$, $M_\alpha^{(A,\mathcal{F})} = M_\alpha^A \cap M^{(A,\mathcal{F})}$.*

Proof. The "only if" direction is obvious, so let f be $\mathcal{F}$-stable with $\mathrm{D}(f) \subseteq M^{(A,\mathcal{F})}$. Since $\mathrm{D}(f)$ is a set in M, a typical and easy argument shows that $\mathrm{D}(f) \subseteq M_\beta^{(A,\mathcal{F})}$ for some ordinal β; hence $f \in M_{\beta+1}^{(A,\mathcal{F})} \subseteq M^{(A,\mathcal{F})}$.

For the second part, we first use induction on the ordinal α to show the inclusion $M_\alpha^{(A,\mathcal{F})} \subseteq M_\alpha^A$. Assuming this for α, take $f \in M_{\alpha+1}^{(A,\mathcal{F})}$. By definition, $\mathrm{D}(f) \subseteq M_\alpha^{(A,\mathcal{F})} \subseteq M_\alpha^A$ and hence $f \in M_{\alpha+1}^A$ by Proposition 14.1.1. The same property for limit ordinals is now obvious. To prove the reverse inclusion, we apply again induction and, as before, we only need to check the property for successor ordinals. So, admit it holds for α and let $f \in M_{\alpha+1}^A \cap M^{(A,\mathcal{F})}$. This entails that $\mathrm{D}(f) \subseteq M_\alpha^A \cap M^{(A,\mathcal{F})} = M_\alpha^{(A,\mathcal{F})}$. Then $f \in M_{\alpha+1}^{(A,\mathcal{F})}$ by Definition 15.4.3. □

Proposition 15.4.5. *With the notations of Definition 15.4.3, if $f \in M_\alpha^{(A,\mathcal{F})}$ and $\varphi \in \mathcal{G}$, then $\varphi(f) \in M_\alpha^{(A,\mathcal{F})}$. Therefore $\varphi[M^{(A,\mathcal{F})}] \subseteq M^{(A,\mathcal{F})}$.*

Proof. We prove the implication $f \in M_\alpha^{(A,\mathcal{F})} \Rightarrow \varphi(f) \in M_\alpha^{(A,\mathcal{F})}$ by induction. Assume the property for elements in M_α^A and let $f \in M_{\alpha+1}^A \cap M^{(A,\mathcal{F})}$. Every $u \in \mathrm{D}(f)$ belongs to $M_\alpha^{(A,\mathcal{F})}$ and, by the inductive hypothesis, $\varphi(u) \in M_\alpha^{(A,\mathcal{F})}$. I.e., $\mathrm{D}(\varphi(f)) = \varphi[\mathrm{D}(f)] \subseteq M_\alpha^{(A,\mathcal{F})}$. According to the definition of $M_{\alpha+1}^{(A,\mathcal{F})}$, it remains to show that $\varphi(f)$ is $\mathcal{F}$-stable.

We know $\mathrm{Stb}(f) \in \mathcal{F}$. If $\pi \in \mathrm{Stb}(f)$, then $(\varphi \circ \pi \circ \varphi^{-1})(\varphi(f)) = \varphi(f)$, hence $\varphi \circ \mathrm{Stb}(f) \circ \varphi^{-1} \subseteq \mathrm{Stb}(\varphi(f))$ and thus $\mathrm{Stb}(\varphi(f)) \in \mathcal{F}$. □

15.4.2 The intermediate generic extension

For a ZFC-universe M, a complete Boolean algebra A, a group $\mathcal{G}$ of automorphisms of A and a filter $\mathcal{F}$ of subgroups of $\mathcal{G}$, we have just constructed a kind of Boolean universe $M^{(A,\mathcal{F})}$ and, due to Proposition 15.3.2, a functional $\Phi : M \to M^{(A,\mathcal{F})}$, the restriction of the functional of Proposition 14.1.3. So we are basically in the same position of Chapter 14 when we defined the hypotheses (GE); only with $M^{(A,\mathcal{F})}$ substituted for M^A. Our next step is also parallel to that of the preceding chapter. We set a new hypothesis extending (GE), and consider a functional from $M^{(A,\mathcal{F})}$ to V, as we then did. As with Φ, this new functional is not really new, just the restriction of the functional Ψ of Proposition 14.1.5.

(1) We assume the hypotheses of (GE), with the ZFC-universes $M \in V$, the algebra $A \in M$, the M-generic filter G of A^*.

(2) We add to those data a group $\mathcal{G} \in M$ of automorphisms of A, and a filter $\mathcal{F} \in M$ of subgroups of $\mathcal{G}$ which is closed under conjugation.

We globally refer to these assumptions as the *(IGE) hypotheses* – IGE for "intermediate generic extensions".

We know that $M[G]$ is the codomain of the functional $\Psi : M^A \to V$ of Proposition 14.1.5. Since $M^{(A,\mathcal{F})} \subseteq M^A$, we define $M[G,\mathcal{F}] = \Psi[M^{(A,\mathcal{F})}]$ so that $M[G,\mathcal{F}] \subseteq M[G]$; and if $x \in M^{(A,\mathcal{F})}$, then $\Psi(x) = x^G \in M[G,\mathcal{F}]$. The goal of the present section is to show that $M[G,\mathcal{F}]$ is a ZF-universe. Note that $M \subseteq M[G,\mathcal{F}] \subseteq M[G]$, because if $x \in M$, then $\check{x} \in M^{(A,\mathcal{F})}$ by Proposition 15.3.2; and $\Psi(\check{x}) = x$ by Equation 14.2.

Proposition 15.4.6. *Under the hypotheses (IGE), $M[G,\mathcal{F}]$ is a transitive class of V which satisfies the axioms of empty set, infinity, extension and regularity.*

Proof. Let $x^G \in M[G,\mathcal{F}]$ with $x \in M^{(A,\mathcal{F})}$. Any element of x^G in V is t^G for $t \in \mathrm{D}(x)$. But $\mathrm{D}(x) \subseteq M^{(A,\mathcal{F})}$ by Lemma 15.4.4, so that $t^G \in M[G,\mathcal{F}]$. This shows that $M[G,\mathcal{F}]$ is transitive in V.

Since $M[G,\mathcal{F}]$ is transitive, it satisfies the extension axiom. Also $\emptyset \in M^{(A,\mathcal{F})}$ so that $\emptyset \in M[G,\mathcal{F}]$. Therefore the axiom of the empty set, as well as the axiom of regularity are satisfied in $M[G,\mathcal{F}]$. Finally, $\omega \in M$ and thus $\omega \in M[G,\mathcal{F}]$ and $\omega \subseteq M[G,\mathcal{F}]$ by the transitivity. □

Proposition 15.4.7. *Assume the hypotheses (IGE). $M[G,\mathcal{F}]$ satisfies the axioms of pairs, unions and power set.*

Proof. Pairs. Let $x^G, y^G \in M[G,\mathcal{F}]$ with $x, y \in M^{(A,\mathcal{F})}$. If $z \in M^A$ has domain $\{x, y\}$ with values 1, then $\mathrm{Stb}(z) \supseteq \mathrm{Stb}(x) \cap \mathrm{Stb}(y)$ and hence $\mathrm{Stb}(z) \in \mathcal{F}$ as x, y are $\mathcal{F}$-stable. So, z is $\mathcal{F}$-stable and hence $z \in M^{(A,\mathcal{F})}$. Since $z^G = \{x^G, y^G\} \in M[G,\mathcal{F}]$, this proves the validity of the axiom of pairs.

Unions. Let $x^G \in M[G,\mathcal{F}]$ with $x \in M^{(A,\mathcal{F})}$. As seen in Proposition 14.2.3, we define $h \in M^A$ with $\mathrm{D}(h) = \bigcup_{y \in D(x)} \mathrm{D}(y)$ and

$h(t) = \sum_{\{y\in D(x)|t\in D(y)\}} y(t)x(y)$; then $h^G = \cup x^G$ (in $M[G]$ and in V); thus it will be enough to show that $h \in M^{(A,\mathcal{F})}$, because then $h^G = \cup x^G$ also in $M[G,\mathcal{F}]$, since $M[G,\mathcal{F}]$ is transitive in V by Proposition 15.4.6.

Each $y \in \mathrm{D}(x)$ and each $t \in \mathrm{D}(y)$ belong to $M^{(A,\mathcal{F})}$ and thus every element of $\mathrm{D}(h)$ belongs to $M^{(A,\mathcal{F})}$ by Lemma 15.4.4. Now, let $\varphi \in \mathrm{Stb}(x)$. $\mathrm{D}(\varphi(h)) = \varphi[\cup_{y\in D(x)} \mathrm{D}(y)] = \cup_{y\in D(x)}\varphi[\mathrm{D}(y)] = \cup_{y\in D(x)} \mathrm{D}(y) = \mathrm{D}(h)$, because φ permutes the elements of $\mathrm{D}(x)$ according to Lemma 15.4.2.

So it will suffice to see that $\varphi(h)(\varphi(t)) = h(\varphi(t))$ for every $\varphi \in \mathrm{Stb}(x)$ and $t \in \mathrm{D}(h)$, to get that $\mathrm{Stb}(x) \subseteq \mathrm{Stb}(h)$ and hence $h \in M^{(A,\mathcal{F})}$. Now, for any $y \in \mathrm{D}(x)$ we have $\varphi(t) \in \mathrm{D}(\varphi(y)) \Leftrightarrow t \in \mathrm{D}(y)$, and it follows that $h(\varphi(t)) = \sum_{\{y\in D(x)|t\in D(y)\}} \varphi(y)(\varphi(t))x(\varphi(y)) = \sum_{\{y\in D(x)|t\in D(y)\}} \varphi(y(t))\varphi(x(y))$, because φ stabilizes x and permutes $\mathrm{D}(x)$. Therefore $h(\varphi(t)) = \varphi(h(t)) = \varphi(h)(\varphi(t))$ and $h = \varphi(h)$.

Power set. Here we follow the proof of Proposition 14.2.6. As there, given the inclusion $x^G \subseteq y^G$ with $x, y \in M^{(A,\mathcal{F})}$, we may define $z : \mathrm{D}(y) \to A$ by $z(u) = B(u, x)$ so that $z^G = x^G$. We see that z is $\mathcal{F}$-stable from the implication $\varphi \in \mathrm{Stb}(x) \cap \mathrm{Stb}(y) \Rightarrow \varphi(z) = z$; this is so because (1) $\varphi[\mathrm{D}(y)] = \mathrm{D}(y)$ as $\varphi \in \mathrm{Stb}(y)$, and (2) $\varphi(z)(\varphi(u)) = \varphi(z(u)) = B(\varphi(u), \varphi(x)) = B(\varphi(u), x) = z(\varphi(u))$, by Proposition 15.3.4 and because $\varphi \in \mathrm{Stb}(x)$. Once we obtain in this way that $z \in M^{(A,\mathcal{F})}$, the rest of the proof is the same of Proposition 14.2.6 with the substitution of $M_\alpha^{(A,\mathcal{F})}$ for M_α^A if $\mathrm{D}(y) \subseteq M_\alpha^{(A,\mathcal{F})}$. The new function $p : M_{\alpha+1}^{(A,\mathcal{F})} \to A$ with $p(x) = I(x, y)$ is easily seen to be $\mathcal{F}$-stable, because $\mathrm{Stb}(y) \subseteq \mathrm{Stb}(p)$ by Propositions 15.4.5 and 15.3.4. Thus $p^G \in M[G,\mathcal{F}]$, and the proof that $p^G = \mathbb{P}(y^G) \cap M[G,\mathcal{F}]$ is analogous to that of the corresponding equality in Proposition 14.2.6. □

To study the axiom of replacement in $M[G,\mathcal{F}]$ we follow the same path of Section 14.2.4, but using functionals of M with domain $M^{(A,\mathcal{F})}$ instead of M^A. For instance, the completion H of a functional M-class $F : M^{(A,\mathcal{F})} \to A$ is defined for $M^{(A,\mathcal{F})}$ and given by $H(x) = \sum_{u\in M^{(A,\mathcal{F})}} F(u)E(x, u)$; the same argument of Lemma 14.2.11 shows that $H^G = F^G \subseteq M[G,\mathcal{F}]$ and $x^G \in H^G \Leftrightarrow H(x) \in G$ for any $x \in M^{(A,\mathcal{F})}$.

We also have a version of Lemma 14.2.7. Let us say that the set $S \subseteq M^{(A,\mathcal{F})} \times A$ is $\mathcal{F}$-stable when there is a subgroup $H \in \mathcal{F}$ such that $\pi \in H \Rightarrow S = \{\langle\pi(u), \pi(a)\rangle | \langle u, a\rangle \in S\}$.

Lemma 15.4.8. *Let $S \subseteq M^{(A,\mathcal{F})} \times A$ be a set of M and call $S^G = \{u^G | \exists a \in G\ (\langle u, a\rangle \in S)\}$. If S is $\mathcal{F}$-stable, then $S^G \in M[G,\mathcal{F}]$.*

Proof. As in the proof of Lemma 14.2.7, we define $f : \mathrm{D}(S) \to A$ by $f(u) = \sum\{a \in A | \langle u, a\rangle \in S\}$. Let H be a subgroup that satisfies the condition above for S being $\mathcal{F}$-stable. Since $f \in M^A$ and every automorphism of A preserves sums (Lemma 15.3.3), it follows that for each $\pi \in H$, $\pi(f) = f$. Since $H \in \mathcal{F}$, we conclude that $f \in M^{(A,\mathcal{F})}$. But also $f^G = S^G$ as in the case of Lemma 14.2.7, and hence $S^G \in M[G,\mathcal{F}]$. □

The analogous result holds for sets included in $M^{(A,\mathcal{F})} \times M^{(A,\mathcal{F})} \times A$, again under a corresponding condition of $\mathcal{F}$-stability.

The functionals S_0, P_0, Q_0, T_0 have been introduced just before Lemma 14.2.8. We adapt these functionals to the current setting by taking their respective domains as $(M^{(A,\mathcal{F})})^2$, $(M^{(A,\mathcal{F})})^3$ or $(M^{(A,\mathcal{F})})^4$. Besides, a functional like Q_0 would be defined now in $(M^{(A,\mathcal{F})})^3$ as $Q_0(x,b,c) = \sum_{u,v\in M^{(A,\mathcal{F})}} P_0(x,u,v)S_0(u,b)P_0(v,b,c)$. The new definition of T_0 is similar, while those for S_0, P_0 may remain as in the mentioned lemma. One may then obtain the analogous result to Proposition 14.2.12.

Proposition 15.4.9. *Let C be a $M[G,\mathcal{F}]$-class. Then there is a functional class F of M with domain $M^{(A,\mathcal{F})}$ and values in A such that $F^G = C$.*

Proof. One must simply follow the steps in the proof of Proposition 14.2.12. For the membership relation, we must take $F : M^{(A,\mathcal{F})} \to A$ with $F(x) = \sum_{u,v\in M^{(A,\mathcal{F})}} Q_0(x,u,v)B(u,v)$. The analogous functional works for equality, and the rest of the cases follow precisely from the arguments in the proof of Proposition 14.2.12. □

It is now time to extend Proposition 15.3.4 as promised.

Proposition 15.4.10. *We assume the hypotheses (IGE). Let $C_1, \ldots, C_n$ be a class sequence of $M[G,\mathcal{F}]$ and let $F_1, \ldots, F_n$ be the corresponding functional M-classes obtained in Proposition 15.4.9. Let φ be an automorphism of the algebra A such that whenever $C_j = x_j$ is introduced by rule 1 of Definition 3.1.2 and $c_j \in M^{(A,\mathcal{F})}$ is chosen with $x_j = c_j^G$, then $\varphi \in Stb(c_j)$. Then for any index i,*

$$\varphi(F_i(x)) = F_i(\varphi(x)) \quad (\forall x \in M^{(A,\mathcal{F})}) \tag{15.2}$$

Proof. We first justify the analogous property for the auxiliary functionals S_0, P_0, Q_0, T_0; the four cases being similar, we just prove the argument for Q_0, under the assumption that the property holds for S_0 and P_0; we also know that it holds for the functionals B and E by Proposition 15.3.4.

$$\varphi(Q_0(x,b,c)) = \sum_{u,v\in M^{(A,\mathcal{F})}} \varphi(P_0(x,u,v))\varphi(S_0(u,b))\varphi(P_0(v,b,c)) =$$
$$= \sum_{u,v\in M^{(A,\mathcal{F})}} P_0(\varphi(x),\varphi(u),\varphi(v))S_0(\varphi(u),\varphi(b))P_0(\varphi(v),\varphi(b),\varphi(c)).$$

and since φ induces a permutation of the elements of $M^{(A,\mathcal{F})}$ (Proposition 15.4.5), this gives $\varphi(Q_0(x,b,c)) = Q_0(\varphi(x),\varphi(b),\varphi(c))$.

We now give the proof of the proposition by following the items in the class sequence and prove the property of the statement successively for each of them, according to the constructions of these items in the proof of Proposition 14.2.12.

For items 2, 3, the result follows from Proposition 15.3.4. For instance, if $F(x) = \sum_{u,v\in M^{(A,\mathcal{F})}} Q_0(x,u,v)B(u,v)$, then

$$\varphi(F(x)) = \sum_{u,v\in M^{(A,\mathcal{F})}} Q_0(\varphi(x),\varphi(u),\varphi(v))B(\varphi(u),\varphi(v))$$

by Lemma 15.3.3, Proposition 15.3.4 and the above observation for Q_0. It follows then $\varphi(F(x)) = F(\varphi(x))$ by Proposition 15.3.1. In item 1, $C = x = c^G$ and $F = c$ and the property holds by the hypothesis of the statement about φ.

Let $C = C_1 \setminus C_2$ with $F^G = C_1$ and $H^G = C_2$ (H with the property of a completion). Then $U(x) = F(x)H(x)^\perp$ for any $x \in M^{(A,\mathcal{F})}$. By the inductive hypothesis, $\varphi(F(x)) = F(\varphi(x))$ and $\varphi(H(x)) = H(\varphi(x))$ with $\varphi(x) \in M^{(A,\mathcal{F})}$ by Proposition 15.4.5. Then $\varphi(U(x)) = \varphi(F(x))\varphi(H(x))^\perp = F(\varphi(x))H(\varphi(x))^\perp = U(\varphi(x))$.

The rest of cases is similar to that one. For instance, let $C = \mathrm{D}(K)$ with $K = F^G$. We obtained U with $U^G = C$ by setting $U(x) = \sum_t(\sum_z F(z)Q_0(z,x,t))$ (we abbreviate the notation $\sum_{t,u\in M^{(A,\mathcal{F})}} L(t,u)$ by writing instead $\sum_{t,u} L(t,u)$; since the field for the variables is always the same, this abbreviation is safe). Now, $\varphi(U(x)) = \sum_t(\sum_z \varphi(F(z))\varphi(Q_0(z,x,t))) = \sum_t(\sum_z F(\varphi(z))Q_0(\varphi(z),\varphi(x),\varphi(t))) = U(\varphi(x))$, again by Proposition 15.3.1 and the hypothesis on F. □

Proposition 15.4.11. *Under the hypotheses (IGE), $M[G,\mathcal{F}]$ is a ZF-universe.*

Proof. By the previous results, it will suffice to prove that each instance of the axiom of replacement holds in the universe $M[G,\mathcal{F}]$. Thus let C be a class of $M[G,\mathcal{F}]$ whose elements are triples, and let $u, s \in M[G,\mathcal{F}]$ be given. We have $v, w \in M^{(A,\mathcal{F})}$ such that $u = w^G$ and $s = v^G$. The hypothesis is that for each $x \in \mathrm{D}(v)$ with $x^G \in v^G$, there is at most one element $y^G \in M[G,\mathcal{F}]$ such that $\langle w^G, x^G, y^G\rangle \in C$. We follow now the proof of Proposition 14.2.13. We keep the names F, L, H for the corresponding functional classes, though they are not exactly the same as in the proof of Proposition 14.2.13, but they are defined in the same form with the substitution of $M^{(A,\mathcal{F})}$ for M^A.

By using Proposition 15.4.9, we take the functional class F defined on $M^{(A,\mathcal{F})}$ and such that $F^G = C$. Then F_0 is defined on $(M^{(A,\mathcal{F})})^3$ with $F_0(x,y,z) = \sum_u F(u)T_0(u,x,y,z)$, and we have that $F_0(x,y,z) \in G \Leftrightarrow \langle x^G, y^G, z^G\rangle \in C$. We then construct the M-class L and the M-class H of triples as in Proposition 14.2.13 – but of course defined over $M^{(A,\mathcal{F})}$ instead of M^A; in particular, when we use the rank of the element $z \in M^{(A,\mathcal{F})}$ for the construction of L, it is the rank as an element of $M^{(A,\mathcal{F})}$ that we consider. By applying the strong form of the axiom of replacement in M for the class H, we obtain the set $t \subseteq M^{(A,\mathcal{F})} \times A$ as in that proof; and it satisfies the equality $t^G = \mathrm{cD}(C|_{\{u\}\times s})$ by the same argument. So it will be enough to show that t is $\mathcal{F}$-stable in the sense of Lemma 15.4.8: by the lemma $t^G \in M[G,\mathcal{F}]$ and we will be done.

Let us consider the subgroup $S \subseteq \mathcal{G}$ of those automorphisms φ of $\mathcal{G}$ which stabilize w and v, as well as the finitely many elements (which belong to $M^{(A,\mathcal{F})}$) introduced by rule 1 in the sequence of Proposition 15.4.10 when we construct F from C. Since this is an intersection of finitely many subgroups belonging to the filter, $S \in \mathcal{F}$.

By Proposition 15.4.10, each $\varphi \in S$ commutes with F, i.e., equation (15.2) holds for F. We see now that φ commutes also with F_0 and L. To check that $\varphi(F_0(x, y, z)) = F_0(\varphi(x), \varphi(y), \varphi(z))$ is routine. To study the commutativity between φ and L, we must see that $\varphi[\mathrm{D}(L)] = \mathrm{D}(L)$. So, suppose that $\langle x, y, z, a\rangle \in L$ and $\langle \varphi(x), \varphi(y), z', \varphi(a)\rangle \in L$ with $z' \in M_\alpha^{(A,\mathcal{F})}$ and $\varphi(z) \notin M_\alpha^{(A,\mathcal{F})}$ – in such hypothesis, $\langle \varphi(x), \varphi(y), \varphi(z), \varphi(a)\rangle \notin L$. By Proposition 15.4.5, $\varphi^{-1}(z') \in M_\alpha^{(A,\mathcal{F})}$ and $z \notin M_\alpha^{(A,\mathcal{F})}$, which is absurd as $\langle x, y, \varphi^{-1}(z'), a\rangle \in F_0$, and z has minimum rank with such property for the elements of F_0. Therefore $\langle x, y, z, a\rangle \in L \Rightarrow \langle \varphi(x), \varphi(y), \varphi(z), \varphi(a)\rangle \in L$.

Next, we check that if $\varphi \in S$, then $t = \{\langle \varphi(z), \varphi(a_0)\rangle | \langle z, a_0\rangle \in t\}$, following the notation in the proof of Proposition 14.2.13. It will then follow from Lemma 15.4.8 that $t^G \in M[G, \mathcal{F}]$.

Now, $\langle z, a_0\rangle \in t \Rightarrow \exists y, a, a'(\langle y, a'\rangle \in v \wedge a_0 = aa' \wedge \langle w, y, z, a\rangle \in L)$. And in turn this implies that $\langle \varphi(y), \varphi(a')\rangle \in v \wedge \langle w, \varphi(y), \varphi(z), \varphi(a)\rangle \in L$. Therefore $\langle \varphi(z), \varphi(a_0)\rangle \in t$. The converse implication is obtained by using φ^{-1}, and this completes the proof. □

15.5 Cohen's Theorem on the Axiom of Choice

15.5.1 Overview

We head now toward the proof that the axiom of choice (AC) cannot be deduced from the theory of ZF-universes (if these are consistent). Thus our objective is the same we had in the second section of this chapter, but now relative to AC in ZF-universes instead of CH in ZFC-universes. Also our strategy will be the same: we postulate that there is a proof of AC from axioms 1-8 and then deduce a contradiction. However, there will be a difference: with the CH, we constructed a map in a ZFC$^+$-universe which directly refuted the CH; now, and still in the frame of a ZFC$^+$-universe, we construct a set which refutes a certain consequence of AC. Namely this:

(MIS) If x is a set which is not equipotent to any finite ordinal, then there is an injective map $f : \omega \to x$ – we denote this assertion (MIS) for "map for infinite sets".

This property holds in every ZFC-universe by Proposition 10.4.1. But we will show that it fails in some ZF-universe of the form $M[G, \mathcal{F}]$. Consequently AC fails in that universe, because AC + ZF implies (MIS).

The proof is developed in three steps. First, working in an arbitrary ZFC-universe M, we construct a particular ordered set P with its completion algebra A; and the group $\mathcal{G}$ of A-automorphisms and the filter $\mathcal{F}$ of subgroups that will later allow the construction of the intermediate universe $M[G, \mathcal{F}]$. Second, working in a ZFC$^+$-universe V with distinguished set M, we find how these data satisfy the hypotheses (IGE) so that we obtain a ZF-universe

$M[G, \mathcal{F}]$; and in this universe we define a certain set Z which, viewed in $M[G]$ is denumerable and hence, there exists an injective $M[G]$-map $\omega \to Z$. Finally, we see that not only that map does not belong to $M[G, \mathcal{F}]$, but also that there is, in that universe, no injective map $\omega \to Z$. This implies that (MIS) fails in $M[G, \mathcal{F}]$. We easily infer from this the metatheoretical consequence that AC is independent from the axioms of the ZF-theory.

15.5.2 The basic data

Let M be a ZFC-universe. We define an ordered set $P \in M$, a group $\mathcal{G}$ of automorphisms of the completion algebra A of P, and a proper filter $\mathcal{F}$ of subgroups of $\mathcal{G}$ which is closed under conjugation.

1) The ordered set P. P will be the set of all finite functional sets with values in $\{0, 1\}$ and domain included in $\omega \times \omega$. The order relation is reverse inclusion: $f \leq g \Leftrightarrow g \subseteq f$, i.e., f extends g. This ordering of P is separative, by the same reason given just before Corollary 15.2.3 for the ordered set P therein. Thus the completion algebra A of P can be seen as consisting of all regular segments of P, and the completion map $i : P \to A$ is injective.

2) The group $\mathcal{G}$. Any permutation π of the set ω (that is, a bijective map $\pi : \omega \to \omega$) determines the following morphism π^* of the ordered set P: for $f \in P, \pi^*(f) = \{\langle \pi(x), y, b\rangle | \langle x, y, b\rangle \in f\}$. It is clear that $\pi^* : P \to P$ preserves inclusions, hence it is indeed a morphism and, in fact, an isomorphism. We then extend π^* to an automorphism π^* of A (hoping that no confusion will arise from this abuse of notation) by applying the isomorphism to any regular segment of P, i.e., $\pi^*(S) = \pi^*[S]$; like this, $\pi^* \circ i = i \circ \pi^*$, i being the preserving map of Theorem 13.3.13. These automorphisms form a group that is clearly isomorphic to the group of the permutations of ω, and we shall identify this group of automorphisms as $\mathcal{G}$.

3) The filter $\mathcal{F}$. We identify $\mathcal{G}$ and the group of permutations of ω. As such, given $E \subseteq \omega$, the set of those permutations π with the property that $\pi(x) = x$ for every $x \in E$ will be denoted as $S(E)$. It is routine to show that $S(E)$ is a subgroup of $\mathcal{G}$. Moreover, the following holds.

Lemma 15.5.1. *Let $\mathcal{G}$ be the group of permutations of ω. The set $\mathcal{F}$ of those subgroups H of $\mathcal{G}$ with the property that there is a finite set $E \subseteq \omega$ such that $S(E) \subseteq H$, is a proper filter of subgroups of $\mathcal{G}$ and is closed under conjugation.*

Proof. We may readily see that $\mathcal{F}$ is a filter; and it is proper because for any finite $E \subseteq \omega$, $S(E)$ contains non-trivial elements. Finally, if $\pi \in \mathcal{G}$, then $\pi \circ S(E) \circ \pi^{-1} = S(\pi(E))$, and this plainly entails that $\mathcal{F}$ is closed under conjugation. □

Let us observe that the above property has been proved assuming that $\mathcal{G}$ is the group of permutations of ω. But, since the group of automorphisms of A is isomorphic to this group, the property about the filter is valid too for this group of A-automorphisms. Our filter $\mathcal{F}$ will consist then of those subgroups $H \subseteq \mathcal{G}$ such that $H \supseteq S(E)$ for some finite subset $E \subseteq \omega$.

Remark 15.5.2. Note that we have implicitly charged π^* with three different meanings. Two of these meanings have already been presented: there is $\pi^* : P \to P$ and there is $\pi^* : A \to A$ which is an A-automorphism. By the way, the first of these maps does not directly translates the permutation π to P: if $\langle n, m\rangle \in \mathrm{D}(f)$ for certain $f \in P$, then we cannot say that $\langle \pi(n), \pi(m)\rangle \in \mathrm{D}(\pi^*(f))$; instead, what we know is that $\langle \pi(n), m\rangle \in \mathrm{D}(\pi^*(f))$.

But Proposition 15.3.1 shows that $\pi^* : A \to A$ induces a transformation of the elements of M^A. This was denoted $\tilde{\pi}^*$ in the beginning, but we ended up by writing also π^* for this permutation of the elements of M^A. So now $\pi^*(x)$ with $x \in M^A_\alpha$ will be another element of M^A_α with domain $\pi^*[\mathrm{D}(x)]$ and values $\pi^*(x)(\pi^*(u)) = \pi^*(x(u))$.

As an example, consider the sets M^A_2, M^A_3 that we discussed in Examples 14.1.2, now in the particular situation we are considering: M^A_2 consisted of all functions of the form $s_a = \{\langle \emptyset, a\rangle | a \in A\}$ plus the empty function. Since $\pi^*(\emptyset) = \emptyset$, we have that $\pi^*(s_a) = \{\langle \emptyset, \pi^*(a)\rangle\}$; and if $a = i(p)$ for $p \in P$, $\pi^*(s_a) = \{\langle \emptyset, \pi^*(p)\}$, by identifying $\pi^*(p)$ and $i(\pi^*(p))$. If $x \in M^A_3$ is the function $\{\langle s_a, c_1\rangle, \langle s_b, c_2\rangle\}$, then $\pi^*(x) = \{\langle \pi^*(s_a), \pi^*(c_1)\rangle, \langle \pi^*(s_b), \pi^*(c_2)\rangle\}$. Let us write in full just the first of these elements: $\langle \{\langle \emptyset, \pi^*(a)\}, \pi^*(c_1)\rangle$. But if $a = i(p)$ and $\{\langle n, m, 0\rangle\} = p$, then $\{\langle \pi(n), m, 0\rangle\} = \pi^*(p)$; and with the natural identification, this is $\pi^*(a)$.

We know from Proposition 15.3.2 that for every $u \in M$, $\check{u}$ belongs to $M^{(A,\mathcal{F})}$. We are now going to identify other elements of $M^{(A,\mathcal{F})}$.

Given $x, y \in \omega$, we write $p_{x,y} = \{\langle x, y, 1\rangle\} \in P$, a function with only one element. For each $k \in \omega$, $f_k \in M^A$ will be the following: $\mathrm{D}(f_k) = \mathrm{D}(\check{\omega})$; and $f_k(\check{n}) = i(p_{k,n})$, where $i : P \to A^*$ is the canonical preserving inclusion of the separative ordered set P into its completion algebra. We will normally identify the elements $p \in P$ and the corresponding $i(p)$, writing simply p as an element of A. Accordingly, $f_k(\check{n}) = p_{k,n}$.

Lemma 15.5.3. *With the notations above, for any $\pi \in \mathcal{G}$, $\pi^*(f_k) = f_{\pi(k)}$. Moreover, each f_k belongs to $M^{(A,\mathcal{F})}$.*

Proof. By Proposition 15.3.2, $\pi^*(\check{\omega}) = \check{\omega}$ and $\pi^*(\check{m}) = \check{m}$ for any $m \in \omega$. Thus $\mathrm{D}(\pi^*(f_k)) = \pi^*[\mathrm{D}(f_k)] = \pi^*[\ \mathrm{D}(\check{\omega})] = \mathrm{D}(\check{\omega}) = \mathrm{D}(f_{\pi(k)})$ so both functions have the same domain. Also, $\pi^*(p_{k,n}) = \{\langle \pi(k), n, 1\rangle\} = p_{\pi(k),n}$. Then, for any $\check{n} \in \mathrm{D}(\check{\omega})$, $\pi^*(f_k)(\pi^*(\check{n})) = \pi^*(f_k(\check{n})) = \pi^*(p_{k,n}) = p_{\pi(k),n} = f_{\pi(k)}(\check{n}) = f_{\pi(k)}(\pi^*(\check{n}))$, so that $\pi^*(f_k) = f_{\pi(k)}$, as we had to show.

For the second assertion, we must only check that f_k is $\mathcal{F}$-stable, because $\mathrm{D}(\check{\omega}) \subseteq M^{(A,\mathcal{F})}$ is immediate from Proposition 15.3.2. Note that if $\pi \in S(\{k\})$ (with the notation $S(-)$ introduced in item (3) of the above list of basic data), then $\pi(k) = k$ and $\pi^*(f_k) = f_k$ by the first part of this proof. Therefore $S(\{k\}) \subseteq \mathrm{Stb}(f_k)$ and $\mathrm{Stb}(f_k) \in \mathcal{F}$. □

Of course, $p_{k,n} \neq p_{t,n}$ if $t \neq k$. Thus we have implicitly defined an injective function with domain ω giving $k \mapsto f_k$. Its codomain $Y = \{f_k | k \in \omega\} \subseteq M^{(A,\mathcal{F})}$ is a denumerable set of the universe M. Moreover, Lemma 15.5.3 shows that every $\pi \in \mathcal{G}$ induces a permutation of Y, $\pi^*[Y] = Y$.

15.5.3 The construction of the set Z

Up to now, we have worked in this section inside a ZFC-universe M. We now work from the axioms of the ZFC$^+$-theory: V is a ZFC-universe, M is a transitive and denumerable set of V with $\omega \in M$; and M is a ZFC-universe too. We consider the M-sets $P, A, \mathcal{G}, \mathcal{F}$ as defined in the preceding section. From Propositions 15.1.2 and 15.1.3 we infer that there exists a generic filter G_0 of P and a generic filter G of A^* such that $G_0 = i^{-1}[G]$. Thus hypotheses (IGE) are satisfied with these data and therefore $M[G, \mathcal{F}]$ is a ZF-universe.

Lemma 15.5.4. *Let us use the notation of Lemma 15.5.3. Then $f_k^G \neq f_t^G$ whenever $t \neq k$ for $t, k \in \omega$; and $Z = \{f_k^G | k \in \omega\}$ is an infinite set of the ZF-universe $M[G, \mathcal{F}]$.*

Proof. Let $n \neq m$ elements of ω, and call $D_{n,m} = \{p \in P | \exists k \in \omega\ (\langle n,k\rangle, \langle m,k\rangle \in \mathrm{D}(p) \wedge p(n,k) \neq p(m,k))\}$. Each of these $D_{n,m}$ is an M-set which is a dense subset of P. To see this, let $q \in P$ and choose k such that $\langle n,k\rangle, \langle m,k\rangle \notin \mathrm{D}(q)$; such k exists because $\mathrm{D}(q)$ is finite. There is an extension $p \leq q$ with $p(n,k) \neq p(m,k)$. Therefore $p \in D_{n,m}$.

Since G_0 is M-generic in P, for $n \neq m \in \omega$ there is some $q \in G_0 \cap D_{n,m}$. Suppose $q(n,k) = 1$ and $q(m,k) = 0$ for a certain $k \in \omega$. Then q extends $p_{n,k}$, that is, $q \leq p_{n,k}$ and $p_{n,k} \in G_0$. Then $f_n(\check{k}) = i(p_{n,k}) \in G$, from which it follows that $(\check{k})^G = k \in f_n^G$. We use RAA to prove that $k \notin f_m^G$ and thus $f_n^G \neq f_m^G$. For suppose we had $k \in f_m^G$; then $f_m(\check{k}) = i(p_{m,k}) \in G$. Since $i(q) \in G$, $i(q) \cdot i(p_{m,k}) \neq 0$, and hence $p_{m,k}$ and q are compatible elements of P because i is preserving, but this is impossible since $p_{m,k}(m,k) = 1$ and $q(m,k) = 0$. Consequently, $f_n^G \neq f_m^G$, showing the first part of the lemma.

Recall that $Y = \{f_k | k \in \omega\} \subseteq M^{(A,\mathcal{F})}$, and define $h \in M^A$ as the function with domain Y and constant value 1. We have shown after the proof of Lemma 15.5.3 that $\pi^*[Y] = Y$ for any $\pi \in \mathcal{G}$. Therefore $\pi^*(h) = h$ because h is constantly 1, so that $h \in M^{(A,\mathcal{F})}$. Clearly $h^G = Z$ so $Z \in M[G, \mathcal{F}]$. Since $f_n^G \neq f_m^G$ for $n \neq m$, any map $k \to Z$ in $M[G, \mathcal{F}]$ (for $k \in \omega$) is not bijective and thus Z is not finite. □

The set Z of Lemma 15.5.4 is denumerable in the universe V and even in $M[G]$, because Y is denumerable and the map $g : Y \to Z$ with $g(f_k) = f_k^G$ is a bijective map. But this does not mean that $g \in M[G, \mathcal{F}]$ so it is unclear whether Z is denumerable as an object of $M[G, \mathcal{F}]$.

15.5.4 A counterexample to AC

Proposition 15.5.5. *Let V be a ZFC$^+$-universe with distinguished set M, and $P, A, \mathcal{G}, \mathcal{F}$ as defined at the beginning of Section 15.5.2 so that $M[G,\mathcal{F}]$ is a ZF-universe. Assertion (MIS) does not hold in $M[G,\mathcal{F}]$; consequently, AC does not hold in $M[G,\mathcal{F}]$.*

Proof. We have defined the $M[G,\mathcal{F}]$-set Z in Lemma 15.5.4 and proved that it is infinite. We now use RAA to prove that there is no injective map $h : \omega \to Z$ in $M[G,\mathcal{F}]$; thus (MIS) fails in $M[G,\mathcal{F}]$.

Assume that such an injective map $h : \omega \to Z$ exists, and let $h = u^G$ for some $u \in M^{(A,\mathcal{F})}$. In the universe M, consider the two-variable map $s : D(\check{\omega}) \times Y \to A$ with $s(a,b) = \sum_{v\in D(u)} u(v)Q_0(v,a,b)$ –$Y = \{f_k | k \in \omega\}$ was defined after the proof of Lemma 15.5.3. The function $s \subseteq M^A \times M^A \times A$ defines in the usual way (see Corollary 14.2.9) the set $s^G \in M[G]$. It is straightforward to check that $s^G = u^G = h$; and in fact $s(a,b) \in G \Leftrightarrow \langle a^G, b^G\rangle \in s^G = h$ for any $\langle a,b\rangle \in \mathrm{D}(s)$. Moreover, if $\pi \in \mathrm{Stb}(u)$, then $\pi^*(s(a,b)) = s(\pi^*(a), \pi^*(b))$; that is, s commutes with the automorphisms of $\mathrm{Stb}(u)$. This follows from Lemma 15.4.2 and Proposition 15.3.4 because Q_0 is defined through sums and products of the functionals B, I, E.

Consider the following elements of the algebra A:

$q_0 = \Pi_{n\in\omega}(\sum_{k\in\omega} s(\check{n}, f_k) \cdot \Pi_{t\in\omega\setminus\{k\}}(s(\check{n}, f_t)^\perp))$ and
$q_1 = \Pi_{n,k,t\in\omega}(s(\check{n}, f_k)^\perp + s(\check{n}, f_t)^\perp + E(f_k, f_t))$

Observe that these elements correspond to conditions that, for each generic filter H of A^*, force s to give that s^H is an injective function with domain ω whenever q_0, q_1 belong to H.

Since $s^G = h$, it follows easily from Lemma 15.5.4 that $q_0 \in G$; and similarly $q_1 \in G$ by the injectivity of h. Let E be a finite subset of ω such that $S(E) \subseteq \mathrm{Stb}(u)$, which exists because $\mathrm{Stb}(u) \in \mathcal{F}$. Thus we may choose some $k \in \omega$ such that $k \notin E$ and $f_k^G \in \mathrm{cD}(h)$, so that $s(\check{n}, f_k) = q_2 \in G$ for some n. Because $q_0, q_1, q_2 \in G$, their product also belongs to G by Proposition 13.4.3. Hence it is $\neq 0$ and, by density of P in A^*, there is some $p \in P$ with $p \le q_0, q_1, q_2$ (writing as we do p instead of $i(p)$ alleviates the notation without risk of confusion). Choose next an index $t \in \omega$ such that $t \notin E$, $t \neq k$ and t does not appear in the elements of the domain of the finite function $p \in P$. Finally, let π be the permutation of ω which is just the transposition $(k\ \ t)$.

From the construction of π we see that p and $\pi^*(p)$ are compatible elements of P: if p contains elements $\langle k, v, b\rangle$ (with $b \in \{0,1\}$), then the corresponding elements of $\pi^*(p)$ are $\langle t, v, b\rangle$, and there is no collision because $\langle t, v\rangle \notin \mathrm{D}(p)$; the rest of elements of p are not changed by π^*. Therefore $p \cdot \pi^*(p) \neq 0$ in A^* and there exists a M-generic filter G_1 of A^* such that $p \cdot \pi^*(p) \in G_1$, by Proposition 15.1.2. Thus $p, \pi^*(p) \in G_1$.

Then $q_0, q_1 \in G_1$, and hence s^{G_1} is an injective function with domain ω. Since $q_2 \in G_1$, we have $s(\check{n}, f_k) \in G_1$ so that $s^{G_1}(n) = f_k^{G_1}$. But we also know that π fixes the elements of E, hence $\pi \in \mathrm{Stb}(u)$ and therefore π^*

commutes with s, as observed in the second paragraph of this proof. Thus $s(\check{n}, f_k) = s(\check{n}, f_{\pi(t)}) = s(\pi^*(\check{n}), \pi^*(f_t))) = \pi^*(s(\check{n}, f_t)) = \pi^*(q_2)$. Now, $q_2 \geq p$ hence $\pi^*(q_2) \geq \pi^*(p) \in G_1$ and thus $s^{G_1}(n) = f_t^{G_1}$. Since s^{G_1} is injective and $f_k^{G_1} \neq f_t^{G_1}$ by Lemma 15.5.4, we have reached a contradiction. □

We may basically repeat the proof of Theorem 15.2.7 from this new result.

Theorem 15.5.6. *(Cohen's theorem on the AC) If the axioms for ZF-universes are consistent, then there is no proof of AC from those axioms.*

Proof. We know from Gödel's theorem (see the first paragraph of Section 15.3.1) that AC is consistent with the ZF-theory. Consequently, if we prove that ZFC is inconsistent, then also ZF is inconsistent. Then, by Theorem 15.1.1, it will be enough to show that if there is a proof of AC from the ZF-axioms, then the ZFC$^+$-theory is inconsistent.

So, suppose that there is a proof of AC for ZF-universes. By Proposition 15.2.1, there exists a neat class sequence $W_1, \ldots, W_k$ such that the AC holds in every ZF*-universe in which the W_i-instances of replacement are satisfied. Working in the theory of ZFC$^+$-universes, we have just constructed the ZF-universe $M[G, \mathcal{F}]$ where AC does not hold. But surely $M[G, \mathcal{F}]$ is a ZF*-universe satisfying the W_i-instances of replacement, so that AC must hold in $M[G, \mathcal{F}]$. This contradiction shows that, under that supposition, the theory of pointed ZFC$^+$-universes is inconsistent, as it was to be proved. □

15.6 Exercises

1. In the proof of Proposition 15.1.2 a sequence of elements $p_0, p_1, \ldots$ of P has been defined in order to construct the generic filter G_0. Give a complete proof of the existence of that sequence and say if it is enough to use the principle of dependent choices (see, e.g., Exercise 16 of Chapter 10) for the proof.

2. Show from some of the results in this chapter, that the axiom of constructibility $V = L$ is independent of the axioms of ZFC-universes.

3. It follows from Proposition 15.2.5 that $\aleph_1 < 2^{\aleph_0}$ in the universe $M[G]$. Let us suppose that the ZFC-universe M satisfies CH, so that $2^{\aleph_0} = \aleph_1$ in M. How can this be, when we know from Corollary 14.3.4 that M and $M[G]$ have the same cardinals?

4. Recall that (under property NNS) every infinite sequence of digits identifies a real number of the interval $[0, 1]$. Explain how, through the construction of the injective function g of Proposition 15.2.5, the elements of $\aleph_2 \in M$ may be seen as real numbers in the universe $M[G]$.

5. Suppose we had defined the ordered set P given at the beginning of Section 15.2.2 by taking the cardinal $\aleph_\omega$ of M instead of $\aleph_2$. Prove that P is still a separative ordered set, and that the construction of the generic filters G_0, G and the conclusion that $\aleph_\omega \leq 2^{\aleph_0}$ are valid for this $M[G]$ too.

6. Let α be a given ordinal. Show that the assertion $2^{\aleph_0} < \aleph_\omega$ cannot be proved in the ZFC-theory.

7. Prove that, if the ZFC$^+$-theory is consistent, there is a ZFC-universe of the form $M[G]$ in which $2^{\aleph_0} = (\aleph_\omega)^{\aleph_0}$ (Hint: Choose some universe $M[G]$ where $\aleph_\omega \leq 2^{\aleph_0}$).

8. For an infinite cardinal κ and an ordered set P, let us say that P satisfies the κ-chain condition (κ-cc) when every antichain of P has cardinal $< \kappa$ (like this, the ccc is the $\aleph_1$-cc). Under the hypotheses of Proposition 14.3.3, but with the κ-cc for A^*, prove that if λ is a regular cardinal of M with $\kappa \leq \lambda$, then λ is also a regular cardinal of $M[G]$.

9. Under the (GE) hypothesis, let κ be an uncountable cardinal of M. Define $P \in M$ as the set of M-functions whose domain is an element of ω, and whose codomain is included in κ. The ordering of P is reverse inclusion. Let G_0 be a M-generic filter of P and $f = \cup G_0$. Show that f gives a map $f : \omega \to \kappa$ in $M[G]$ and that $\mathrm{cD}(f) = \kappa$. Thus the cardinal in $M[G]$ of the ordinal $\kappa \in M[G]$ is $\aleph_0$. One says that κ *collapses* to $\aleph_0$ in $M[G]$. Why is it that this does not contradict the property about M and $M[G]$ having the same cardinals?

10. Under the (GE) hypotheses, let $\rho < \kappa$ uncountable cardinals of M. Let $P \in M$ be the ordered set consisting of all functions f with $\mathrm{D}(f) \in \rho$, $\mathrm{cD}(f) \subseteq \kappa$, with the usual reverse ordering. Let G_0 be a M-generic filter of P with the corresponding A^*-filter G, and let $f = \cup G_0$. Show, as in the previous exercise, that there exists a surjective $M[G]$-map $\rho \to \kappa$ (thus ρ and κ have the same cardinal in $M[G]$).

11. Let P be an ordered set in a ZFC-universe M. A subset $D \subseteq P$ is *open dense* if it is dense and the following property holds: $\forall x \in P\ ((\exists y \in D\ (x \leq y)) \Rightarrow (x \in D))$. Prove that a subcollection $G \subseteq P$ (G may belong to M or not) is a generic filter of P if and only if it is a filter and $G \cap D \neq \emptyset$ for every open dense subset $D \subseteq P$.

12. Let M be a ZFC-universe, ρ, κ infinite cardinals of M with $\kappa > \rho$ and ρ regular. Let $P \in M$ be the ordered set consisting of all functions f with $\mathrm{D}(f) \subseteq \rho$, $\mathrm{cD}(f) \subseteq \kappa$ and $|\mathrm{D}(f)| < \rho$, with the usual reverse ordering. Let $\lambda < \rho$ an ordinal of M. Prove that if

$\{D_\alpha | \alpha \in \lambda\}$ is a family of open dense subsets of P, then $\bigcap_{\alpha \in \lambda} D_\alpha$ is dense in P.

13. M, ρ, κ and P are as in the preceding exercise. Now, we assume that M is a transitive class of a ZFC-universe V, and $G_0 \in V$ is a generic filter of P, with G being the corresponding generic filter of A^*. Suppose that $\lambda < \rho$ is an ordinal of M, and that there is a function in $M[G]$, $f : \lambda \to \rho$. Prove that the function f belongs to M (Hint: Show that $f = z^G$ for some function in M, $z : \mathrm{D}(\check{\lambda}) \times \mathrm{D}(\check{\rho}) \to A$. By forcing z to be a function with domain λ as in the proof of Proposition 15.5.5 and applying the preceding exercise, get some element $q \in G_0$ so that for each $x \in \lambda$, there is $y \in \rho$ with $q \leq z(\check{x}, \check{y})$. Prove that the function constructed in M by using these data equals f).

14. Prove the claim in the proof of Proposition 15.3.1.

15. Let A be a complete Boolean algebra of the ZFC-universe M. Show that the equation 15.1 is valid for arbitrary automorphisms ψ, φ of A; i.e., $\widetilde{\psi \circ \varphi} = \tilde{\psi} \circ \tilde{\varphi}$.

16. Let P be a separative ordered set, $i : P \to A$ its completion algebra. From results of Chapter 13, it may be inferred that an isomorphism of ordered sets $h : P \to P$ induces a unique automorphism $g : A \to A$ such that $i \circ h = g \circ i$ (it is easy to see this for the canonical completion first). Let now P be the ordered set of the beginning of Section 15.5.2. Show with detail that any permutation π of ω induces the isomorphism which we have called π^*, of the ordered set P. Infer from this that the permutation π of ω induces the automorphism π^* of the algebra A. Show also that the map $\pi \mapsto \pi^*$ from the group of permutations of ω to the group of automorphisms of A, is a group homomorphism.

17. Let M be a ZFC-universe which is a transitive class of the ZFC-universe V, and A a complete Boolean algebra of the universe M. Let $\mathcal{G}$ be a group of automorphisms of A, and $\mathcal{F}_1 \subseteq \mathcal{F}_2$ filters of subgroups of $\mathcal{G}$ both closed under conjugation. Prove that $M^{(A,\mathcal{F}_1)} \subseteq M^{(A,\mathcal{F}_2)}$. Deduce that if G is a M-generic filter of A^*, then $M(G, \mathcal{F}_1) \subseteq M(G, \mathcal{F}_2)$.

18. Let $M, V, P, A, \mathcal{G}$ as in the preceding exercise and $\mathcal{F}$ a filter of subgroups of $\mathcal{G}$ which is closed under conjugation. Prove that $M[G, \mathcal{F}] = M[G]$ if and only if there exists in M a map $h : A \to A$ with the following two properties: (1) $\forall a \in A\ (a \in G \Leftrightarrow h(a) \in G)$; and (2) the subgroup $H = \{\pi \in \mathcal{G} | \pi \circ h = h\}$ belongs to the filter $\mathcal{F}$ (Hint: For the "if" part, use induction to prove that there are maps $g : M_\alpha^A \to M_\alpha^{(A,\mathcal{F})}$ such that for every $x \in M_\alpha^A$, $(g(x))^G = x^G$ and every element of the subgroup H stabilizes $g(x)$).

19. Explain why the proof of Proposition 14.2.18 cannot be carried over to the universe $M[G, \mathcal{F}]$.

20. Consider the sets $P, A, \mathcal{G}, \mathcal{F}$ of the beginning of Section 15.5.2. Try to describe (at least, some of) the elements of $M_2^{(A,\mathcal{F})}$ in this example; is it possible to extend the description to the elements of $M_3^{(A,\mathcal{F})}$?.

21. For the functions u, s introduced at the beginning of the proof of Proposition 15.5.5, prove the equation $\pi^*(s(a,b)) = s(\pi^*(a), \pi^*(b))$ for $\pi \in \mathrm{Stb}(u)$.

22. In the situation of Lemma 15.5.4, we have constructed $Y \subseteq M^{(A,\mathcal{F})} \subseteq M^A$ as the set of the functions f_k; consider the restriction of the functional Ψ of Proposition 14.1.5 to Y, giving the V-map $g : Y \to Z$ with $g(f_k) = f_k^G$. Justify that $g \in M[G]$ by finding some $s \subseteq M^A \times M^A \times A$ such that $s^G = g$ (it is possible to take s a constant function). With that choice of s, show directly, without using Proposition 15.5.5, that s is not $\mathcal{F}$-stable – in the sense that corresponds to that of Lemma 15.4.8.

23. A 6-universe that satisfies the axiom of regularity (asserting that if s is a non-empty set then some $y \in s$ does not contain any element of s) will be called a *ZFA-universe*[1]. That is, it is a ZF-universe except for the axiom of extension (but, of course, the principle of extension for sets is still assumed). It is also assumed in ZFA-universes that an object which is not a set is an *atom*, i.e., it does not contain any object of the universe. Prove that it follows from Proposition 7.2.5 that the class of pure sets of a ZFA-universe is a ZF-universe (it is called the *kernel* of the ZFA-universe). Define ordinals so that every ordinal is a pure set.

24. Suppose that in the ZFA-universe V, the class of all atoms A is an infinite set. Define a hierarchy of objects of the universe V analogous to the von Neumann hierarchy in ZF-universes, and check that this gives all the objects of the universe V (Hint: Start with $V_0 = A$. You will have to revise the content of Section 9.2, especially 9.2.3; and some care with the proof of the result corresponding to Proposition 9.2.10 is advisable).

25. Let V be a ZFA-universe with an infinite set A of atoms. Choose a group $\mathcal{G}$ of permutations of the set A, and a proper filter $\mathcal{F}$ of subgroups of A which is closed under conjugation of subgroups. Use the von Neumann hierarchy of the preceding exercise to show how each element of $\mathcal{G}$ induces a permutation of the elements of V. Let us identify such permutations with the elements of $\mathcal{G}$ and prove that if $s \in V$ is a pure set and $\pi \in \mathcal{G}$, then $\pi(s) = s$.

[1]Not to be confused with the ZFA$^-$-universes of the exercises of Chapter 9.

26. Let $V, A, \mathcal{G}, \mathcal{F}$ as in the preceding exercise. Given $x \in V$, we say that x is *symmetric* when the subgroup $\mathrm{Stb}(x)$ belongs to $\mathcal{F}$. Write $\mathcal{S}(\mathcal{F})$ for the class of all symmetric elements of V, and suppose moreover that $A \subseteq \mathcal{S}(\mathcal{F})$. Prove that the class $N(\mathcal{F}) = \{x \in V | \operatorname{tc}(x) \cup \{x\} \subseteq \mathcal{S}(\mathcal{F})\}$ is transitive, closed under the Gödel operations and a big class; and deduce from the exercises on Gödelian classes of Chapter 12 that $N(\mathcal{F})$ is a ZFA-universe.

27. With the same hypotheses, let $\mathcal{G}$ be the group of all permutations of A and for any $E \subseteq A$, write $S(E) = \{\pi \in \mathcal{G} | \pi(a) = a \ (\forall a \in E)\}$. Let $\mathcal{F}$ be the set of those subgroups H of $\mathcal{G}$ that include some $S(E)$ with E finite. Show that $\mathcal{F}$ is a filter which is closed under conjugation. Consider then the ZFA-universe $N(\mathcal{F})$ as in the preceding exercise (check that the condition $A \subseteq S(\mathcal{F})$ is satisfied) and prove that there is no injective function $f \in N(\mathcal{F})$ with $\mathrm{D}(f) = \omega$ and $\mathrm{cD}(f) \subseteq A$. Show that this implies that the axiom of choice does not hold in the ZFA-universe $N(\mathcal{F})$ (Hint: Assume f exists and let $E \subseteq A$ finite with $S(E) \subseteq \mathrm{Stb}(f)$. Choose $\pi \in S(E)$ and $a \in A \setminus E$ such that $\pi(a) \neq a$ and $a \in f[\omega]$. Then find a contradiction).

Part III

Appendices

A

The NBG Theory

A.1 Introduction

NBG is an alternative to ZFC-theory. As it happens with the latter, it may be constructed as a symbolic first-order theory, but as we have done with ZFC-theory, we will present it as an intuitive theory. The version that we essentially follow is that in [16, Chapter 4]. This version is an elaboration on the theory that was first developed by von Neumann, then by Bernays and subsequently modified by Gödel.

The main difference between NBG and ZFC-theories lies in their treatment of the idea of *class*. In our presentation of the ZFC-theory, classes are not objects of the universe. Instead, they are a product of our observation and study of the universe. The universe itself is not altered by these observations of our mind, it is forever unchangeable, but classes arise from the universe, are part of it in a sense, thus creating like a second level of objects – not belonging to the universe but stemming from it. The existence of these two levels may create a certain unrest in the theory, since we are always forced to distinguish between objects of the real world, so to say, and objects that we introduce and which are therefore subject to different rules than those of the first level.

The solution of NBG-theory to this dichotomy is radical and simple: classes will be objects of the universe. Like this, all finiteness limitations on classes simply disappear: they are full-right objects, we may speak of all classes if we so want and there is no need to introduce in our consideration a subjective level. As we shall see, however, there is a price to be paid for this: since classes will now be data and not our constructions, we have not a concrete description of which are the classes, contrary to what happens in the ZFC-theory.

A.2 NBG-Universes

We present now an intuitive view of NBG-theory. Like in ZF-theory, in NBG-theory we postulate the existence of a universe $\mathcal{U}$ of objects. And like in ZF, all

DOI: 10.1201/9781003449911-A

objects of $\mathcal{U}$ are collections of objects of $\mathcal{U}$. The objects of $\mathcal{U}$ are called *classes*. So far, the only difference between NBG and ZF is the collective name for the objects of the universe. The substantial difference is that classes may be formed only with objects of a certain subcollection of the universe.

Definition A.2.1. If $\mathcal{U}$ is a NBG-universe, a class $X \in \mathcal{U}$ is called a *set* when there exists $B \in \mathcal{U}$ such that $X \in B$.

Like this, the concept of set is a defined concept in NBG-theory, as it was for ZF-theory, though the definitions are not the same. However, in both theories the collection $\mathcal{S}$ of all sets of $\mathcal{U}$ will be proved to be a class.

We introduce now the axioms of the theory, by dividing them into three groups.

- Axioms of formation of sets. Most of the axioms of ZFC are also axioms of NBG, but specifying that they are given only for sets, not for all objects of the universe. This is the case for the axioms of the empty set, pairs, unions, power set, infinity and choice. For instance, the axiom of pairs states that if x, y are two sets, then there is a set $\{x, y\}$ having precisely these elements; the axiom of the power set asserts that to any set x there is a set having as elements precisely all the sets included in x; the axiom of choice admits the existence of a choice function (thus a set) for each set x when $\emptyset \notin x$; and similarly for the other axioms mentioned.

- Axioms of formation of classes. We have mentioned that all the objects of the universe $\mathcal{U}$ are classes, that is, collections of sets. Several axioms establish that some constructions or operations with arbitrary classes give classes. These can be reduced to the following two.

  - In $\mathcal{U}$, the collections $\{\langle a, b\rangle | a \in b \wedge (a, b \in \mathcal{S})\}$ and $\{\langle a, b\rangle | a = b \wedge (a, b \in \mathcal{S})\}$ are classes[1].
  - Classes are closed under the operations of rules 3 to 7 of Definition 3.1.2; i.e., difference, cartesian product, domain and the two transformations of triples of rules 6 and 7 in that definition.

- Axioms involving classes and sets. The regularity axiom is applied for any non-empty class C, asserting that some element $x \in C$ satisfies $x \cap C = \emptyset$. The axiom of replacement is now just one axiom:

  If a class C is functional and $s \in \mathcal{S}$, then $\mathrm{cD}(C|_s)$ is a set.

  Finally, we include in this group (a bit improperly) the axiom of extension; it says that two classes that have the same elements are equal.

[1]One can define ordered pairs exactly as in ZF theory: a, b have to be sets, then $\langle a, b\rangle = \{\{a\}, \{a, b\}\}$ is a set by the axiom of pairs.

An immediate consequence is that $\mathcal{S}$ is a class because it is the domain of the membership class. But $\mathcal{S}$ is not a set, because this would entail that $\mathcal{S} \in \mathcal{S}$ and $C = \{\mathcal{S}\}$ is a non-empty class (and set), which contradicts the axiom of regularity. Thus $\mathcal{S}$ is a proper class, and clearly a transitive $\mathcal{U}$-class which is a 2-universe in the sense of the first part of this book.

As happens with sets, classes appear both in NBG and ZFC-universes, but their meaning is not the same. Nevertheless, there is a direct relationship. To state it, let us call $\mathcal{S}$-classes to the classes of the universe $\mathcal{S}$, as considered in the first part of this book.

Lemma A.2.2. *If C is an $\mathcal{S}$-class, then it is an element of $\mathcal{U}$.*

Proof. The $\mathcal{S}$-classes $\mathcal{M}, \mathcal{E}$ belong to $\mathcal{U}$ by the first of the axioms of formation of classes. Also, any set belongs to $\mathcal{U}$, i.e., it is a $\mathcal{U}$-class. The fact that C is given through a class sequence of $\mathcal{S}$, and that the rules for the formation of class sequences give classes from classes by the second axiom on formation of classes, entail, by applying a finite induction process, that C is an object of $\mathcal{U}$. □

A.3 NBG vs ZF

We may now study the relationship between NBG- and ZFC-theories.

Proposition A.3.1. *Let $\mathcal{U}$ be a NBG-universe, and $\mathcal{S}$ the class of all sets of $\mathcal{U}$. Then $\mathcal{S}$ is a ZFC-universe.*

Proof. Except replacement, it has been assumed that all the other axioms for sets in ZF-universes hold in $\mathcal{S}$ when $\mathcal{U}$ is an NBG-universe. Thus we must only consider the axioms of replacement in $\mathcal{S}$.

Let C be a class of $\mathcal{S}$; by Lemma A.2.2 $C \in \mathcal{U}$. Let $u, s \in \mathcal{S}$ such that $C|_{\{u\}\times s}$ is functional – obviously, both as a $\mathcal{U}$-class or a $\mathcal{S}$-class. In $\mathcal{U}$, the class of all pseudo-triples $\langle x, \langle u, x\rangle\rangle$ with $x \in s$ is functional, hence $\{u\} \times s$ is a set in that universe by the axiom of replacement. On the other hand, $C|_{\{u\}\times s} = C \cap (\{u\} \times s \times \mathcal{S})$ is again a $\mathcal{U}$-class since $\mathcal{U}$-classes are closed under intersections by the same reason as that given in Proposition 3.1.5. Therefore, we can apply the axiom of replacement in $\mathcal{U}$ to that $\mathcal{U}$-class, and hence its codomain $\mathrm{cD}(C|_{\{u\}\times s})$ will be a set of $\mathcal{U}$, that is, an element of $\mathcal{S}$. □

In all, the axioms for NBG-universes consist basically of the axioms that state that $\mathcal{S}$ is a ZFC-universe, plus the two statements which give the axioms of classes (that ensure in particular that all $\mathcal{S}$-classes in the sense of ZFC-theory are elements of $\mathcal{U}$). Note however, that the axiom of replacement in NBG has stronger value, as it can be applied to arbitrary elements of $\mathcal{U}$, not only to $\mathcal{S}$-classes.

Proposition A.3.1 has some metamathematical consequences. Let (P) be an assertion which holds in every ZFC-universe and refers only to sets. If $\mathcal{U}$ is any NBG-universe, assertion (P) may be seen as referring to the objects of $\mathcal{S}$ and thus (P) holds in $\mathcal{S}$ due to Proposition A.3.1, so that (P) holds in $\mathcal{U}$. That is, any assertion (P) which refers to sets and which holds in ZFC-universes, does hold in NBG-universes as well – of course, referred to the elements of $\mathcal{S}$.

We may look for a certain converse to that relationship. Let V be a ZFC-universe; we know that inside ZFC-theory, we cannot consider the collection of all possible V-classes as a finished entity. But if we are willing to transcend this frame, we may imagine the universe $\mathcal{U}$ whose objects are all the V-classes (which includes, by definition, all elements of V)[2].

Proposition A.3.2. *Let V be a ZFC-universe and let $\mathcal{U}$ be the collection of all V-classes. Then $\mathcal{U}$ is a NBG-universe.*

Proof. By introducing the proper V-classes as elements of $\mathcal{U}$, no new membership relation $x \in A$ is added. Therefore the sets of $\mathcal{U}$ are the elements of V, according to Definition A.2.1, thus $\mathcal{S} = V$. The axioms of NBG-universes for sets are automatically valid in $\mathcal{U}$ because they only depend on the class $\mathcal{S}$; and the same holds with the axioms on classes, in view of Definition 3.1.2. Since extension is assumed, only the axioms of regularity and replacement remain to be proved. As for replacement, let C be a $\mathcal{U}$-class (i.e., a V-class) which is functional; and let $s \in V = \mathcal{S}$. Then $C|_s$ is a functional class of V and therefore its codomain is a set by Proposition 5.1.4; and $\mathcal{U}$ satisfies replacement.

Let us now prove regularity in $\mathcal{U}$ by RAA. Suppose C is a $\mathcal{U}$-class hence a V-class such that $x \cap C \neq \emptyset$ for every $x \in C$. Let us choose $u \in C$ and $b = C \cap \operatorname{tc}(u)$, which will be a non-empty set by the principle of separation in V and because $u \subseteq \operatorname{tc}(u)$. Define $f(x) = x \cap C$ for every $x \in b$, so that $\emptyset \notin \operatorname{cD}(f)$ – and also each $f(x)$ is a set. Then $\operatorname{cD}(f)$ is a set by replacement (in V) and it has a choice function g; this gives a map $g \circ f = h : b \to b$ and $h(x) \in x$. Finally, we may define by natural recursion a function $\varphi : \omega \to b$ by setting $\varphi(0) = u_0$ (for some $u_0 \in b$) and $\varphi(s(\alpha)) = h(\varphi(\alpha))$. This gives a denumerable chain of elements of V,

$$u_0 = \varphi(0) \ni \varphi(1) \cdots \ni \varphi(\alpha) \ni \varphi(s(\alpha)) \ni \dots$$

which contradicts Proposition 9.2.15. This contradiction shows that $\mathcal{U}$ satisfies regularity and it is a NBG-universe. □

Again, let (P) be a statement which refers only to sets and which is valid for the class of sets of any NBG-universe. Then, if V is a ZFC-universe, we may construct the NBG-universe $\mathcal{U}$ as just shown, and thus (P) will be valid

[2] By throwing this universe into being, we are necessarily out of the theory. What follows is only an informal but plausible argument. Its consequence below, however, may be proved in a correct way by other means.

in $\mathcal{U}$, hence in $\mathcal{S} = V$. That is, for an statement (P) referring to sets in a universe, (P) is valid for all ZFC-universes if and only if it is valid in all NBG-universes. This shows that NBG-theory and ZFC-theory are not different in the results they produce about sets.

B

Logic and Set Theory

B.1 Introduction

The desire of constructing mathematics (and even other fields of knowledge) in a completely symbolic way is now centuries old. Certainly, many symbols are employed when writing mathematics, but mathematics is normally developed by using at the same time some natural language, which conveys and transmits arguments and thinking about the objects in study. Since set theory is our concern, we will examine here a symbolic or *formalized* version of set theory which, though not unique, is generally agreed upon as a natural and reasonable presentation of formal set theory. This theory will consist entirely of well specified symbols and of the manipulation of those symbols, which is done through precise mechanical rules.

Before going into the matter, some questions seem to be of interest in this connection. First, assuming that it is possible to give a symbolic version of set theory, how does this version relate to the intuitive version? Is there an easy-to-describe correspondence between both? And if so, do both forms of the theory reach the same results? Second, what is the idea of that symbolic version? More concretely, which are the advantages of studying first-order set theory with respect to the intuitive study we have carried so far?

The question about the possibility of obtaining the symbolic version has an affirmative answer and is explained in many books produced in the last century or so. We will give a brief account in the next section. As for the relationship between symbolic and intuitive theory, we will mention that there is indeed a perfect correlation about the results found through both approaches. This is a consequence of the celebrated *completeness theorem* of Güdel[1]. About the last question, before speaking about the objectives that mathematicians have pursued with the symbolization of set theory, let us give a brief description of that symbolic set theory, the *first-order set theory.*

A word of caution is necessary. First-order theories are normally presented as independent from set theory. However, the description of first-order theories uses, from its very beginning, sets and even infinite sets. As Hilbert once put it: "We observe that the principles of Logic, as they are usually presented,

[1] In their book *Grundzüge der theoretischen Logik*, Hilbert and Ackermann state that the result was also included in the work of Herbrand in the same year 1930.

DOI: 10.1201/9781003449911-B

suppose certain arithmetic notions, for example the notion of Set and, to some extent, the notion of Number. Like this, we are caught in a circle and this is why, for avoiding all paradoxes, I find it necessary to develop simultaneously the principles of Logic and those of the Arithmetic"[2] (as can be deduced from the text, Arithmetic here comprises set theory as well). Since the simultaneous development of logic and set theory (including arithmetic) proposed here by Hilbert is far from our possibilities and intentions in this book, we prefer to describe first-order set theory as a piece of set theory.

This may be shocking to those who have already some knowledge of logic. But there are some reasons. In usual presentations the necessary concepts of set or countability appear to the unadvised reader as clear, intuitive concepts, not requiring any knowledge of set theory. But we know that countability is not living in nature, so to speak: a set may be countable in one universe and uncountable in another one. Even the non-technical concept of set is not free of traps, as we have been taught by the antinomies and more importantly by the discussions fuelled by them. On the other hand, the identification of first-order logic inside set theory does not change anything essential nor important, it is just an added precision to the normal presentations. It has nevertheless a weakness: it requires that the elements of the symbolic theory belong to a concrete universe $\mathcal{U}$. But we shall see that this universe plays basically no role in the theory, and that all the properties and arguments are exactly the same in whatever universe $\mathcal{U}$ we choose.

B.2 First-Order Set Theory

B.2.1 Language of set theory

In intuitive set theory (from now on, IST for the sake of brevity; in opposition, we will abbreviate "symbolic set theory" as FOST, for "first-order set theory") we postulate a universe $\mathcal{U}$, and we now want to describe FOST inside the universe $\mathcal{U}$. Though the choice of $\mathcal{U}$ is not relevant, there is one point to be made here. In explaining the symbolic theory, the concept of finiteness has to be frequently used and we want this concept to have its usual intuitive meaning; for this, we will require that ω corresponds to the ordinary natural numbers. Therefore, we suppose that $\mathcal{U}$ is a 6-universe[3] that satisfies the property NNS: look Section 6.2.2.

The *first-order alphabet* for the language of set theory will be a set which is the union of two disjoint sets:

[2]Cited in H. Poincaré: Les Mathématiques et la Logique, Revue de Métaphysique et de Morale, Nov. 1905, 815-835.

[3]Though this is not strictly necessary for the development of the basics of first-order logic; just the first two axioms, plus infinity and natural recursion would be enough.

(1) A denumerable set $v \in \mathcal{U}$, whose elements we call *variables.* A bijection $\omega \to v$ is also specified, and thus we represent the elements of v as $x_1, x_2, \ldots$.

(2) A finite set $s \in \mathcal{U}$ of 11 elements which we call the *logical symbols*[4]. We will represent these elements by the symbols $\neg, \wedge, \vee, \to, \leftrightarrow, \in, =,), (, \forall, \exists$. We also ask that $\emptyset$ is not an element of v or s, and that no element of $v \cup s$ is a sequence of elements of this same set.

Any non-empty finite sequence of elements of the alphabet will be called an *expression*, and $E \subseteq \mathcal{U}$ will denote the set of all expressions. A variable t *appears* in an expression f when $t \in \mathrm{cD}(f)$.

Besides the alphabet, the language of set theory is constructed by a *syntax.* The syntax of the language selects a subset of the set of all expressions of the alphabet. To define the chosen subset of expressions, we first obtain sets of ordered pairs $\langle f, \langle a, b \rangle \rangle$ with $f \in E$ and $a, b \subseteq v$ such that $a \cap b = \emptyset$.

Before giving the actual definition, let us explain its motivation. In the pairs $\langle f, \langle a, b \rangle \rangle$ to be defined, it is intended that f be a formula (basically in the same sense that we have been using formulas since Chapter 3), while a, b are sets of variables that appear in the formula f, with a being the set of *free variables*, b including the *bounded variables.*

We now define the sets K_n by natural recursion. K_0 consists of all pairs $\langle f, \langle a, b \rangle \rangle$ where f is an expression represented by one of the forms $t_1 \in t_2$ or $t_1 = t_2$, with t_1, t_2 variables; and $a = \{t_1, t_2\}$, $b = \emptyset$. Note that, for all we know, we might have $1, 2 \in \mathcal{U}$ as variables and 3 as a logical symbol, and thus an expression could be $\{\langle 0, 1 \rangle, \langle 1, 3 \rangle, \langle 2, 2 \rangle\}$, i.e., the sequence $1, 3, 2$. If $1, 2, 3$ are represented respectively by $x_2, x_5, \in$, then the above expression would be represented as $x_2 \in x_5$, so it belongs to K_0. Without further mention, we shall usually speak of these representations as being the expressions themselves – or, what is the same, we assume that the elements of the alphabet **are** $x_1, x_2, \exists, \wedge$, etc.

Given K_n, we set K_{n+1} as the union of K_n and the set of ordered pairs of one of the following forms:

- For any $\langle P, \langle a, b \rangle \rangle \in K_n$, the pair $\langle \neg(P), \langle a, b \rangle \rangle$.
- For $\langle P, \langle a, b \rangle \rangle \in K_n$ with $x_i \in a$, the pair $\langle \forall x_i(P), \langle a \setminus \{x_i\}, b \cup \{x_i\} \rangle \rangle$; and the pair $\langle \exists x_i(P), \langle a \setminus \{x_i\}, b \cup \{x_i\} \rangle \rangle$.
- For two elements of K_n, $\langle P, \langle a_1, b_1 \rangle \rangle$ and $\langle Q, \langle a_2, b_2 \rangle \rangle$, such that $a_1 \cap b_2 = a_2 \cap b_1 = \emptyset$, the pair $\langle (P) \wedge (Q), \langle a_1 \cup a_2, b_1 \cup b_2 \rangle \rangle$.
- The constructions analogous to the last one, but with $\vee, \to, \leftrightarrow$ substituted for $\wedge$.

The set F of *formulas* is defined as $\bigcup_{n \in \omega} \mathrm{D}(K_n)$ (and we set $F_n = \mathrm{D}(K_n)$). In particular, $\mathrm{D}(K_0) = F_0$ is called the set of *atomic formulas.* This definition

[4]The symbol $\in$ is not properly a logical symbol; it is what is normally called a *predicate* symbol. But this difference is not of interest to us here.

of formula is basically the same as the one explained in Section 3.2 (and atomic formulas correspond to our old simple formulas) with just two exceptions: names for elements of $\mathcal{U}$ do not appear in formulas in the present sense; and the requirement in Proposition 3.2.6 that the formula P (with which $\forall x(P)$ or $\exists x(P)$ are formed) has at least two free variables is softened to at least one free variable – in the case of Proposition 3.2.6 our formulas must always have free variables, as they are used therein to describe classes, but now this is no longer necessary. Despite this strong similarity, we have preferred to give a complete definition of formulas in the current setting to emphasize that the construction is made inside IST, so that formulas can be seen as objects of the universe $\mathcal{U}$. In particular, K_n, F_n are sets in $\mathcal{U}$, and so is F.

By using the so called *unique reading lemma* (which essentially asserts that if a formula is, say, a disjunction, i.e., of the form $(P)\vee(Q)$, it cannot belong to any of the other types, negation, conjunction, conditional, quantified, atomic; and the same for the other kinds of formulas) one may show that $\cup_{n\in\omega}K_n$ is a function; that is, a formula P determines uniquely the pair $\langle a, b\rangle$ of sets of variables such that $\langle P, \langle a, b\rangle\rangle \in \cup_{n\in\omega}K_n$. The elements of a are the *free variables* of the formula P, those of b are the *bound variables* of P. When the set of free variables of P is empty, we say that P is a *closed formula* or a *sentence.* The closed formulas form a set, that we shall denote as C.

Formulas with free variables have been used in this text to describe classes. Closed formulas of the language of set theory may be similarly interpreted as describing properties that the universe $\mathcal{U}$ may have (or not). For instance,

$$\forall x_1(\forall x_2((\forall x_3((x_3 \in x_1) \leftrightarrow (x_3 \in x_2))) \rightarrow (x_1 = x_2)))$$

can naturally be seen as asserting that any two objects containing the same elements are equal, i.e., as asserting the axiom of extension.

B.2.2 Deduction

The most characteristic ingredient in the developing of intuitive set theory is the process of deduction, by which properties of $\mathcal{U}$ are inferred from the axioms. First-order logic constructs a version of this process in terms of the symbolic language we have presented. But the version of deduction in first-order logic is much more rigid and mechanical than the usual process of intuitive deduction.

In IST, our objective is to obtain properties that are true in universes that satisfy a given collection of axioms; and we use any argument that can assure that a property holds in any such universe, provided that previously proven properties (and, of course, the axioms) hold in that universe. Like this, the inspiration for deductions comes mainly from the consideration of the truth or falsehood of statements in each context. On the other hand, deduction in first-order logic is defined without mention of any characterization of formulas as true or false. The process of deduction has explicit rules that make it almost automatic; and, in fact, this process could be carried out by a computer[5].

Formal deduction contains two fundamental items: premises and rules of inference. Premises are formulas, and from these formulas, the application of the rules of inference give other formulas, the conclusions of the deduction. Specifically, a *deduction* from a set A of formulas in the language of set theory is a finite sequence of ordered pairs $\langle P, t\rangle$ where P is a formula and t can be 0, a positive integer or a pair $\langle i, j\rangle$ of positive integers. t is a code that identifies the reason that makes P a consequence of previously obtained formulas in the sequence; $t = 0$ when P is simply one of the elements of A; in any other case, P is obtained from previous formulas of the sequence by a rule of inference, and t identifies which is the rule of inference employed and to which previous formulas has that rule been applied. Explicitly, for the sequence to be a deduction, each item of the sequence must be of one of the following forms:

- $\langle P, 0\rangle$, where $P \in A$.
- $\langle P, i\rangle$ is the k-th item of the sequence with $i < k$, the formula in the i-th item of the sequence is Q and $P = \forall x_h(Q)$ – by the rules of formation of formulas, x_h must be a free variable of Q. This is called the *rule of generalization.*
- $\langle P, \langle i, j\rangle\rangle$ is the k-th item of the sequence with $i, j < k$, the formula in the i-th item of the sequence is Q, the formula in the j-th item is $Q \to P$. This is called the *modus ponens*, MP.

The elements of the set A are called the *premises* of the deduction; generalization and MP are called the *rules of inference* and they are common to all deductions. Each item of the deduction is usually called a *line*. Each formula that is the first component of a line is a *conclusion* of the deduction. When P is a conclusion of a deduction with the set A of premises, we say also that P is deduced or *proved* from A (written $A \vdash P$); and the deduction itself is a *proof* of P (relative to the given set A).

As mentioned above, deduction in IST is quite different and apparently richer in resources than the picture of formal deduction we have just given. For instance, assertions that can be formally represented by the formula $(x_1 = x_2 \wedge x_2 = x_3) \to x_1 = x_3$[6] are safe to use in an intuitive reasoning; as is a formula like $P \to (Q \to P)$ for whatever formulas P, Q. Like these, there are several other formulas, the *logical axioms* as they are known, that reflect convincing assertions in our normal intuitive reasoning. To approach deductions in IST and deductions in FOST, these logical axioms are also accepted as premises in formal deductions. In this sense, they belong to the *extended set of axioms* of set theory.

[5]Provided that the possible premises of any deduction can be mechanically produced.

[6]The reader should not be disturbed by this not being a formula by the strict rules of construction of formulas given before. $((x_1 = x_2) \wedge (x_2 = x_3)) \to (x_1 = x_3)$ would be the formula, but as long as, through some conventions, the expression identifies the formula, everything is in order.

For presenting the rest of the logical axioms, let us write $P(z_1, z_2, \ldots, z_n)$ $(n \geq 1)$ to denote a formula where $z_1, z_2, \ldots, z_n$ are free variables – perhaps not the only free variables; note that this convention is different than that we followed in Chapter 3, see the sentences before Example 3.2.3. We also use this writing in the form $P(z, z)$ for a formula which has several appearances of the free variable z. The following is another writing convention. When $P = P(z)$ is a formula and y is a variable that is not a bounded variable of P, then $P(y)$ denotes the formula obtained from P by substituting y for any appearance of z in the formula P. Likewise, if y is not a bounded variable of $Q = Q(x, x)$, then $Q(x, y)$ denotes (ambiguously) any formula obtained from Q by substituting y for some, but possibly not all, of the appearances of the free variable x in Q. We now complete the list of the other types of logical axioms.

(1) For any formula P where x is a free variable, the formula $(\exists x(P)) \leftrightarrow (\neg(\forall x(\neg(P))))$ is a logical axiom.

(2) If x does not appear in a formula P but is a free variable in the formula Q, then $(\forall x(P \to Q)) \to (P \to (\forall x(Q)))$ is a logical axiom.

(3) For a formula P, $(\forall x(P(x))) \to P(y)$ is a logical axiom. Of course, according with the previous conventions, y is not a bounded variable of $P(x)$. Note that this includes formulas of the form $(\forall x(P(x))) \to P(x)$.

(4) For any variable x, $x = x$ is a logical axiom. Also, given the formula $P(x, x)$, the formula $x = y \to ((P(x, x)) \to (P(x, y)))$ is a logical axiom.

(5) For arbitrary formulas P, Q, R, the following are also logical axioms:

(5.1) $P \to (Q \to P)$
(5.2) $(P \to (Q \to R)) \to ((P \to Q) \to (P \to R))$
(5.3) $((\neg P) \to (\neg Q)) \to (((\neg P) \to Q) \to P$
(5.4) $(P \vee Q) \to ((\neg P) \to Q)$
(5.5) $((\neg P) \to Q) \to (P \vee Q)$
(5.6) $(P \wedge Q) \to \neg((\neg P) \vee (\neg Q))$
(5.7) $\neg((\neg P) \vee (\neg Q)) \to (P \wedge Q)$
(5.8) $(P \leftrightarrow Q) \to ((P \to Q) \wedge (Q \to P))$
(5.9) $((P \to Q) \wedge (Q \to P)) \to (P \leftrightarrow Q)$

These types of formulas are known as *tautologies*. The term "tautology" includes many other formulas, but all of them have a deduction from these nine forms, and using only the rule of inference MP.

Given a set A of formulas of the language of set theory, the *theory* of that set A consists of the set[7] $T(A)$ of all the formulas P such that there is a proof of P from A and the logical axioms. As said above, it is this use of logical axioms what approaches deductions in FOST and usual deductions in IST; otherwise, using just the two rules of inference for making deductions would be clearly insufficient. Note that, by the definition of deduction, $A \subseteq T(A)$.

[7]It can be seen to be a set by separation.

For theories, the formulas of the set A are called the *axioms* of the theory, and the formulas of $T(A)$ are the *theorems.* A theory is called *consistent* if $T(A) \neq F$; that is, not every formula is a theorem. When $T(A) = F$ the theory is *inconsistent.*

A final observation about the axioms of a theory: we may suppose that all the axioms are closed formulas (so $A \subseteq C$); and the theorems are also considered to be those closed formulas for which there exists a proof[8]; that is, we consider that the set of theorems is $T(A) \cap C$. For example, there are closed formulas that translate the axioms of ZFC-universes as presented in this book – and these are considered to be the non-logical axioms of the formal ZFC-theory. We have already given the formula that translates the axiom of extension; we give below closed formulas for the other axioms of the ZFC-theory, and leave the reader the task of identifying each of these axioms.

- $\exists x_1((\forall x_2((x_2 \in x_1) \rightarrow (\forall x_3((\forall x_4((x_4 \in x_3) \leftrightarrow ((x_4 \in x_2) \vee (x_4 = x_2)))) \rightarrow (x_3 \in x_1))))) \wedge (\forall x_5((\forall x_6(\neg(x_6 \in x_5))) \rightarrow (x_5 \in x_1))))$.
- $\forall x_1(\exists x_2(\forall x_3((x_3 \in x_2) \leftrightarrow (\exists x_4((x_4 \in x_1) \wedge (x_3 \in x_4))))))$.
- $\forall x_1((\exists x_2(x_2 \in x_1)) \rightarrow (\exists x_3((x_3 \in x_1) \wedge (\forall x_4((x_4 \in x_3) \rightarrow (\neg(x_4 \in x_1)))))))$.
- $\exists x_1(\forall x_2(\neg(x_2 \in x_1)))$.
- $\forall x_1((\forall x_2((x_2 \in x_1) \rightarrow (\exists x_3(x_3 \in x_2)))) \rightarrow (\exists x_4((\forall x_5((x_5 \in x_4) \rightarrow (\exists x_6(\exists x_7((x_5 = \langle x_6, x_7\rangle) \wedge ((x_6 \in x_1) \wedge (x_7 \in x_6))))))) \wedge ((\forall x_6((x_6 \in x_1) \rightarrow (\exists x_5(\exists x_7((x_5 \in x_4) \wedge (x_5 = \langle x_6, x_7\rangle)))))) \wedge (\forall x_5(\forall x_6(\forall x_7(\forall x_8(\forall x_9(((x_5 = \langle x_6, x_7\rangle) \wedge ((x_5 \in x_4) \wedge ((x_8 = \langle x_6, x_9\rangle) \wedge (x_8 \in x_4)))) \rightarrow (x_7 = x_9)))))))))))$

  where $x_5 = \langle x_6, x_7\rangle$ and similar formulas are abbreviations of formulas that the reader can find without trouble.
- $\forall x_1(\forall x_2(\exists x_3(\forall x_4((x_4 \in x_3) \leftrightarrow ((x_4 = x_1) \vee (x_4 = x_2))))))$.
- $\forall x_1(\exists x_2(\forall x_3((x_3 \in x_2) \leftrightarrow (\forall x_4((x_4 \in x_3) \rightarrow (x_4 \in x_1))))))$.
- Given any formula P with three free variables (we write $P = P(x_1, x_2, x_3)$), we have the following axiom: $\forall x_1(\forall x_2((\forall x_3((x_3 \in x_2) \rightarrow (\forall x_4(\forall x_5(((P(x_1, x_3, x_4)) \wedge (P(x_1, x_3, x_5))) \rightarrow (x_4 = x_5)))))) \rightarrow (\exists x_6(\forall x_7((x_7 \in x_6) \leftrightarrow (\exists x_8((x_8 \in x_2) \wedge (P(x_1, x_8, x_7)))))))))$.

  Here $P(x_1, x_8, x_7)$, for instance, means the formula that is obtained when we replace in the formula $P = P(x_1, x_2, x_3)$, all the appearances of x_2 or x_3 with x_8 and x_7, respectively.

In this form, the axiom of replacement has still infinitely many instances, but we may consider them as an actual infinity because these instances form a set, a subset of the set F of formulas.

[8]That this reduction is possible is a consequence of the rule of generalization and the scheme of logical axioms of the form $\forall x(P) \rightarrow P$, number (3) of the list above.

B.3 FOST and IST

In the foregoing paragraphs, the question of the relationship between IST (intuitive set theory) and FOST (first-order set theory) has shown up now and then. The examples just seen of the non-logical axioms of ZFC-theory are instances of this relationship, and we may see from them how both directions of the translation work: the symbolic version of an intuitive statement is found by restricting our powers of expression to the use of connectives and quantifiers, and reducing assertions to minimal components that express an equality or a membership relation. In the opposite direction, the proof of the unique reading lemma shows how any formula can be effectively reduced to a structured composition of atomic formulas and thus interpreted intuitively. Trying to precise how the translation of formulas as assertions of IST works, we rely on the fact that our interpretation of a symbolic formula is, if not coincident with, completely identified by the determination of its truth or falsehood in every imaginable context. To that end, we introduce now another ingredient of mathematical logic, the *semantics* of the language.

B.3.1 Evaluation of formulas

The semantics of the language of FOST is the way in which we attribute truth values (1 for true sentences, 0 for false ones) to the closed formulas of the language. In order to achieve this, we need also to assign values to arbitrary non-closed formulas, but since these formulas have free variables, those values will be dependent on replacing the free variables with concrete elements of some set or class. Three steps are necessary to the definition of that assignement.

(1) *The interpretation.* To attribute a truth value to a closed formula we need a domain or universe where the interpretation of the formula may be true or not. In principle, this could be any universe, but we shall assume that this universe or *domain of interpretation* is a class M of the same universe $\mathcal{U}$ where the alphabet belongs. This is because we need to define maps which involve formulas (hence elements of $\mathcal{U}$) and elements of the domain of interpretation M.

Given the domain M, an *M-interpretation* is a relation $L \subseteq M \times M$. The *standard M-interpretation* is the membership relation $\mathcal{M} \cap (M \times M)$.

(2) *Evaluation of variables.* Let $M \subseteq \mathcal{U}$ the domain and $v \in \mathcal{U}$ the set of variables. An *M-evaluation of variables* (or simply *M-evaluation*) is any function ϕ with $\mathrm{D}(\phi) = v$ and $\mathrm{cD}(\phi) \subseteq M$. With an M-evaluation, we assign a value in M to any variable. There is a class $V(M)$ which contains all the M-evaluations. There is also a notation for M-evaluation which agree on every variable except possibly one: if $\phi \in V(M)$, $x \in v$ and $u \in M$, we write ϕ^x_u to

denote the M-evaluation with $\phi_u^x(x) = u$ and $\phi_u^x(y) = \phi(y)$ whenever $y \neq x$. ϕ_u^x is called a *variation* of ϕ in the variable x.

(3) *Evaluation of formulas.* Let L be an M-interpretation. Recall that $F_n = \mathrm{D}(K_n)$ is the set of formulas obtained in $\leq n$ steps from the atomic formulas by the process described in Section B.2.1. For each n, we may define a functional class $e_n : V(M) \times F_n \to \{0, 1\}$ by using a finite process of recursion:

First, let $\langle \phi, P \rangle \in V(M) \times F_0$. Then

If P is $x \in y$, then $e_0(\phi, P) = 1$ when $\langle \phi(x), \phi(y) \rangle \in L$; and 0 otherwise.

If P is $x = y$, then $e_0(\phi, P) = 1$ when $\phi(x) = \phi(y)$; and 0 otherwise.

That is: the formula $x \in y$, when we give the variables x, y the values $a, b \in M$, is true in the standard interpretation when $\langle a, b \rangle \in \mathcal{M}$ (i.e., $a \in b$ in the universe M); or when $\langle a, b \rangle \in L$ in the general case. In turn, $x = y$ is true when $a = b$, both in standard or general interpretations.

Let now $\langle \phi, P \rangle \in V(M) \times F_{n+1}$. Then

If P is obtained from formulas in F_n through any of the connectives $\neg, \wedge, \vee, \to, \leftrightarrow$, the value $e_{n+1}(\phi, P)$ is just the value we showed for these composed formulas as they were introduced in Chapter 1 (depending on the values $e_n(\phi, Q)$ of their components), considering that 1 is the value for true and 0 is the value for false. For instance, $e_{n+1}(\phi, \neg(P)) = 1 - e_n(\phi, P)$; $e_{n+1}(\phi, Q \wedge R) = e_n(\phi, Q) \cdot e_n(\phi, R)$; or $e_{n+1}(\phi, P \to Q) = 1 - e_n(\phi, P) + e_n(\phi, P) \cdot e_n(\phi, Q)$.

If $P = \forall x(Q)$ with $Q \in F_n$, then $e_{n+1}(\phi, P) = 1$ if and only if $e_n(\phi_u^x, Q) = 1$ for every $u \in M$.

If $P = \exists x(Q)$ with $Q \in F_n$, then $e_{n+1}(\phi, P) = 1$ if and only if $e_n(\phi_u^x, Q) = 1$ for some $u \in M$.

If P is atomic, then $e_{n+1}(\phi, P) = e_0(\phi, P)$.

A couple of observations is in order. First, one may see by finite induction that if $P \in F_k$ and $k < n$, then $e_n(\phi, P) = e_k(\phi, P)$ because the component formulas of P are independent of the values n, k we might consider. Consequently, we write $e(\phi, P)$ to denote any of the values $e_n(\phi, P)$ with $P \in F_n$. Second, if $x_1 \ldots, x_n$ are the free variables of the formula P, $\phi, \psi \in V(M)$ and $\phi(x_i) = \psi(x_i)$ for $i = 1, 2, \ldots, n$, then $e(\phi, P) = e(\psi, P)$; that is, the evaluation of P for the M-evaluation ϕ of variables depends only on the values assigned by ϕ to the free variables of P. This is obvious for atomic formulas and is also easy to extend inductively. In particular, if P is a closed formula, then $e(\phi, P)$ is the same for whatever ϕ, and thus we write $e(P)$ to denote this well determined value.

We say that the formula P is (M, L)-*true* when $e(\phi, P) = 1$ for every M-evaluation ϕ. Therefore, if P is closed then P is (M, L)-true if and only if $e(P) = 1$. In this way, the "meaning" of the sentences (in the sense that they give true or false statements for a particular relation $L \subseteq M \times M$) is determined.

When L is the standard M-interpretation, the (M, L)-evaluation of formulas is called the *standard evaluation*. It is important to observe that the concept of M-evaluation has been so defined that the value of a sentence P for the standard M-interpretation agrees with the truth or falsity of the statement which translates P into an intuitive assertion about the universe M.

B.3.2 Models of theories

Recall that for a set A of sentences (which we will assume that includes the logical axioms), the theory $T(A)$ of A consists of all the sentences that appear in some deduction from the formulas of A. The elements of $T(A)$ are called the *theorems* of the theory. The following is a key concept that relates deduction and evaluation.

Definition B.3.1. Let T be a theory, for some set of sentences A; and L be an M-interpretation. We say that M (with the interpretation L) is a *model* of the theory T when $e(P) = 1$ for every sentence $P \in T$. If L is the standard M-interpretation, then M is called a *standard model.*

Probably the most usual and useful way of viewing models is suggested by the following result.

Proposition B.3.2. *Let A be a set of sentences, A' the union of A with the set of all the logical axioms, and T the theory of A'. Then the class $M \subseteq \mathcal{U}$ with the M-interpretation L is a model of T if and only if $e(P) = 1$ for every sentence $P \in A$.*

Proof. (Sketch of) Obviously, if M is a model, then $P \in A \Rightarrow e(P) = 1$ because $A \subseteq T$. Suppose now that $P \in A \Rightarrow e(P) = 1$. The first element we need is the observation that deductions preserve truth. That is, in any deduction, if the premises have value 1 for a given M-interpretation, then the corresponding value of the conclusion is also 1. This is clear when MP is applied. As for generalization, assume that a formula $P(x)$ has value 1; i.e., $e(\phi, P(x)) = 1$ for any $\phi \in V(M)$. Then $e(\forall x\ (P(x))) = 1$ by the rule for evaluation of quantified formulas.

The second ingredient is the proof that $e(P) = 1$ for any logical axiom P. This has to be shown directly for each axiom. For instance, consider a logical axiom of the form $P = (\forall x(Q \to R)) \to (Q \to (\forall x(R)))$. If we had $e(\phi, P) = 0$, then $e(\phi, \forall x(Q \to R)) = 1$, $e(\phi, Q) = 1$ and $e(\phi, \forall x(R)) = 0$. This entails that there is $u \in M$ with $e(\phi^x_u(R)) = 0$ while $e(\phi^x_u(Q)) = 1$ because x does not appear in Q. Then $e(\phi^x_u, (Q \to R)) = 0$ and $e(\phi, (\forall x(Q \to R))) = 0$, which is a contradiction.

This having been done, let the sentence P belong to T. There is a deduction of P where the premises belong to A or are logical axioms; and in each of these two cases they have value 1, and so does P because deduction preserves truth. □

So, a standard model for a given set A of sentences is basically a set or class M which, through the translation from FOST to IST, satisfies those sentences. A standard model for the ZF-theory, for instance, is nothing else than a class of $\mathcal{U}$ which is a ZF-universe itself.

B.4 The Completeness Theorem and Consequences

B.4.1 The theorem

We have already observed the possibility of translation between FOST and IST, and we now focus on the relationship between the theorems obtained in each of them, a problem which is solved by the completeness theorem. Though this theorem has a much larger scope, we will limit ourselves to the language of set theory. Like this, ZF or ZFC-theories will fall under the results that follow.

One direction of that relationship is easy. Though the usual way of making proofs in IST is not through the precise rules of inference of formal deduction, it is certain that arguments are considered valid in IST when they show that the consequences are necessarily true provided that the premises are true. Now, since formal deduction preserves truth, we must accept that formal deduction is a valid method of proof in IST too because of the correspondence between formal truth for formulas and truth of their translation in IST. Consequently, if A is a set of axioms and T is the theory of A, then each theorem of T has a translation in IST which is also a theorem.

It is for the other direction that Gödel's completeness theorem may be fruitfully applied. The shortest statement for the completeness theorem (in what is of interest to us here) is the following.

Given a set A of sentences of the language of set theory which includes the logical axioms, if the set A is consistent then there is a countable set $M \in \mathcal{U}$ with an M-interpretation L which is a model for the theory of A. Furthermore, when A contains the axioms for the ZF-theory, then there is a countable standard model M.

The second part of this statement is a direct consequence of the first. Because if M with the interpretation L is a model for an extension of ZF(C)-theory, this entails that the relation L is extensional and well-founded. It follows by Mostowski's theorem (Theorem 9.3.9) that there is a transitive countable set $M' \in \mathcal{U}$ with an isomorphism $(M, L) \cong (M', \in)$ which gives that M' is a standard model for the formulas of A.

Corollary B.4.1. *Consider both versions, intuitive and symbolic, of the ZF or ZFC-theories. If some sentence P is the symbolic translation of a theorem of the IST, then P is a theorem of FOST.*

The proof is easy from the above form of the completeness theorem. We may assume that the symbolic ZF(C)-theory is consistent, otherwise the conclusion is obvious. Suppose then that P is not a theorem, i.e., there is no proof of P from the axioms of the corresponding FOST. In FOST there is also a version of RAA, from which we may deduce from the above assumption that ZF(C) + $\neg(P)$ is consistent (otherwise we would have a proof for P by RAA). By the completeness theorem, there is a standard model M for that theory; therefore M will be a standard model for the ZF(C)-theory which satisfies $\neg(P)$. In terms of IST, M is a ZF(C)-universe where the intuitive form of P fails. This is absurd since we have a proof in IST of that form of P; therefore the ZF(C)-theory is inconsistent when developed as an intuitive theory. But this is impossible because we have just seen that there exists a universe M which satisfies the axioms of the ZF(C)-theory.

One might think of an objection to the last inference in the preceding argument. The existence of the universe M assumes that the starting universe $\mathcal{U}$ is a 6-universe. But if intuitive ZF(C)-theory is inconsistent, then the theory of 6-universes is inconsistent too, so $\mathcal{U}$ simply cannot exist and M would not have been constructed.

The fact is that the hypothesis that $\mathcal{U}$ is a 6-universe is just a convenient one; but, as we mentioned, it is not really necessary for developing the language, syntax and semantics of FOST. All we need is: the existence of an infinite set (and more conveniently, the property NNS to deal with the ordinary notion of finiteness), that $\mathcal{U}$ be a 2-universe so that we may speak of classes in studying semantics, and the principle of natural recursion. We may be reasonably confident that these principles do not involve a contradiction; and under those hypotheses, there is an effective construction of the standard model M, hence we may safely (or almost) arrive to the above conclusion. But, of course, the mere assumption of the existence of an infinite set may be questionable.

B.4.2 Final remarks

As we have just seen, all the results that we may obtain with our intuitive ZF(C)-theory can be also obtained in the symbolic theory, as long as they can be translated into sentences. In this sense, both ways of dealing with set theory are, in principle, equally powerful. This brings us to the second question stated at the beginning of this appendix: what is the aim or the advantage of introducing the symbolic theory?

Since FOST and IST produce basically the same theorems, any difference between both lies necessarily in the metatheory. In this connection, the crucial fact can be explained easily from the point of view presented here, that sees FOST as a construction inside IST. A typical question in the metatheory could be whether CH can be proved or not. For IST this is a question *about the theory* and may be treated by supposing that there is a proof and reasoning from that assumption. But in FOST, this metatheoretical question on the provability of CH is also a question *inside the theory*. Because there is a particular formula (let us write it "CH") that translates the assertion CH, hence that formula is an object of the universe $\mathcal{U}$. And what is asked is whether "CH" $\in T$, T

being the set of (symbolic) theorems. In this way, besides the metatheoretical meaning of the provability of CH, the question "does "CH" belong to T?" has also the meaning about membership between a certain sequence of known elements of $\mathcal{U}$ and a set that has been precisely defined, though somewhat artificially from the point of view of set theory, the set of theorems.

The reduction of the metatheory to the theory itself allows FOST to boast of some superiority, relatively to IST. Recall how we proved after Proposition 15.2.1 the consistency of CH relative to ZFC: assuming that there is a proof of $\neg$(CH), we had to rely on the property that such proof uses only finitely many axioms, and then take the constructible universe L as a counterexample, but we could not use directly that L is a ZFC-universe, because we had no proof of that fact, only the (infinitely many) proofs that L satisfies each of the axioms. FOST has, in principle, the same problem: there is no finite proof giving that L is a ZFC-universe because the infinite proofs of the axiom of replacement cannot be reduced to one proof. But the metatheoretical statement "L satisfies all the axioms of replacement" is translated into a sentence of FOST because the instances of the axioms form, when translated in FOST, a set. Like this, there is a sentence of FOST which corresponds to "L satisfies all the axioms of replacement" and which surely is a quite weird and, in itself, dry and meaningless statement about some peculiarly defined objects of the universe, in which it is difficult to recognize the metatheoretical assertion. But it follows from the completeness theorem that such sentence must be provable because otherwise some instance would be false for some ZFC-universe M, and this cannot happen due to the value of the statement as a result of the metatheory. Briefly, while in our intuitive metatheory we cannot consider, for instance, all theorems as a completed entity, the metatheory of FOST is reflected in the theory, and thus the collection of all the theorems is represented as a set. The metatheory of IST has the limitations of human finiteness, the metatheory of FOST allows the consideration of actual infinity because it can be represented as a piece of intuitive set theory.

Historically, there have been a variety of reasons for advocating the symbolization of set theory, and even of other theories, mathematical or not. Ancient forerunners of these attempts, like Leibniz, saw this as a way of avoiding ambiguities and instilling rigor and accuracy in the theory; by this move, that theory will be developed in some mechanical and indisputable form that would free it from the endless debates that plague other (mainly, philosophical, but sometimes also scientific) theories. However, this objective can hardly be taken literally. If we tried to write down in symbolic form any significant part of the theory, it would produce thousands and thousands of pages and the intended meaning of a segment of that symbolic text would be impossible to decipher in practice; not to mention that it would also be humanly impossible to construct that symbolic version in the first place. True, it could be in principle feasible for a machine to carry out this job. But still, if the machine simply prints out that symbolic version of the theory, it would remain unattainable for humans to understand the outcome. It is not impossible, however, that the progress in computation could lead machines to be able to make some triaging and communicate only theorems that are really interesting from a human point of view, and also present them in some understandable form (perhaps IA could become useful in this regard). But this is not the situation right now.

For the "logicistic school" of Russell and Whitehead, the purpose of symbolization was to give support to their main thesis: that mathematics is a part of Logic; and especially that natural numbers can be produced as logical concepts without any particular intuition justifying them. But this is no more a tenable thesis currently, so that, in spite of the great influence that the school had around one hundred years ago, and the tools and ideas that their work has provided to the development of mathematics, this cannot be seen now as one of the accomplishments of symbolization.

In those same years 1920s, the consideration of set theory as a symbolic first-order theory, proposed by Skolem among others, was starting to become a winner. This proposal was entangled with and justified by the discussions about Zermelo's separation axiom. Zermelo's axiomatic system was intended to be a formal axiomatic system, free of intuitive notions, but he was forced to leave a loose end. His *Axiom der Aussonderung* stated that given a set x, there is a set which consists of all those elements of x that satisfy a "definite class-assertion" as he called, something that we may oversimplify as a "property". Though he tried to explain this notion as accurately as possible, this remained as an appeal to intuition and raised criticism. The tries for clarifying the axiom were mixed with observations on the necessity of, at the same time, extend somehow the axiom; which in the end resulted in the axiom of replacement. The solution proposed by Skolem (and basically accepted in the long run by the mathematicians) was to symbolize the theory so that the intuitive idea of "property" was replaced with "satisfaction of a formula". This approach and the use of mathematical logic is responsible for fundamental advances in the theory, like constructibility or forcing. But at the elementary level, FOST is not so different in practice from IST: though the assumed object of study is the symbolic theory itself, it is impossible to develop set theory completely in the symbolic form, as mentioned before. Hence, what is usually done is to implicitly accept that the symbolic theory reflects intuitive properties of sets, and carry out the study in a way which is much similar to IST. The main difference at this level between IST as we have presented it, and FOST is our assumptions in IST about the natural numbers: in contrast, FOST does not need those assumptions, but cannot identify the collection of ordinary natural numbers as an object of the universe.

Hilbert's reasons for the symbolization of the theory were closer to the idea at the beginning of this section, of viewing problems in the metatheory as questions in set theory itself; though for him it was more to view the fundamental metatheoretical problem, the consistency of arithmetic, inside the symbolic study of arithmetic. Hilbert had observed that the unprovability of a formula like $1 = 2$ is equivalent to consistency (note that also the formula $C \neq T$ is equivalent to consistency in set theory). Unprovability of that formula is, in our terms, the set theoretic property $'1 = 2' \notin T$. Of course, Hilbert does not use this representation, but understands that proving the unprovability of that formula depends on a combinatorial analysis (hopefully finite, to be absolutely convincing) of the symbols used for formulas, and a theory of deduction. In his paper *Über das Unendliche* (Math. Ann. 95 (1926), 161-190), Hilbert fostered the use of the symbolic theory in the search for a proof of consistency of arithmetic, because he hoped that, since the result to be proved (the unprovability of a particular formula) refers to objects which are manifestly finite, namely formulas and deductions, such proof could be reached by finitary arguments, that is, without using the idea of infinity of any kind. But in the end, this idea was not rewarded with success: Gödel proved that the consistency of arithmetic (and this also holds for set theory) cannot be proved within arithmetic (respectively, within set theory) through an embedding of the metatheory inside the theory; unless, of course, if arithmetic is inconsistent.

Anyway, motivations for the symbolization of set theory have been numerous, and its value is beyond doubt. Let us add a final remark. As we have presented it, symbolic set theory is constructed inside a particular universe $\mathcal{U}$. But its development and properties are in fact independent of the universe. Certainly, the alphabet is a set in $\mathcal{U}$, but the representation of the symbols of the alphabet can be taken the same in every universe. The syntax gives then equivalent sets of formulas in any universe; and the deduction through logical axioms and rules of inference is again independent of $\mathcal{U}$; and so is then the concept of a theory, consistency, etc. As for the semantics, the formulas have values in a certain set $M \in \mathcal{U}$ (with a given relation L); but, since the formulas are the same for every universe, and the evaluation is also defined in the same way, any formula can be supposed to belong to any possible $\mathcal{U}$ and can therefore be evaluated in any suitable universe. Then if a sentence P is true in every set $M \in \mathcal{U}$ which is a model for a given collection A of axioms, then it is deducible from A and hence it is also true in every possible model of A in whatever universe. That is, the collection of the theorems (or, better, of their symbolic representation) is the same whatever the universe $\mathcal{U}$ we take at the start.

C

Real Numbers Revisited

C.1 Only One Set of Real Numbers?

At the start of this journey through set theory, we set out to study the universe of mathematical objects, but lacking a precise idea of what it was, we followed the way of the axiomatization: try to identify that universe through its assumed properties. Have we been successful?

Partially, yes. We have obtained inside set theory approximate copies of the basic number sets, natural, integers, rationals, real, complex; and from this we are able to describe and study functions, structures, spaces, and most of the mathematical objects we use. We have also developed a measure of a kind of size of the sets formed with those mathematical objects, thus starting an analysis of different infinities.

But if we ask about the objective of identifying the universe of mathematical objects, the answer to the above question is clearly no. If we may consistently imagine some ZF-universe for example, then there are many more ZF-universes available to our imagination, and many properties may be different in one or another of those universes. And, of course, there are good reasons for this. Our axioms are far from telling us which are the objects of the universe. They simply assert that those objects, the sets, are closed under taking pairs, unions, powers and functional images; and while these closure properties agree with mathematical practice and allow (along with the axiom of infinity) the construction of the classical mathematical objects, they obviously fall short of positively identifying the universe. Certainly, there is also the principle of separation, but it only makes possible the construction of subsets of any previously given set; and, for infinite sets, only of some of those subsets, leaving us helpless for any attempt at describing all the subsets of the given infinite set. Finally, in ZFC-theory there is also the axiom of choice, but again it works on previously defined sets and it just asserts the existence of some set related to that starting set. It is highly useful for some purposes, but it does not tell much about the whole universe.

Even the task of identifying the subsets of the set of natural numbers is more difficult than it could seem. As we know, a much more modest objective, that of evaluating the size of $\mathbb{P}(\omega)$ is also unattainable in a sense. Because the size of that set is precisely the size (if we identify size and cardinal) of the

DOI: 10.1201/9781003449911-C

set $\mathbb{R}$ of the real numbers, $2^{\aleph_0}$ (see Exercise 15 of Chapter 6); and since the continuum hypothesis is independent in the ZFC-theory, we cannot know from the ZFC axioms which is this cardinal in the series of the alephs; we do not even know whether there exists some infinite non-countable subset of $\mathbb{R}$ with a cardinal strictly smaller that the cardinal of $\mathbb{R}$ – which would precisely be the case if $\aleph_1 < 2^{\aleph_0}$, since some subset of $\mathbb{R}$ would have, in that hypothesis, cardinal $\aleph_1$.

This impossibility has shocked many mathematicians. Because a common idea would run like this: the set $\mathbb{R}$ of real numbers is well-known, hence its subsets are well determined; so to speak, they exist somehow. Therefore the problem whether some subset of $\mathbb{R}$ is non-countable and with cardinal $< |\mathbb{R}|$, must have a definite answer *in the real world.* Since the axioms are incapable of finding the answer, it must be because they do not capture the truth about the subsets of real numbers.

This is a sound argument as long as we accept its premises. And certainly we know (and have remarked above) that the axioms are far from describing the true nature of the universe of sets, if such a universe exists at all. In particular, even if the set $\mathbb{R}$ is well-known, knowledge of its subsets is not guaranteed: since they are those objects of the universe that happen to be included in $\mathbb{R}$, they depend both on $\mathbb{R}$ and on the universe itself. Since our axioms do not give us the knowledge of the universe, it is not surprising that they cannot determine all the subsets of $\mathbb{R}$, neither.

But we may question the premises of the argument, too; namely, the assumption that the set $\mathbb{R}$ is well-known or well determined. Let us observe that $\mathbb{R}$ may be different in different universes, provided the ZFC-theory is consistent. By Theorem 15.1.1, the ZFC$^+$-theory is also consistent, so let us pick a ZFC$^+$-universe V, and let $M \in V$ be a transitive and countable ZFC-universe with $\omega \in M$. Let $\mathbb{R}$ be the set of real numbers of V, as defined in Section 6.8, and write $\mathbb{R}_M$ for the set of real numbers of the universe M, which will also be an element of V. We have then:

Proposition C.1.1. *With the above notations,* $\mathbb{R}_M \subseteq \mathbb{R}$.

Proof. Since $\omega \in M$, both universes have the same finite ordinals and same ω, and the basic operations $+$ and $\cdot$ of ω also belong to both universes. M is closed under ordered pairs and thus $\omega \times \omega$ is the same in both universes; it follows that the equivalence relation of Proposition 6.7.2 is also a member of M and again $\mathbb{Z}$ is the same in V and in M. By the same reason, $\mathbb{Q}$ is the same in V and in M.

If $C \in \mathbb{R}_M$, then C is a Dedekind cut of the rationals by definition, but then it is also a Dedekind cut of $\mathbb{Q}$ in the universe V and thus $C \in \mathbb{R}$. □

So $\mathbb{R}_M \subseteq \mathbb{R}$, but, viewed in V, $\mathbb{R}_M$ is countable and $\mathbb{R}$ is not, whence there are elements of $\mathbb{R}$, hence D-cuts of $\mathbb{Q}$, which do not belong to $\mathbb{R}_M$; neither do they belong to M, because if they did, then they would also be D-cuts of $\mathbb{Q}$ in M. Therefore we have two versions of the real numbers in two ZFC-universes,

V and M, which perfectly fill the role of the complete field of real numbers (each in its own universe), but which are different, with one of them properly including the other.

If the set of real numbers is well determined, how is it that it depends upon the ZFC-universe that we consider? Is it again that the ZFC axioms are insufficient to identify not only the subsets of $\mathbb{R}$ but even $\mathbb{R}$ itself, so that additional axioms would be necessary to identify the one and only set $\mathbb{R}$? There are some reasons in favor of this view: first, the real numbers are those used to measure magnitudes, for instance, lengths; and lengths seem to have a reality independent of set theory conventions or universes; second, real numbers correspond to the points in a line, and again a line seems to be a perfectly well determined object, at least in our imagination.

However, this does not settle the question. As for lengths, they are not measured in practice by real numbers, but always by rational numbers. It is only ideally that real numbers appear as the limit of the process of indefinitely improving the measure of a length, but in reality, matter does not seem to be indefinitely divisible. Even if we trust our imagination and accept the existence of that ideal measure, what we may see as a real number is an infinite sequence of approximations, i..e., of rational numbers. And the concept of an infinite sequence of rational numbers is not more clear than that of a Dedekind cut of $\mathbb{Q}$. So, we are back in our definition of real numbers which has shown to be relative to the universe.

We still have the correspondence between real numbers and the points of a line. However, while the idea of a line seems to be reasonably concrete (as an ideal Platonic object), that of the collection of all the points of the line is not as clear. In fact, Proposition C.1.1 suggests that the points in a line when we think of the universe M are just a small subcollection of the points of the same line for the universe V. In the end, it is doubtful that our intuition of *the totality* of the points in a line is clearer than other intuitions about infinity.

We get a different approach to this same problem by considering the following universe, which gives what is perhaps the prevalent interpretation of set theory. We choose the universe $\mathcal{U}$ consisting of **all** collections C of objects of $\mathcal{U}$, except those for which there is a bijective functional between C and the universe $\mathcal{U}$ itself. $\mathcal{U}$ will be a ZFC-universe if we assume that it satifies the main ZFC-axioms: unions, power set, infinity and choice; the rest of the axioms are either conventions (empty set and limitation to pure well founded sets) or immediate consequences of the assumptions (pairs, replacement). Moreover, if one accepts the simple intuition about natural numbers, the objects $s^n(\emptyset)$ form obviously a (small) collection, hence a set. Like this, the number sets $\mathbb{N}, \mathbb{Z}, \mathbb{Q}$ are uniquely identified as the familiar ones, and $\mathbb{R}$ should also have a definite meaning in this interpretation. Briefly, if you believe in the reality of the collection of all subsets of $\mathbb{Q}$ (or the collection of subsets of $\mathbb{N}$ for that matter) and see it as something determined and absolute, then you will also accept that the collection of real numbers is well determined and absolute. On the contrary, if you feel that one cannot give an absolute meaning to the collection of those subsets, then you may also feel that the idea of the "true" real numbers has no definite meaning neither.

As the reader may see, the arguments about the question of the reality of real numbers are inconclusive, and the mathematicians have not reached a definite agreement on it. Be it as it may, the result, in its current state, of the set theory quest for the mathematical objects leads to different universes. Most of mathematics can be developed from set theory, but each mathematical theory based on set theory is constructed, explicitly or not, upon some of the possible universes. Of course, most of the results obtained in those particular theories are valid in every universe. But there are also results (surely in set theory, but also in practically any domain) which are dependent on the universe where the theory is supposed to be founded.

Bibliography

[1] G. Cantor: Beiträge zur Begründung der transfiniten Mengenlehre. I. Math. Annalen 46 (1895), 481–512.

[2] P. J. Cohen: Set Theory and the Continuum Hypothesis, Benjamin/Cummings Publ. Co., 1966.

[3] K. Devlin: The Joy of Sets, Springer-Verlag, 1993.

[4] J. Ferreirós: Labyrinth of Thought. A History of Set Theory and its Role in Modern Mathematics, 2nd Ed., Birkhäuser, 2007.

[5] A. A. Fraenkel: Historical introduction, in: P. Bernays, Axiomatic Set Theory, North-Holland, 1958.

[6] A. Fraenkel, Y. Bar-Hillel, A. Levy: Foundations of Set Theory, 2nd Ed., North-Holland, 1973.

[7] K. Gödel: Consistency-proof for the Generalized Continuum Hypothesis, Proc. Nat. Acad. Sci. USA 25 (1939), 220–224.

[8] P. Halmos: Naive Set Theory, Van Nostrand, 1960.

[9] K. Hrbacek, T. Jech: Introduction to Set Theory, 3rd Ed., Marcel Dekker, 1999.

[10] T. Jech: Set Theory. The Third Millenium Edition, Springer, 2006.

[11] W. Just, M. Weese: Discovering Modern Set Theory. I. American Mathematical Society, 1996.

[12] E. Kamke: Théorie des ensembles, Dunod 1964.

[13] K. Kunen: Set Theory. An Introduction to Independence Proofs, Elsevier 1980.

[14] A. Levy: Basic Set Theory, Springer-Verlag, 1979.

[15] J. P. Mayberry: The Foundations of Mathematics in the Theory of Sets, Cambridge University Press, 2000.

[16] E. Mendelson: Introduction to Mathematical Logic, Van Nostrand, 1979.

[17] Y. N. Moschovakis, Notes on Set Theory, Springer-Verlag, 1994.

[18] M. Potter: Set Theory and its Philosophy, Oxford University Press, 2004.

[19] J. Roitman: Introduction to Modern Set Theory, Wiley and Sons, 1990.

[20] E. Schimmerling: A Course on Set Theory, Cambridge University Press, 2011.

[21] R. Smullyan, M. Fitting: Set Theory and the Continuum Problem, Clarendon Press, 1996.

[22] P. Suppes: Axiomatic Set Theory, Van Nostrand, 1960.

[23] D. van Dalen, H. C. Doets, H. de Swart: Sets: Naive, Axiomatic and Applied, Pergamon Press, 1978.

[24] N. Weaver: Forcing for Mathematicians, World Scientific, 2014.

[25] E. Zermelo: Untersuchungen über die Grundlagen der Mengenlehre. I. Math. Annalen 65 (1908), 261–281.

Index

9781032581200